Hypercompositional Algebra and Applications

Hypercompositional Algebra and Applications

Editor

Christos G. Massouros

MDPI • Basel • Beijing • Wuhan • Barcelona • Belgrade • Manchester • Tokyo • Cluj • Tianjin

Editor
Christos G. Massouros
Core Department
National and Kapodistrian
University of Athens
Athens
Greece

Editorial Office
MDPI
St. Alban-Anlage 66
4052 Basel, Switzerland

This is a reprint of articles from the Special Issue published online in the open access journal *Mathematics* (ISSN 2227-7390) (available at: www.mdpi.com/journal/mathematics/special_issues/ Hypercompositional_Algebra_Applications).

For citation purposes, cite each article independently as indicated on the article page online and as indicated below:

LastName, A.A.; LastName, B.B.; LastName, C.C. Article Title. *Journal Name* **Year**, *Volume Number*, Page Range.

ISBN 978-3-0365-2104-6 (Hbk)
ISBN 978-3-0365-2103-9 (PDF)

© 2021 by the authors. Articles in this book are Open Access and distributed under the Creative Commons Attribution (CC BY) license, which allows users to download, copy and build upon published articles, as long as the author and publisher are properly credited, which ensures maximum dissemination and a wider impact of our publications.

The book as a whole is distributed by MDPI under the terms and conditions of the Creative Commons license CC BY-NC-ND.

Contents

About the Editor vii

Preface to "Hypercompositional Algebra and Applications" ix

Christos Massouros and Gerasimos Massouros
An Overview of the Foundations of the Hypergroup Theory
Reprinted from: *Mathematics* **2021**, *9*, 1014, doi:10.3390/math9091014 1

Gerasimos Massouros and Christos Massouros
Hypercompositional Algebra, Computer Science and Geometry
Reprinted from: *Mathematics* **2020**, *8*, 1338, doi:10.3390/math8081338 43

Milica Kankaras and Irina Cristea
Fuzzy Reduced Hypergroups
Reprinted from: *Mathematics* **2020**, *8*, 263, doi:10.3390/math8020263 75

Hashem Bordbar and Irina Cristea
Regular Parameter Elements and Regular Local Hyperrings
Reprinted from: *Mathematics* **2021**, *9*, 243, doi:10.3390/math9030243 87

Sarka Hoskova-Mayerova, Madeline Al Tahan and Bijan Davvaz
Fuzzy Multi-Hypergroups
Reprinted from: *Mathematics* **2020**, *8*, 244, doi:10.3390/math8020244 101

Jan Chvalina, Michal Novák, Bedřich Smetana and David Staněk
Sequences of Groups, Hypergroups and Automata of Linear Ordinary Differential Operators
Reprinted from: *Mathematics* **2021**, *9*, 319, doi:10.3390/math9040319 115

Dariush Heidari and Irina Cristea
On Factorizable Semihypergroups
Reprinted from: *Mathematics* **2020**, *8*, 1064, doi:10.3390/math8071064 131

Mario De Salvo, Dario Fasino, Domenico Freni and Giovanni Lo Faro
1-Hypergroups of Small Sizes
Reprinted from: *Mathematics* **2021**, *9*, 108, doi:10.3390/math9020108 139

Štěpán Křehlík
n-Ary Cartesian Composition of Multiautomata with Internal Link for Autonomous Control of Lane Shifting
Reprinted from: *Mathematics* **2020**, *8*, 835, doi:10.3390/math8050835 157

Vahid Vahedi, Morteza Jafarpour, Sarka Hoskova-Mayerova, Hossein Aghabozorgi, Violeta Leoreanu-Fotea and Svajone Bekesiene
Derived Hyperstructures from Hyperconics
Reprinted from: *Mathematics* **2020**, *8*, 429, doi:10.3390/math8030429 175

Christos G. Massouros and Naveed Yaqoob
On the Theory of Left/Right Almost Groups and Hypergroups with their Relevant Enumerations
Reprinted from: *Mathematics* **2021**, *9*, 1828, doi:10.3390/math9151828 191

About the Editor

Christos G. Massouros

Christos G. Massouros is a Full Professor of Mathematics at the National and Kapodistrian University of Athens. Christos G. Massouros is a graduate of the National and Kapodistrian University of Athens, Department of Mathematics. He did graduate work in theoretical physics and mathematics at the National Center for Scientific Research "Democritus" of Greece and he received his Ph.D. from the School of Applied Mathematical and Physical Sciences of the National Technical University of Athens. His main research interests are in the theory of hypercompositional structures and their applications. He is the author of a number of research, review, and survey papers on the subject and has participated in many relevant research programs. He is a reviewer for several scientific journals and member of their editorial boards.

Preface to "Hypercompositional Algebra and Applications"

This Special Issue is about Hypercompositional Algebra, which is a recent branch of Abstract Algebra.

As it is known, Algebra is a generalization of Arithmetic, where Arithmetic (from the Greek word —arithmós—meaning "number") is the area of mathematics that deals with numbers. During the classical period, Greek mathematicians created a Geometric Algebra where terms were represented by sides of geometric objects, while the Alexandrian School that followed (which was founded during the Hellenistic era) changed this approach dramatically. Especially, Diophantus made the first fundamental step towards Symbolic Algebra, as he developed a mathematical notation in order to write and solve algebraic equations.

In the Middle Ages, the stage of mathematics shifted from the Greek world to the Arabic, and in AD 830 the Persian mathematician Muammad ibn Mūsā al-Khwārizmī wrote his famous treatise al-Kitāb al-mukhtaar fī isāb al-jabr wa'l-muqābala which contains the first systematic solution of linear and quadratic equations. The word "Algebra" derives from this book's title (al-jabr). After the 15th century AD, the stage of mathematics changed again and went back to Europe, leaving the Islamic world which was in decline, while the European world was ascending as the Renaissance began to spread in Western Europe. However, Algebra essentially remained Arithmetic with non-numerical mathematical objects until the 19th century, when the algebraic mathematical thought changed radically due to the work of two young mathematicians: the Norwegian Niels Henrik Abel (1802–1829) and the Frenchman Evariste Galois (1811–1832), on the algebraic equations. It was then understood that the same processes could be applied to various objects or sets of entities other than numbers. Therefore, Abstract Algebra was born.

The early years of the next (20th) century brought the end of determinism and certainty to science. The uncertainty affected Algebra as well and was expressed in the work of the young French mathematician, Frederic Marty (1911–1940), who introduced an algebraic structure in which the rule of synthesizing elements gives a set of elements instead of one element only. He called this structure "hypergroup", and he presented it during the 8th congress of Scandinavian Mathematicians, held in Stockholm in 1934.

Unfortunately, Marty was killed at the age of 29, when his airplane was hit over the Baltic Sea while he was in military duty during World War II. His mathematical heritage on hypergroups is three papers only. However, other mathematicians such as H. Wall, M. Dresher, O. Ore, M. Krasner, and M. Kuntzmann started working on hypergroups shortly thereafter. Thus, the Hypercompositional Algebra came into being as a branch of Abstract Algebra that deals with structures endowed with multi-valued operations. Such operations, which are also called "hyperoperations" or "hypercompositions", are laws of synthesis of the elements of a nonempty set, which associate a set of elements, instead of a single element, with every pair of elements.

Nowadays, this theory is characterized by huge diversity of character and content, and can present results in mathematics and other sciences such as computer science, information technologies, physics, chemistry, biology, social sciences, etc.

This Special Issue, titled "Hypercompositional Algebra and Applications" contains 11 peer-reviewed papers written by 24 outstanding experts in the theory of hypercompositional structures. The names of the authors are in alphabetical order:

Hossein Aghabozorgi, Madeline Al Tahan, Svajone Bekesiene, Hashem Bordbar, Jan Chvalina, Irina Cristea, Bijan Davvaz, Mario De Salvo, Dario Fasino, Domenico Freni, Dariush Heidari, Sarka Hoskova-Mayerova, Morteza Jafarpour, Milica Kankaras, Štěpán Křehlík, Violeta Leoreanu-Fotea, Giovanni Lo Faro, Christos G. Massouros, Gerasimos Massouros, Michal Novák, Bedřich Smetana, David Staněk, Vahid Vahedi, and Naveed Yaqoob.

The introductory paper in this Issue is a review paper on the foundations of the hypergroup theory that focuses on the presentation of the essential principles of the hypergroup, which is the prominent structure of the Hypercompositional Algebra. The 20th century was not only the period when the hypergroup came into being, but it was also characterized by the multi-volume and extremely valuable work Éléments de mathématique by Nicolas Bourbaki, where the notion of the mathematical structure, an idea related to the broader and interdisciplinary concept of structuralism, was introduced. Towards this direction, the paper generalizes the notion of the magma, which was introduced in Algèbre by Nicolas Bourbaki (1943, by Hermann, Paris), in order to include the hypercompositional structures and thereafter it reveals the structural relation between the fundamental entity of the Abstract Algebra, the group with the hypergroup, as it shows that both structures satisfy exactly the same axioms but under different synthesizing laws.

I hope that this Special Issue will be useful not only to the mathematicians who work on the Hypercompositinal Algebra but to all scientists from different disciplines who study the Abstract Algebra and seek its applications.

Christos G. Massouros
Editor

Review

An Overview of the Foundations of the Hypergroup Theory

Christos Massouros [1,*] and Gerasimos Massouros [2]

1. Core Department, Euripus Campus, National and Kapodistrian University of Athens, GR 34400 Euboia, Greece
2. School of Social Sciences, Hellenic Open University, Aristotelous 18, GR 26335 Patra, Greece; germasouros@gmail.com
* Correspondence: ChrMas@uoa.gr or Ch.Massouros@gmail.com

Abstract: This paper is written in the framework of the Special Issue of *Mathematics* entitled "Hypercompositional Algebra and Applications", and focuses on the presentation of the essential principles of the hypergroup, which is the prominent structure of hypercompositional algebra. In the beginning, it reveals the structural relation between two fundamental entities of abstract algebra, the group and the hypergroup. Next, it presents the several types of hypergroups, which derive from the enrichment of the hypergroup with additional axioms besides the ones it was initially equipped with, along with their fundamental properties. Furthermore, it analyzes and studies the various subhypergroups that can be defined in hypergroups in combination with their ability to decompose the hypergroups into cosets. The exploration of this far-reaching concept highlights the particularity of the hypergroup theory versus the abstract group theory, and demonstrates the different techniques and special tools that must be developed in order to achieve results on hypercompositional algebra.

Keywords: group; hypergroup; subhypergroup; cosets

Citation: Massouros, C.; Massouros, G. An Overview of the Foundations of the Hypergroup Theory. *Mathematics* **2021**, *9*, 1014. https://doi.org/10.3390/math9091014

Academic Editor: Michael Voskoglou

Received: 21 February 2021
Accepted: 16 March 2021
Published: 30 April 2021

Publisher's Note: MDPI stays neutral with regard to jurisdictional claims in published maps and institutional affiliations.

Copyright: © 2021 by the authors. Licensee MDPI, Basel, Switzerland. This article is an open access article distributed under the terms and conditions of the Creative Commons Attribution (CC BY) license (https://creativecommons.org/licenses/by/4.0/).

1. Introduction

The early years of the 20th century brought the end of determinism and certainty to science. The emergence of quantum mechanics rocked the well-being of classical mechanics, which was founded by Isaac Newton in *Philosophiæ Naturalis Principia Mathematica*. In 1927, Werner Heisenberg developed his uncertainty principle while working on the mathematical foundations of quantum mechanics. On the other hand, in 1931 Kurt Gödel published his two incompleteness theorems, thus giving an end to David Hilbert's mathematical dreams and to the attempts that are culminating with *Principia Mathematica* of Bertrand Russell. In 1933, Andrey Kolmogorov published his book *Foundations of the Theory of Probability*, establishing the modern axiomatic foundations of probability theory. In the same decade the uncertainty invaded algebra as well. A young French mathematician, Frédéric Marty (1911–1940), during the 8th Congress of Scandinavian Mathematicians, held in Stockholm in 1934, introduced an algebraic structure in which the rule of synthesizing elements results to a set of elements instead of a single element. He called this structure hypergroup. Marty was killed at the age of 29, when his airplane was hit over the Baltic Sea, while he was in the military during World War II. His mathematical heritage on hypergroups was only three papers [1–3]. However, other mathematicians such as M. Krasner [4–8], J. Kuntzmann [8–10], H. Wall [11], O. Ore [12–14], M. Dresher [13], E. J. Eaton [14,15], and L. W. Griffiths [16] gradually started working on hypergroups shortly thereafter (see the classical book [17] for further bibliography). Thus, hypercompositional algebra came into being.

Hypercompositional algebra is the branch of abstract algebra that deals with structures equipped with multivalued operations. Multivalued operations, also called hyperoperations or hypercompositions, are laws of synthesis of the elements of a nonempty set, which associates a set of elements, instead of a single element, to every pair of elements.

The fundamental structure of hypercompositional algebra is the hypergroup. This paper enlightens the structural relation between the groups and the hypergroups. The study of such relationships is at the heart of structuralism. Structuralism is based on the idea that the elements of a system under study are not important, and only the relationships and structures among them are significant. As it is proved in this paper, the axioms of groups and hypergroups are the same, while these algebraic entities' difference is based on the relationship between their elements, which is created by the law of synthesis. In groups, the law of synthesis of any two elements is a composition, i.e., a single element, while in hypergroups it is a hypercomposition, that is, a set of elements.

The next section of this paper generalizes the notion of magma, which was introduced in *Éléments de Mathématique, Algèbre* [18] by Nicolas Bourbaki, and so will include algebraic structures with hypercomposition. The third section presents a unified definition of the group and the hypergroup. This definition of the group is not included in any group theory book, and its equivalence to the already-known ones is proved in the fourth section. The fifth section presents another, equivalent definition of the hypergroup, while certain of its fundamental properties are proved. As these properties derive directly from the axioms of the hypergroup, they outline the strength of these axioms. So, for instance it is shown that the dominant in the bibliography definition of the hypercomposition includes redundant assumptions. The restriction that a hypercomposition is a mapping from $E \times E$ into the family of nonempty subsets of E is needless, since, in the hypergroups, the result of the hypercomposition is proved to be always a nonvoid set. The sixth section deals with different types of hypergroups. The law of synthesis imposes a generality on the hypergroup, which allows its enrichment with more axioms. This creates a multitude of special hypergroups with many and interesting properties and applications. The join space is one of them. It was introduced by W. Prenowitz in order to study geometry with the tools of hypercompositional algebra, and many other researchers adopted this approach [19–34]. Another one is the fortified join hypergroup, which was introduced by G. Massouros in his study of the theory of formal languages and automata [35–42], and he was followed by other authors who continued in this direction e.g., [42–52]. One more is the canonical hypergroup, which is the additive part of the hyperfield that was used by M. Krasner as the proper algebraic tool in order to define a certain approximation of complete valued fields by sequences of such fields [53]. This hypergroup was used in the study of geometry as well e.g., [32,33,54–60]. Moreover, the canonical hypergroup became part of other hypercompositional structures like the hypermodule [61] and the vector hyperspace [62]. In [61], it is shown that analytic projective geometries and Euclidean spherical geometries can be considered as special hypermodules. Furthermore, the hyperfields were connected to the conic sections via a number of papers [63–65], where the definition of an elliptic curve over a field F was naturally extended to the definition of an elliptic hypercurve over a quotient Krasner's hyperfield. The conclusions obtained in [63–65] can be applied to cryptography as well. Moreover, D. Freni in [66] extended the use of the hypergroup in more general geometric structures, called geometric spaces; [67] contains a detailed presentation of the above. Also, hypergroups are used in many other research areas, like the ones mentioned in [68], and recently, in social sciences [69–73] and in an algebraization of logical systems [74,75]. The seventh section refers to subhypergroups. A far-reaching concept of abstract group theory is the idea of the decomposition of a group into cosets by any of its subgroups. This concept becomes much more complicated in the case of hypergroups. The decomposition of the hypergroups cannot be dealt with in a similar uniform way as in the groups. So, in this section, and depending on its specific type, the decomposition of a hypergroup to cosets is treated with the use of invertible, closed, reflexive, or symmetric subhypergroups.

Special notation: In the following, in addition to the typical algebraic notations, we use Krasner's notation for the complement and difference. So, we denote with $A \cdot \cdot B$ the set of elements that are in the set A, but not in the set B.

2. Magma

In *Éléments de Mathématique, Algèbre* [18], Nicolas Bourbaki used the Greek word *magma*, which comes from the verb μάσσω (= "knead"), to indicate a set with a law of composition. The following definition extends this notion in order to incorporate more general laws of synthesizing the elements in a set.

Definition 1. *Let E be a nonvoid set. A mapping from $E \times E$ into E is called a composition on E and a mapping from $E \times E$ into the power set $P(E)$ of E is called a hypercomposition on E. A set with a composition or a hypercomposition is called a magma.*

The notation $(E, \perp)$, where $\perp$ is the composition or the hypercomposition, is used when it is required to write the law of synthesis in a magma. The image $\perp(x,y)$ of (x,y) is written $x \perp y$. The symbols $+$ and $\cdot$ are the most commonly used instead of $\perp$. A law of synthesis denoted by the symbol $+$ is called *addition* and $x + y$ is called the *sum* of x and y if the synthesis is a composition, and the *hypersum* of x and y if the synthesis is a hypercomposition. A law of synthesis denoted by the symbol $\cdot$ is called *multiplication*, and $x \cdot y$ is called the *product* of x and y if the synthesis is a composition and the *hyperproduct* of x and y if the synthesis is a hypercomposition; when there is no likelihood of confusion, the symbol $\cdot$ can be omitted and we write xy instead of $x \cdot y$.

Example 1. *The power set $P(E)$ of a set $E \neq \emptyset$ is a magma if*

$$(X,Y) \to X \cup Y \text{ or } (X,Y) \to X \cap Y.$$

Example 2. *The set $\mathbb{N}$ of natural numbers is a magma under addition or multiplication.*

Example 3. *A nonvoid set E becomes a magma under the following law of synthesis:*

$$(x,y) \to \{x,y\}$$

The above law of synthesis is called b-hypercomposition. E also becomes a magma if

$$(x,y) \to E.$$

This law of synthesis is called total hypercomposition.

Two significant types of hypercompositions are the closed and the open ones. A hypercomposition is called *closed* [76] (or *containing* [77], or *extensive* [78]) if the two participating elements always belong to the result of the hypercomposition, while it is called *open* if the result of the hypercomposition of any two different elements does not contain these two elements.

Example 4. *Let S be the set of points of a Euclidian geometry. In S we define the following law of synthesis "·": if a and b are distinct points of S, then $a \cdot b$ is the set of all elements of the segment ab; while $a \cdot a$ is taken to be the point a, for any point a of S (Figure 1). Then, the set S of the points of the Euclidian geometry becomes a magma. Usually, $a \cdot b$ is written simply as ab and it is called the join of a and b. It is worth noting that we can actually define two laws of synthesis: an open and a closed hypercomposition, depending on whether we consider the open or the closed segment ab.*

Figure 1. Example 4.

Any finite magma E can be explicitly defined by its synthesis table. If E consists of n elements, then the synthesis table is a $n \times n$ square array heading both to the left and above by a list of the n elements of E. In this table (Cayley table), the entry in the row headed by x and the column headed by y is the synthesis $x \perp y$.

Example 5. *Suppose that $E = \{1, 2, 3, 4\}$. Then the law of the synthesis is a composition in the first table and a hypercomposition in the second.*

$\perp$	1	2	3	4
1	1	3	2	3
2	3	2	1	4
3	1	2	1	1
4	2	3	4	1

$\perp$	1	2	3	4
1	{1,2,3}	{1,3}	{2,4}	{1,3,4}
2	{3}	{1,2,3,4}	{1,3}	{2,4}
3	{1,2,3}	{1,2}	{1,2,3,4}	{1}
4	{2,3}	{2,3}	{3,4}	{1,3}

Let $(E, \perp)$ be a magma. Given any two nonvoid subsets X, Y of E, then

$$X \perp Y = \{x \perp y \in E \mid x \in X, y \in Y\}, \quad \text{if } \perp \text{ is a composition}$$

and

$$X \perp Y = \bigcup_{x \in X, y \in Y} (x \perp y), \quad \text{if } \perp \text{ is a hypercomposition.}$$

If X or Y is empty, then $X \perp Y$ is empty. If $a \in E$ we usually write $a \perp Y$ instead of $\{a\} \perp Y$ and $X \perp a$ instead of $X \perp \{a\}$. In general, the singleton $\{a\}$ is identified with its member a. Sometimes it is convenient to use the relational notation $A \approx B$ to assert that subsets A and B have a nonvoid intersection. Then, as the singleton $\{a\}$ is identified with its member a, the notation $a \approx A$ or $A \approx a$ is used as a substitute for $a \in A$. The relation $\approx$ may be considered as a weak generalization of equality, since, if A and B are singletons and $A \approx B$, then $A = B$. Thus, $a \approx b \perp c$ means $a = b \perp c$, if the synthesis is a composition and $a \in b \perp c$, if the synthesis is a hypercomposition.

Definition 2. *Let $(E, \perp)$ be a magma. The law of synthesis*

$$(x, y) \to x \perp^{op} y = y \perp x$$

is called the opposite of $\perp$. The magma $(E, \perp^{op})$ is called the opposite magma of $(E, \perp)$. When $\perp^{op} = \perp$, the law of synthesis is called commutative and the magma is called commutative magma.

Definition 3. *Let E be a magma and X a subset of E. The set of elements of E that commute with each one of the elements of X is called the centralizer of X. The centralizer of E is called the center of E. An element of the center of E is called the central element of E.*

Every law of synthesis in a magma induces two new laws of synthesis. If the law of synthesis is written multiplicatively, then the two induced laws are:

$$a/b = \{x \in E \mid a \approx xb\} \quad \text{and} \quad b \backslash a = \{x \in E \mid a \approx bx\}.$$

Thus, $x \approx a/b$ if and only if $a \approx xb$ and $x \approx b \backslash a$ if and only if $a \approx bx$. In the case of a multiplicative magma, the two induced laws are named *inverse laws* and they are called the *right* and *left division*, respectively. It is obvious that if the magma is commutative, then the right and left divisions coincide.

Proposition 1. *If the law of synthesis in a magma $(E, \cdot)$ is an open hypercomposition, then $a/a = a\backslash a = a$ for all a in E, while $a/a = a\backslash a = E$ for all a in E, when the law of synthesis is a closed (containing) hypercomposition.*

Example 6. *On the set $\mathbb{Q}$ of rational numbers, multiplication is a commutative law of synthesis. The inverse law is as follows:*

$$a/b = \begin{cases} \frac{a}{b} & if\ b \neq 0 \\ \varnothing & if\ b = 0,\ a \neq 0 \\ \mathbb{Q} & if\ b = 0,\ a = 0 \end{cases}$$

Here, the law of synthesis is a composition and the inverse law is a hypercomposition.

Example 7. *In Example 4, a law of synthesis was defined on the set S of the points of a Euclidian geometry. If we consider the open hypercomposition, then the inverse law is the following: If a and b are two distinct points in S, then a/b is the set of the points of the open halfline with endpoint a, that is opposite to point b, while $a/a = a$, for any point a of S (Figure 2). Usually, this law is called the extension of a over b, or "a from b".*

Figure 2. Example 7.

Example 8. *Let $(E, \cdot)$ be a magma and let $/$ and $\backslash$ be the right and left division. A new law of synthesis, called the extensive enlargement of "$\cdot$", can be defined on E as follows:*

$$a \odot b = a \cdot b \cup \{a, b\}, for\ all\ a, b \in E$$

Denoting the two induced laws of synthesis by $_\odot /$ and $\backslash_\odot$, it is immediate that:

$$a_\odot/b = \begin{cases} a/b \cup \{a\}, & if\ a \neq b \\ E, & if\ a = b \end{cases} \quad and \quad b\backslash_\odot a = \begin{cases} b\backslash a \cup \{a\}, & if\ a \neq b \\ E, & if\ a = b \end{cases}$$

Obviously, the extensive enlargement is a closed (containing) hypercomposition.

Definition 4. *An element 0 of a magma $(E, \perp)$ is called left absorbing (resp. right absorbing) if $0\perp x = 0$ (resp. $x\perp 0 = 0$) for all $x \in E$. An element of a magma is called absorbing if it is a bilaterally absorbing element.*

A direct consequence of the above definition is Proposition 2:

Proposition 2. *If a magma has a left (resp. right) absorbing element, then the relevant induced law of synthesis is a hypercomposition.*

Definition 5. *A law of synthesis $(x, y) \to x\perp y$ on a set E is called associative if the property*

$$(x\perp y)\perp z = x\perp(y\perp z)$$

is valid, for all elements x, y, z in E. A magma whose law of synthesis is associative is called an associative magma.

Example 9. *If the law of synthesis is the b-hypercomposition denoted by +, then*

$$(x+y)+z = \{x,y\}+z = (x+z) \cup (y+z) = \{x,z\} \cup \{y,z\} = \{x, y, z\}$$

and $\quad x+(y+z) = x+\{y,z\} = (x+y) \cup (x+z) = \{x,y\} \cup \{x,z\} = \{x, y, z\}$

Hence $(x+y)+z = x+(y+z)$, *for all* x, y, z *in E. Thus, the b-hypercomposition is associative.*

Example 10. *The law of synthesis defined in Example 4 on the set S of the points of a Euclidian geometry is associative. For the verification, it is required to consider many cases. The following Figure 3 presents the general case for the open hypercomposition, in which the points a, b, c are not collinear. The result of both $a(bc)$ and $(ab)c$ is the interior of the triangle abc.*

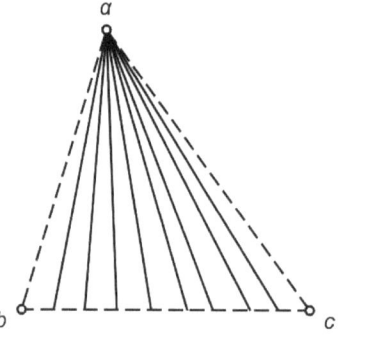 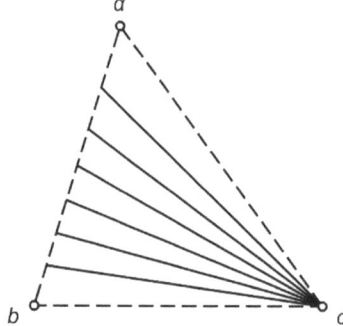

Figure 3. Example 10.

The interaction of the law of synthesis with the two induced laws in an associative magma gives the mixed associativity.

Proposition 3. *In an associative magma $(E, \cdot)$ the properties*

$$\begin{array}{ll} (a/b)/c = a/(cb) & \text{right mixed associativity} \\ c \backslash (b \backslash a) = (bc) \backslash a & \text{left mixed associativity} \\ (b \backslash a)/c = b \backslash (a/c) & \end{array}$$

are valid, for all $a, b, c \in E$.

Proof. Let $x \approx (a/b)/c$. Then we have the following sequence of equivalent statements:

$$x \approx (a/b)/c \Leftrightarrow xc \approx a/b \Leftrightarrow a \approx (xc)b \Leftrightarrow a \approx x(cb) \Leftrightarrow x \approx a/cb,$$

therefore $(a/b)/c = a/(cb)$. Similar is the proof of the left mixed associativity. Next, let $x \approx (b \backslash a)/c$. Then we have the sequence of implications:

$$x \approx (b \backslash a)/c \Leftrightarrow xc \approx b \backslash a \Leftrightarrow a \approx b(xc) \Leftrightarrow a \approx (bx)c \Leftrightarrow bx \approx a/c \Leftrightarrow x \approx b \backslash (a/c),$$

hence $(b \backslash a)/c = b \backslash (a/c)$. □

Corollary 1. *In an associative magma $(E, \cdot)$ the equalities*

$$\begin{array}{l} (A/B)/C = A/(CB) \\ C \backslash (B \backslash A) = (BC) \backslash A \\ (B \backslash A)/C = B \backslash (A/C) \end{array}$$

are valid for all nonvoid subsets A, B, C of E.

Proposition 4. *In an associative magma $(E, \cdot)$ it holds that*

$$b \approx (a/b)\backslash a \quad \text{and} \quad b \approx a/(b\backslash a)$$

for all $a, b \in E$.

Proof. Let $x \approx a/b$. Then, $a \approx xb$, therefore $a \approx (a/b)b$, hence $b \approx (a/b)\backslash a$. The proof of the second relation is similar. □

Corollary 2. *In an associative magma $(E, \cdot)$ the inclusions*

$$B \subseteq (A/B)\backslash A \quad \text{and} \quad B \subseteq A/(B\backslash A)$$

are valid for all nonvoid subsets A, B of E.

Definition 6. *A hypercomposition $(x, y) \to x \perp y$ on a set E is called weakly associative if for all elements x, y, z in E,*

$$(x \perp y) \perp z \approx x \perp (y \perp z).$$

A magma whose law of synthesis is weakly associative is called a weakly associative magma.

Example 11. *Suppose that the law of synthesis on a magma E, with more than three elements, is the following one:*

$$\begin{aligned} x \perp y &= E \cdot \cdot \{x, y\} & \text{for every } x, y \in E \text{ with } x \neq y \\ x \perp x &= x & \text{for all } x \in E \end{aligned}$$

Then E is not an associative magma, because

$$(x \perp x) \perp y = x \perp y = E \cdot \cdot \{x, y\}, \text{ while, } x \perp (x \perp y) = x \perp [E \cdot \cdot \{x, y\}] = E \cdot \cdot \{x\}.$$

However, E is a weakly associative magma, since

$$(x \perp y) \perp z \cap x \perp (y \perp z) \neq \emptyset, \text{ for all } x, y, z \text{ in } E.$$

Proposition 5. *The result of the hypercomposition in a weakly associative magma E is always a nonempty set.*

Proof. Suppose that $x \perp y = \emptyset$ for some $x, y \in E$. Then, $(x \perp y) \perp z = \emptyset$ for any $z \in E$. Therefore, $(x \perp y) \perp z \cap x \perp (y \perp z) = \emptyset$, which is absurd. Hence, $x \perp y$ is nonvoid. □

Definition 7. *A hypercomposition $(x, y) \to x \perp y$ on a set E is called weakly commutative if, for all elements x, y in E,*

$$x \perp y \approx y \perp x.$$

A magma whose law of synthesis is weakly commutative is called a weakly commutative magma.

Example 12. *Let $(E, \cdot)$ be a magma and let $\odot$ be the extensive enlargement of the law of synthesis. Then $(E, \odot)$ is a weakly commutative magma, since*

$$\{x, y\} \subseteq x \odot y \cap y \odot x$$

for all elements x, y in E.

Two statements of magma theory are dual statements if each one results from the other by interchanging the order of the law of synthesis $\perp$, that is, interchanging $a \perp b$ with $b \perp a$. Observe that the axiom of associativity is self-dual. The two induced laws of synthesis $\perp/$

and $\setminus_\perp$ have dual definitions, hence they must be interchanged in a construction of a dual statement. Therefore, the following principle of duality holds:

> Given a theorem, the dual statement, which results from the interchange of the order of the synthesis $\perp$ (and necessarily interchange $_\perp/$ and $\setminus_\perp$), is also a theorem.

A direct consequence of the principle of duality is the following proposition:

Proposition 6. *The opposite law of an associative law of synthesis is associative.*

Proposition 7. *The extensive enlargement of an associative law of synthesis is associative.*

Proof. Let $\odot$ denote the extensive enlargement of an associative law of synthesis, which is written multiplicatively, then:

$$(a \odot b) \odot c = [\{a,b\} \cup ab] \odot c = a \odot c \cup b \odot c \cup (ab) \odot c =$$
$$= \{a,c\} \cup ac \cup \{b,c\} \cup bc \cup ab \cup \{c\} \cup (ab)c =$$
$$= \{a,b,c\} \cup ac \cup bc \cup ab \cup a(bc) = a \odot (b \odot c) \quad \square$$

Proposition 8. *Let $(E, \perp)$ be an associative magma. If an element a of E commutes with the elements b and c of E, then it commutes with their synthesis as well.*

Proof. $a \perp (b \perp c) = (a \perp b) \perp c = (b \perp a) \perp c = b \perp (a \perp c) = b \perp (c \perp a) = (b \perp c) \perp a.$ $\square$

Definition 8. *A law of synthesis $(x, y) \to x \perp y$ on a set E is called reproductive if the equality*

$$x \perp E = E \perp x = E$$

is valid for all elements x in E. A magma whose law of synthesis is reproductive is called a reproductive magma.

Example 13. *The following laws of synthesis in $E = \{1, 2, 3\}$ are reproductive:*

$\perp$	1	2	3
1	1	2	3
2	2	3	1
3	3	1	2

$\perp$	1	2	3
1	{1,3}	{3}	{2}
2	{2}	{1,2,3}	{2}
3	{2}	{1}	{1,2,3}

3. Groups and Hypergroups

Definition 9. *An associative and reproductive magma is called a group if the law of synthesis on the magma is a composition, while it is called a hypergroup if the law of synthesis is a hypercomposition.*

We observe that the two algebraic structures we just defined are equipped with the same operating rules, that is, the axioms of associativity and reproductivity, but with different synthesizing laws for their elements. Of course, if the singletons are identified with their members, then the groups are special cases of the hypergroups. The term *group* was first used in a technical sense by the French mathematician Évariste Galois (25 October 1811–31 May 1832). His brilliant paper "Une Mémoire sur les Conditions de Résolubilité des Equations par Radicaux" was submitted in January 1831, for the third time, to the French Academy of Sciences. It was followed by his famous letter describing his discoveries, written the night before he was killed in a duel at the age of 21, in May 1832. Galois's manuscripts, with annotations by Joseph Liouville, were published in 1846 in the *Journal de Mathématiques Pures et Appliquées*. Évariste Galois discovered the notion of normal subgroups and realized their importance. Another young French mathematician,

Frédéric Marty (23 June 1911–14 June 1940), who was born a hundred years later, while working on cosets determined by not normal subgroups, introduced the *hypergroup*. His mathematical heritage on hypergroups was only three papers [1–3], as he was killed at the age of 29 during World War II, in the Gulf of Finland, while serving in the French Air Force as a lieutenant. For the record, Frédéric Marty's father, Joseph Marty, was also a mathematician who was killed in 1914, when Frédéric was only 3 years old, during World War I.

Example 14. *Let $E = \{a\}$. There is only one associative and reproductive magma, whose law of synthesis is defined by the following table:*

$\perp$	a
a	a

Example 15. *Let $E = \{a,b\}$. There is only one group with two elements. The following table shows this group.*

$\perp$	a	b
a	a	b
b	b	a

On the other hand, there exist eight nonisomorphic hypergroups with two elements. The following tables display these hypergroups. Observe that the first one is the above group.

$\perp$	a	b
a	a	b
b	b	a

$\perp$	a	b
a	a	b
b	b	$\{a,b\}$

$\perp$	a	b
a	a	b
b	$\{a,b\}$	$\{a,b\}$

$\perp$	a	b
a	a	$\{a,b\}$
b	b	$\{a,b\}$

$\perp$	a	b
a	a	$\{a,b\}$
b	$\{a,b\}$	b

$\perp$	a	b
a	a	$\{a,b\}$
b	$\{a,b\}$	$\{a,b\}$

$\perp$	a	b
a	b	$\{a,b\}$
b	$\{a,b\}$	$\{a,b\}$

$\perp$	a	b
a	$\{a,b\}$	$\{a,b\}$
b	$\{a,b\}$	$\{a,b\}$

Example 16. *Let $E = \{a,b,c\}$. There is only one group with three elements. The following table displays this group.*

$\perp$	a	b	c
a	a	b	c
b	b	c	a
c	c	a	b

However, there are 3999 nonisomorphic hypergroups with three elements; [79] presents a software that can calculate and print all these 3999 hypergroups.

Example 17. *A magma endowed with the b-hypercomposition is a hypergroup, called b-hypergroup.*

Example 18. *A magma endowed with the total hypercomposition is a hypergroup, called total hypergroup.*

Theorem 1. *Let $(G, \cdot)$ be a group or a hypergroup. If "$\odot$" is the extensive enlargement of "$\cdot$", then $(G, \odot)$ is a hypergroup.*

Proof. By Proposition 7, the extensive enlargement of $\cdot$ is associative. Moreover:

$$x \odot H = H \odot x = \{x\} \cup H = H$$

and the theorem holds. $\square$

4. The Reproductive Axiom in Groups

Recall the definition of the group that is mentioned in the previous section:

Definition 10. (FIRST DEFINITION OF A GROUP). *An associative and reproductive magma is called a group, if the law of synthesis on the magma is a composition.*

In other words, a group is a set of elements equipped with a law of composition that is associative and reproductive. The next theorems give some important properties of the group structure. In this section, unless otherwise indicated, the law of synthesis is a composition that will be written multiplicatively, and G will denote a multiplicative group.

Theorem 2. *Let G be a group. Then:*
i. *There exists an element $e \in G$ such that $ea = a = ae$, for all $a \in G$.*
ii. *For each element $a \in G$ there exists an element $a' \in G$ such that $a'a = e = aa'$.*

Proof. (i) Let $x \in G$. By reproductive axiom, $x \in xG$. Consequently, there exists $e \in G$, for which $xe = x$. Next, let y be an arbitrary element in G. Since the composition is reproductive $y \in Gx$, therefore, there exists $z \in G$ such that $y = zx$. Consequently, $ye = (zx)e = z(xe) = zx = y$. In an analogous way, there exists an element $\hat{e}$ such that $\hat{e}y = y$ for all $y \in G$. Then the equality $e = \hat{e}e = \hat{e}$ is valid. Therefore, there exists $e \in G$ such that $ea = a = ae$, for all $a \in G$.

(ii) Let $a \in G$. By reproductive axiom, $e \in a G$. Thus, there exists $a' \in G$, such that $e = aa'$. Also by reproductive axiom, $e \in Ga$. Therefore, there exists $a'' \in G$, such that $e = a''a$. However:

$$a' = ea' = (a''a)a' = a''(aa') = a''e = a''.$$

Hence a' and a'' coincide. Therefore $aa' = e = a'a$. $\square$

The element e is called the *neutral element* of G or the *identity* of G. Moreover, a' is called the *symmetric* of a in G. If the composition is written multiplicatively, then e is called the *unit* of G and it is denoted by 1. Furthermore, a' is called the *inverse* of a and it is denoted by a^{-1}. If the composition is written additively, then e is called the *zero* of G and it is denoted by 0. Also a' is called the *opposite* of a and it is denoted by $-a$.

Corollary 3. *For each $a, b \in G$,*

$$a/b = ab^{-1} \text{ and } b\backslash a = b^{-1}a.$$

Corollary 4. *Let e be the identity of a group G. Then:*

$$e/b = b^{-1} \text{ and } b\backslash e = b^{-1}$$

for all $b \in G$.

Corollary 5. *Let e be the identity of a group G. Then:*

$$b/b = e = b\backslash b \text{ and } (b/b)/b = b^{-1} = b\backslash(b\backslash b)$$

for all $b \in G$.

Theorem 3. *Let G be an associative magma whose law of synthesis is a composition. Then G is a group if the following two postulates are fulfilled:*
i. *There exists an element $e \in G$ such that $ea = a = ae$, for all $a \in G$.*
ii. *For each element $a \in G$, there exists an element $a' \in G$ such that $a'a = e = aa'$.*

Proof. It must be proved that the reproductive axiom is valid for G. Since $aG \subseteq G$, for all $a \in G$, it has to be proved that $G \subseteq aG$. Suppose that $x \in G$. Then:

$$x = ex = (aa')x = a(a'x).$$

The product $a'x$ is an element of G, thus $x \in aG$. Hence $G \subseteq aG$, therefore $aG = G$. Similarly, $Ga = G$. □

Theorems 2 and 3 lead to another definition of the group.

Definition 11. (SECOND DEFINITION OF A GROUP). *An associative magma G in which the law of synthesis is a composition is called a group if:*
i. *There exists an element $e \in G$ such that $ea = a = ae$, for all $a \in G$.*
ii. *For each element $a \in G$, there exists an element $a' \in G$ such that $a'a = e = aa'$.*

Yet, one-half of the above definition's postulates (i) and (ii) can be omitted by the following dual propositions:

Proposition 9. *The postulates (i) and (ii) of Definition 11 are equivalent to:*
i*. *There exists an element $e \in G$ with $ea = a$, for all $a \in G$.*
ii*. *For each element $a \in G$, there exists an element $a' \in G$ such that $a'a = e$.*

Proposition 10. *The postulates (i) and (ii) of Definition 11 are equivalent to:*
i**. *There exists an element $e \in G$ with $ae = a$, for all $a \in G$.*
ii**. *For each element $a \in G$, there exists an element $a' \in G$ such that $aa' = e$.*

We quote the following well known and important Theorems 4, 5, 6, 7, 8, and 9, which can be easily proved with the use of the second definition of the group, because we want to show in the next sections that similar theorems can be proved only in very specific types of hypergroups.

Theorem 4. *The neutral element of a group is unique.*

Theorem 5. *The symmetric of each element of a group is unique.*

Theorem 6. *The symmetric of the neutral element is the neutral element itself.*

Theorem 7. *For each $a \in G$,*
$$\left(a^{-1}\right)^{-1} = a.$$

Theorem 8. *For each $a, b \in G$,*
$$(ab)^{-1} = b^{-1}a^{-1}.$$

Theorem 9. *The cancellation law:*

$$ac = bc \text{ implies } a = b$$
$$ca = cb \text{ implies } a = b$$

Theorem 10. *A finite associative magma G is a group if the law of synthesis on the magma is a composition in which the cancellation law holds.*

Proof. Let $G = \{a_1, \cdots, a_n\}$ and let a be an arbitrary element in G. Then:

$$aG = \{aa_1, \cdots, aa_n\} \subseteq G.$$

From the cancellation law, it follows that the n elements of aG are all distinct. Therefore aG and G have the same cardinality. Consequently, since G is finite, $aG = G$ is valid. Duality gives $Ga = G$ and so the theorem. □

Theorem 11. *An associative magma G whose law of synthesis is a composition is a group if and only if the inverse laws are compositions.*

Proof. Let G be a group. We will prove that b/a and $a\backslash b$ are elements of G, for all pairs of elements a, b of G. By reproduction, $Ga = G$ for all $a \in G$. Consequently, for every $b \in G$ there exists $x \in G$, such that $b = xa$. Thus, $x = b/a$. Dually, $a\backslash b$ is an element of G. Conversely now: suppose that the right quotient b/a exists for all pairs of elements a, b of G. Thus, for each $a, b \in G$, there is an element x in G such that $b = xa$. Therefore $G \subseteq Ga$ for all $a \in G$. Next, since $Ga \subseteq G$, for all $a \in G$, it follows that $Ga = G$, for all $a \in G$. In a similar way, the existence of the left quotient $a\backslash b$ for all pairs of elements a, b of G, yields $aG = G$, for all $a \in G$. Thus, the reproductive law is valid and so G is a group. □

Corollary 6. *An associative magma G is a group if and only if the equations*

$$xa = b \text{ and } ay = b$$

are solvable for all pairs of elements a, b of G.

Proof. By Theorem 11, G is a group if and only if b/a and $a\backslash b$ are elements of G, for all a, b in G, which equivalently implies that the equations $xa = b$ and $ay = b$, respectively, are solvable for all pairs of elements a, b of G. □

Having proved the above, another definition can be given for the group.

Definition 12. (THIRD DEFINITION OF A GROUP). *An associative magma G in which the law of synthesis is a composition is called a group if the right quotient b/a and the left quotient $a\backslash b$ result in a single element of G, for all $a, b \in G$.*

Or else:

An associative magma G in which the law of synthesis is a composition is called a group if the equations

$$xa = b \text{ and } ay = b$$

are solvable for all pairs of elements a, b of G.

Definition 13. *A group that has the additional property that for every pair of its elements*

$$a\, b = b\, a$$

is called an Abelian (after N.H. Abel, 1802-29) or commutative group.

In abstract algebra, we also consider structures that do not satisfy all the axioms of a group.

Definition 14. *An associative magma in which the law of synthesis is a composition is called a semigroup. A semigroup with an identity is called a monoid.*

Definition 15. *A reproductive magma in which the law of synthesis is a composition is called a quasigroup. A quasigroup with an identity is called loop.*

Definition 16. *A magma that is the union of a group with an absorbing element is called almost-group.*

5. Fundamental Properties of Hypergroups.

Recall the earlier mentioned definition for the hypergroup:

Definition 17. (FIRST DEFINITION OF A HYPERGROUP). *An associative and reproductive magma is called a hypergroup if the law of synthesis on the magma is a hypercomposition.*

In this section, unless otherwise indicated, the law of synthesis is a hypercomposition that will be written multiplicatively, and H will denote a multiplicative hypergroup.

Theorem 12. $ab \neq \varnothing$ *is valid for all the elements a, b of a hypergroup H.*

Proof. Suppose that $ab = \varnothing$ for some $a, b \in H$. By the reproductive axiom, $aH = H$ and $bH = H$. Hence:
$$H = aH = a(bH) = (ab)H = \varnothing H = \varnothing$$
which is absurd. □

Theorem 13. $a/b \neq \varnothing$ *and* $b \backslash a \neq \varnothing$ *is valid for all the elements a, b of a reproductive magma E.*

Proof. By the reproductive axiom, $Eb = E$ for all $b \in E$. Hence, for every $a \in E$ there exists $x \in E$, such that $a \in xb$. Thus, $x \in a/b$ and therefore $a/b \neq \varnothing$. Dually, $b \backslash a \neq \varnothing$. □

Theorem 14. *If $a/b \neq \varnothing$ and $b \backslash a \neq \varnothing$ for all pairs of elements a, b of a magma E, then E is a reproductive magma.*

Proof. Suppose that $x/a \neq \varnothing$ for all $a, x \in E$. Thus, there exists $y \in E$, such that $x \in ya$. Therefore $x \in Ea$ for all $x \in E$, and so $E \subseteq Ea$. Next, since $Ea \subseteq E$ for all $a \in E$, it follows that $E = Ea$. By duality, $E = aE$. □

Following Theorem 14, another definition of the hypergroup can be given:

Definition 18. (SECOND DEFINITION OF A HYPERGROUP). *An associative magma is called a hypergroup if the law of synthesis is a hypercomposition and the result of each one of the two inverse hypercompositions is nonvoid for all pairs of elements of the magma.*

Theorem 15. *In a hypergroup H, the equalities*
i. $H = H/a = a/H$ *and*
ii. $H = a \backslash H = H \backslash a$
are valid for all a in H.

Proof. (i) By Theorem 12, the result of the hypercomposition in H is always a nonempty set. Thus, for every $x \in H$ there exists $y \in H$, such that $y \in xa$, which implies that $x \in y/a$. Hence $H \subseteq H/a$. Moreover, $H/a \subseteq H$. Therefore $H = H/a$. Next, let $x \in H$. Since

$H = xH$, there exists $y \in H$, such that $a \in xy$, which implies that $x \in a/y$. Hence $H \subseteq a/H$. Moreover, $a/H \subseteq H$. Therefore $H = a/H$. (ii) follows by duality. □

Likewise to the groups, certain axioms were removed from the hypergroup, thus revealing the following weaker structures.

Definition 19. *A magma in which the law of synthesis is a hypercomposition is called a hypergoupoid if for every two of its elements a, b it holds that $ab \neq \emptyset$, otherwise it is called a partial hypergroupoid.*

In the case of finite hypergroupoids, the ratio of the number of hypergroups over the number of hypergroupoids is exceptionally small. For instance, we come across one 3-element hypergroup in every 1740 hypergroupoids of three elements [79].

Definition 20. *An associative hypergoupoid is called a semihypergroup, while a reproductive hypergoupoid is called a quasihypergroup.*

Definition 21. *The magma which is the union of a hypergroup with an absorbing element is called almost-hypergroup.*

Definition 22. *A reproductive magma in which the law of synthesis is weakly associative is called H_V-group [80].*

Because of Proposition 5, the result of the hypercomposition in an H_V-group is always nonvoid.

Many papers have been written on the construction of examples of the above algebraic structures. Among them are the papers by P. Corsini [81,82], P. Corsini and V. Leoreanu [83], B. Davvaz and V. Leoreanu [84], I. Rosenberg [85], I. Cristea et al. [86–90], M. De Salvo and G. Lo Faro [91,92], C. Pelea and I. Purdea [93,94], C.G. Massouros and C.G. Tsitouras [95,96], and S. Hoskova-Mayarova and A. Maturo [70], in which hypercompositional structures, defined in terms of binary relations, are presented and studied.

It is worth mentioning that a generalization of the hypergroup is the fuzzy hypergroup, which was studied by a multitude of researchers [97–127]. An extensive bibliography on this subject can be found in [124]. It can be proved that similar fundamental properties as the aforementioned ones are valid in the fuzzy hypergroups as well [125,126]. For instance, $a \circ b \neq 0_H$ is valid for any pair of elements a, b in a fuzzy hypergroup $(H, \circ)$ [125]. Generalizations of the fuzzy hypergroups are the mimic fuzzy hypergroups [125,126] and the fuzzy multihypergroups [127].

6. Types of Hypergroups

The hypergroup being a very general structure, was equipped with further axioms, which are more or less powerful and lead to a significant number of special hypergroups. One such important axiom is the *transposition axiom*. Initially this axiom was introduced by W. Prenowitz in a commutative hypergroup, all the elements of which also satisfy the properties $aa = a$ and $a/a = a$. He named this hypergroup *join space* and used it in the study of geometry [19–24]. The transposition axiom in a commutative hypergroup is:

$$a/b \cap c/d \neq \emptyset \quad \text{implies} \quad ad \cap bc \neq \emptyset$$

A commutative hypergroup that satisfies the transposition axiom is called a *join hypergroup*. Later on, J. Jantosciak generalized the transposition axiom in an arbitrary hypergroup H as follows:

$$b \backslash a \cap c/d \neq \emptyset \quad \text{implies} \quad ad \cap bc \neq \emptyset \quad \text{for all} \quad a, b, c, d \in H.$$

A hypergroup equipped with the transposition axiom is called *transposition hypergroup* [128].

These hypergroups attracted the interest of a large number of researchers, including I. Cristea et al. [102,107–112], P. Corsini [100–103,129,130], V. Leoreanu-Fortea [101,103–105,131,132], I. Rosenberg [85,132], S. Hoskova-Mayerova, [133–137], J. Chvalina [135–138], P. Rackova [135,136], Ch.G. Massouros [139–150], G.G. Massouros [144–152], J. Nieminen [153,154], A. Kehagias [115], R. Ameri [117,118,155], M. M Zahedi [117], and G. Chowdhury [123].

Proposition 11. [150] *The following are true in any transposition hypergroup:*

i. $a(b/c) \subseteq ab/c$ and $(c\backslash b)a \subseteq c\backslash ba$,
ii. $a/(c/b) \subseteq ab/c$ and $(b\backslash c)\backslash a \subseteq c\backslash ba$,
iii. $(b\backslash a)(c/d) \subseteq (b\backslash ac)/d = b\backslash(ac/d)$,
iv. $(b\backslash a)/(c/d) \subseteq (b\backslash ad)/c = b\backslash(ad/c)$,
v. $(b\backslash a)\backslash(c/d) \subseteq (a\backslash bc)/d = a\backslash(bc/d)$.

The hypergroups are much more general algebraic structures than the groups, to the extent that a theorem similar to Theorem 2 cannot be proved for the hypergroups. In fact, a hypergroup does not necessarily have an identity element. Moreover, in hypergroups there exist different types of identities [149,150,156]. In general, an element e of a hypergroup H is called *right identity*, if $x \in x \cdot e$ for all x in H. If $x \in e \cdot x$ for all x in H, then x is called *left identity*, while x is called *identity* if it is both a right and a left identity; i.e., if $x \in xe \cap ex$ for all $x \in H$. If the equality $e = ee$ is valid for an identity e, then e is called *idempotent identity*. A hypergroup H is called *semiregular* if every $x \in H$ has at least one right and one left identity. An identity is called *scalar* if $a = ae = ea$ for all $a \in H$, while it is called *strong* if $ae = ea \subseteq \{e, a\}$ for all $a \in H$. More generally, an element $s \in H$ is called *scalar* if the result of the hypercomposition of this element with any element in H is a singleton; that is, if $sa \in H$ and $as \in H$ for all $a \in H$. If only the first membership relation is valid, then s is called *left scalar*, while if only the second relation is valid, then s is called *right scalar*. When a scalar identity exists in H, then it is unique but the strong identity is not necessarily unique. Both scalar and strong identities are idempotent identities.

Remark 1. *If a hypercomposition has a scalar identity e, then it is neither open nor closed (containing) because $e \notin ex$ and $x \approx ex$.*

Proposition 12. *If e is a strong identity in H and $x \neq e$, then $x/e = e\backslash x = x$.*

Proof. Let $t \in x/e$. Then $x \in te \subseteq \{t, e\}$. Since $x \neq e$, it follows that $t = x$. Thus $x/e = x$. Similarly, it can be proven that $e\backslash x = x$. □

Corollary 7. *If e is a strong identity in H and X is a nonempty subset of H, not containing e, then $X/e = e\backslash X = X$.*

Proposition 13. *If e is a scalar identity in H, then $x/e = e\backslash x = x$.*

Corollary 8. *If X is a non-empty subset of H and if e is a scalar identity in H, then $X/e = e\backslash X = X$.*

Theorem 16. i. *If a hypergroup H contains a scalar element, then it contains a scalar identity e as well.*
ii. *The set U of the scalar elements of a hypergroup H is a group.*

Proof. (i) Let s be a scalar element. Then, per reproductivity, there exists an element $e \in H$ such that $se = s$. Also, because of the reproductive axiom, any element $y \in H$ can be written as $y = xs$, $x \in H$. Hence $ye = (xs)e = x(se) = xs = y$. Similarly, $ey = y$.

(ii) Let $s \in U$. Then, per reproductivity, there exists an element $s' \in H$ such that $ss' = e$. If $y \in H$ then, because of the reproductive axiom, $y = xs$, $x \in H$. Hence $ys' = (xs)s' = x(ss') = xe = x$, and therefore s' is a right scalar element. Similarly, there exists a left scalar element s'' such that $s''s = e$. But s' is equal to s'' since $s'' = s''e = s''(ss') = \left(s''s\right)s' = es' = s'$. Consequently, U is a subgroup of the hypergroup H. □

The group of the scalars was named the *nucleus* of H by Wall [11].

An element x' is called *right e-symmetric* of x, or *right e-inverse* in the multiplicative case, if there exists a right identity $e \neq x'$ such that $e \in x \cdot x'$. The definition of the *left e-symmetric* or *left e-inverse* is analogous to the above, while x' is called the *e-inverse* or *e-symmetric* of x, if it is both right and left inverse with regard to the same identity e. If e is an identity in a multiplicative hypergroup H, then the set of the left inverses of $x \in H$, with regard to e, will be denoted by $S_{el}(x)$, while $S_{er}(x)$ will denote the set of the right inverses of $x \in H$ with regard to e. The intersection $S_{el}(x) \cap S_{er}(x)$ will be denoted by $S_e(x)$. A semiregular hypergroup H is called *regular* if it has at least one identity e and if each element has at least one right and one left e-inverse. H is called *strictly e-regular* if for the identity e the equality $S_{el}(x) = S_{er}(x)$ is valid for all $x \in H$. In a strictly e-regular hypergroup, the inverses of x are denoted by $S_e(x)$ and, when there is no likelihood of confusion, e can be omitted, and the notation $S(x)$ is used for the inverses of x. We say that H has *semistrict e-regular structure* if $S_{el}(x) \approx S_{er}(x)$ is valid for any $x \in H$. Obviously, in the commutative hypergroups there exist only the strict e-regular structures.

Proposition 14. *If e is an identity in a hypergroup H, then $S_{el}(x) = (e/x) \cdot \cdot \{e\}$ and $S_{er}(x) = (x \backslash e) \cdot \cdot \{e\}$.*

Corollary 9. *If $S_{el}(x) \cap S_{er}(x) \neq \varnothing$, then $x \backslash e \cap e/x \neq \varnothing$.*

Definition 23. *A regular hypergroup is called reversible if it satisfies the following conditions:*
i. $z \in xy \Rightarrow x \in zy'$, for some $y' \in S(y)$,
ii. $z \in xy \Rightarrow y \in x'z$, for some $x' \in S(x)$.

The enrichment of a hypergroup with an identity creates different types of hypergroups, depending on the type of the identity.

Proposition 15. *If H is a transposition hypergroup with a scalar identity e, then, for any x in H, the quotients e/x and $x \backslash e$ are singletons and equal to each other.*

Proof. Obviously $e/e = e \backslash e = e$. Let $x \neq e$. Because of reproduction, there exist x' and x'', such that $e \in x'x$ and $e \in xx''$. Thus $x \in x' \backslash e$ and $x \in e/x''$. Hence $x' \backslash e \approx e/x''$. Therefore, because of transposition, $ex'' \approx x'e$ is valid. Since e is a scalar identity, the following is true: $x'' = ex''$ and $x'e = x'$. Thus $x' = x''$. However, $x' \in e/x$ and $x'' \in x \backslash e$. Therefore e/x and $x \backslash e$ are equal, and since this argument applies to any $y', y'' \in H$, such that $e \in y'x$ and $e \in xy''$, it follows that e/x and $x \backslash e$ are singletons. □

Definition 24. *A transposition hypergroup that has a scalar identity e is called quasicanonical hypergroup [157,158] or polygroup [159–161].*

The connection of quasicanonical hypergroups with color schemes, relation algebras, and finite permutation groups, as well as with weak cogroups, produces a lot of examples of quasicanonical hypergroups (e.g., see [160]). In [162], quasicanonical hypergroups appear as Pasch geometries. In a Pasch geometry (A,Δ,e), A becomes a quasicanonical hypergroup with scalar identity e and $a^{-1} = a^{\#}$, when $ab = \{x | (a,b,x^{\#}) \in \Delta\}$ (see also [128]). The following example from [158] shows the structural relation of groups with the quasicanonical hypergroups.

Example 19. Let $(G, \cdot)$ be a group and e its identity. The following hypercomposition is defined on G:
$$x \circ y = \{x, y, x \cdot y\}, \quad \text{if } y \neq x^{-1}, x, y \neq e$$
$$x \circ x^{-1} = x^{-1} \circ x = G, \quad \text{if } x \neq e$$
$$x \circ e = e \circ x = x, \quad \text{for all } x \in G$$

Then $(G, \circ)$ becomes a quasicanonical hypergroup. Note that this construction can be used to produce new quasicanonical hypergroups from other quasicanonical hypergroups.

In the quasicanonical hypergroups, there exist properties analogous to (i) and (ii) of Theorem 2, which are valid in the groups:

Theorem 17. [128,141] *If $(Q, \cdot, e)$ is a quasicanonical hypergroup, then:*
i. *For each $x \in Q$ there exists one and only one $x' \in Q$ such that $e \in xx' = x'x$.*
ii. $z \in xy \Rightarrow x \in zy' \Rightarrow y \in x'z$.

Corollary 10. *A quasicanonical hypergroup is a reversible hypergroup.*

The inverse of Theorem 17 is also true:

Theorem 18. [128,141] *If a hypergroup Q has a scalar identity e and:*
i. *For each $x \in Q$ there exists one and only one $x' \in Q$ such that $e \in xx' = x'x$*
ii. $z \in xy \Rightarrow x \in zy' \Rightarrow y \in x'z$
then the transposition axiom is valid in Q.

When the hypercomposition is written multiplicatively, x' is denoted by x^{-1} and it is called the *inverse* of x, while, if the hypercomposition is written additively, the identity is denoted by 0 and the unique element x' is called *opposite* or *negative*, and it is denoted by $-x$.

Proposition 16. *In a quasicanonical hypergroup Q,*
i. $x^{-1} = e/x = x \backslash e$ and $x = e/x^{-1} = x^{-1} \backslash e$ for all $x \in Q$.
ii. $x/y = xy^{-1}$ and $y \backslash x = y^{-1}x$ for all $x, y \in Q$.

Proof. (i) is a direct consequence of Proposition 15. Next, for (ii), applying (i), Proposition 11, and Corollary 8, we have:

$$x/y = x/\left(e/y^{-1}\right) \subseteq xy^{-1}/e = xy^{-1}$$

and

$$xy^{-1} = x(e/y) \subseteq xe/y = x/y.$$

Therefore, the first equality is proved. The second one arises from duality. □

The aforementioned Theorems 4, 5, 6, 7, and 8, which are valid in the groups, are also valid in the quasicanonical hypergroups. The cancellation law (Theorem 9), though, is valid in the quasicanonical hypergroups as follows:

$$ac = bc \text{ implies } b^{-1}a \cap cc^{-1} \neq \emptyset$$
$$ca = cb \text{ implies } ba^{-1} \cap cc^{-1} \neq \emptyset$$

More generally the following theorem is valid:

Theorem 19. *If Q is a quasicanonical hypergroup, then $ab \approx cd$ implies that $bd^{-1} \approx a^{-1}c$ and $c^{-1}a \approx db^{-1}$.*

Proof. As it is mentioned above, Theorem 8 is valid in the quasicanonical hypergroups as well. Indeed, the equality $(ab)^{-1} = b^{-1}a^{-1}$ follows from the sequence of implications:

$$x \in ab \,;\, a \in xb^{-1} \,;\, aa^{-1} \subseteq xb^{-1}a^{-1} \,;\, e \in xb^{-1}a^{-1} \,;\, x^{-1} \in b^{-1}a^{-1}.$$

Next, let $x \in ab \cap cd$. Then $xx^{-1} \in ab(cd)^{-1} = abd^{-1}c^{-1}$. Hence $e \in abd^{-1}c^{-1}$, thus $c \in abd^{-1}$. Therefore, there exists $y \in bd^{-1}$ such that $c \in ay$. Reversibility implies that $y \in a^{-1}c$. Consequently, $bd^{-1} \approx a^{-1}c$. By duality, $c^{-1}a \approx db^{-1}$. □

A commutative quasicanonical hypergroup is called *canonical hypergroup*. The canonical hypergroup owes its name to J. Mittas [163,164] while it was first used by M. Krasner for the construction of the hyperfield, which is a hypercompositional structure that he introduced in order to define a certain approximation of complete valued fields by sequences of such fields [53]. The hyperfields, which were constructed afterward, contain interesting examples of canonical hypergroups [165–170]. An example of such a canonical hypergroup is presented in the following construction by J. Mittas [171].

Example 20. *Let E be a totally ordered set and 0 its minimum element. The following hypercomposition is defined on E:*

$$x + y = \begin{cases} \max\{x, y\} & \text{if } x \neq y \\ \{z \in E \mid z \leq x\} & \text{if } x = y \end{cases}$$

Then (E,+) is a canonical hypergroup.

We cite the above example because the hyperfield, which J. Mittas constructed based on this canonical hypergroup, is now called a tropical hyperfield (see, e.g., [54–58]) and it is used in the development of the tropical geometry. Example 19 also gives a canonical hypergroup, when $(G, \cdot)$ is an abelian group. The hyperfield that is constructed based on this canonical hypergroup leads to open problems in both hyperfield and field theories [168,169].

J. Mittas studied the canonical hypergroup in depth [163,164,172–180]. Also motivated by the valuated hyperfield theory, he introduced ultrametric distances to the canonical hypergroups, thus defining the valuated and the hypervaluated canonical hypergroups. Next, he proved that the necessary and sufficient condition for a canonical hypergroup to be valuated or hypervaluated is the validity of certain additional properties of a purely algebraic type; that is, properties that are expressed without the intervention of the valuation or the hypervaluation, respectively. Thus, three special canonical hypergroups came into being:

(a) The *strongly canonical hypergroup*, which also satisfies the axioms:
S_1: If $x \in x + a$, then $x + a = x$, for all $x, a \in H$.
S_2: If $(x + y) \cap (z + w) \neq \emptyset$, then either $x + y \subseteq z + w$ or $z + w \subseteq x + y$.

(b) The *almost strongly canonical hypergroup*, which also satisfies the above axiom S_2 and the axiom:
AS: If $x \neq y$, then either $(x - x) \cap (y - x) = \emptyset$ or $(y - y) \cap (y - x) = \emptyset$.

(c) The *superiorly canonical*, which is a strongly canonical hypergroup that also satisfies the axioms:
S_3: If $z, w \in x + y$ and $x \neq y$, then $z - z = w - w$.
S_4: If $x \in z - z$ and $y \notin z - z$, then $x - x \subseteq y - y$.

J. Mittas has presented a very deep and extensive study on these hypergroups, with a great number and variety of results, among which we mention the following theorem [178]:

Theorem 20. *The necessary and sufficient condition for a canonical hypergroup to be hypervaluated is to be superiorly canonical.*

In all the above cases, the neutral element is scalar. Let us now consider hypergroups that are equipped with a strong identity.

Definition 25. *A fortified transposition hypergroup (FTH) is a transposition hypergroup T with a unique strong identity e, which satisfies the axiom:*

For every $x \in T \cdot \cdot \{e\}$ there exists one and only one element $x' \in T \cdot \cdot \{e\}$, the symmetric of x, such that: $e \in xx'$ and furthermore, for x' it holds that $e \in x'x$.

If the hypercomposition is commutative, the hypergroup is called a fortified join hypergroup (FJH).

The fortified join hypergroup was introduced for the study of languages and automata with tools of hypercompositional algebra [35–43].

It has been proved that every fortified transposition hypergroup consists of two types of elements, the *canonical* (c-elements) and the *attractive* (a-elements) [141,144,149]. An element x is called canonical if $ex = xe$ is the singleton $\{x\}$, while it is called attractive if $ex = xe = \{e, x\}$. A denotes the set of the attractive elements (a-elements) and C denotes the set of the canonical elements (c-elements). By convention, $e \in A$.

Proposition 17. *If $(T, \cdot, e)$ is a fortified transposition hypergroup, then the following are valid:*

i. *If $x \neq e$, then $x/e = e\backslash x = x$*
ii. *$e/e = e\backslash e = A$*
iii. *If $a \in A$, then $a/a = a\backslash a = A$*
iv. *If x, y are attractive elements, then $\{x, y\} \subseteq xy$*
v. *If x, y are attractive elements, then $x \in x/y$ and $x \in y\backslash x$*
vi. *If $a \in A$ and $c \in C$, then $ac = ca = c$*
vii. *If $a \in A$ and $a \neq e$, then $e/a = ea^{-1} = \{e, a\} = a^{-1}e = a\backslash e$*
viii. *If x, y are attractive elements, then $xy^{-1} = x/y \cup \{y^{-1}\}$ and $y^{-1}x = y\backslash x \cup \{y^{-1}\}$.*

A detailed and thorough study of the properties of the a-elements and c-elements is presented in [141,144,149]. Theorems 4, 5, 6, and 7, which are valid for the groups, are also valid for the fortified transposition hypergroups (FTH). Theorem 8, though, is not valid in the FTH, as generally $(aa^{-1})^{-1} \neq aa^{-1}$ (or $-(a - a) \neq a - a$ in the additive case). This led to the definition of two types of elements: those that satisfy the equality $(aa^{-1})^{-1} = aa^{-1}$ (or $-(a - a) = a - a$ in the additive case), which are called *normal* and for which Theorem 8 is valid, and the rest, which are called *abnormal* [141,144,149].

Theorem 21. *If $(Q, \cdot)$ is a quasicanonical hypergroup and "$\odot$" is the extensive enlargement of "$\cdot$", then $(Q, \odot)$ is a fortified transposition hypergroup consisting of attractive elements only.*

Proof. Per Proposition 7, the extensive enlargement of an associative law of synthesis is also associative. Next, for the proof of the transposition axiom, we observe that:

$$b\backslash_\odot a \cap c_\odot/d = [b\backslash a \cup \{a\}] \cap [c/d \cup \{c\}] \text{ and}$$
$$a \odot d \cap b \odot c = [ad \cup \{a, d\}] \cap [bc \cup \{b, c\}]$$

Therefore, if $b\backslash_\odot a \approx c_\odot/d$, we distinguish the cases:
if $b\backslash a \approx c/d$, then $ad \approx bc$, thus $a \odot d \approx b \odot c$;
if $c \in b\backslash a$, then $a \in bc$, thus $a \in a \odot d \cap b \odot c$, consequently $a \odot d \approx b \odot c$;
if $a \in c/d$, then $c \in ad$, thus $c \in a \odot d \cap b \odot c$, consequently $a \odot d \approx b \odot c$;
if $a = c$, then $a \odot d \approx b \odot c$.

19

Finally, if e is the neutral element of the quasicanonical hypergroup, then:
$$a \odot e = ae \cup \{a, e\} = \{a, e\}, \text{ for all } a \in Q. \ \square$$

Example 21. *Let H be a totally ordered set, dense and symmetric around a center denoted by $0 \in H$. The partition $H = H^- \cup \{0\} \cup H^+$ is defined with regard to this center and according to it, for every $x \in H^-$ and $y \in H^+$ it is $x < 0 < y$; and for every $x, y \in H$, $x \leq y \Rightarrow -y \leq -x$, where $-x$ is the symmetric of x with regard to 0. Then H, equipped with the hypercomposition:*

$$x + y = \{x, y\}, \text{ if } y \neq -x$$

and

$$x + (-x) = [0, |x|] \cup \{-|x|\}$$

becomes an FJH in which $x - x \neq -(x - x)$, for every $x \neq 0$. So, all the elements of $(H, +)$ are abnormal.

The fortified transposition hypergroup is closely related to the quasicanonical hypergroup as per the following structure theorem:

Theorem 22. [141] *A transposition hypergroup H containing a strong identity e is isomorphic to the expansion of the quasicanonical hypergroup $C \cup \{e\}$ by the transposition hypergroup A of all attractive elements with regard to the identity e.*

Definition 26. *A transposition polysymmetrical hypergroup (TPH) is a transposition hypergroup $(P, \cdot)$ with an idempotent identity e, which satisfies the axioms:*
i. $x \in xe = ex$
ii. *For every $x \in P \cdot \cdot \{e\}$ there exists $x' \in P \cdot \cdot \{e\}$, the symmetric of x, such that $e \in xx'$, and furthermore, x' satisfies $e \in x'x$.*

The set of the symmetric elements of x is denoted by $S(x)$. A commutative transposition polysymmetrical hypergroup is called a join polysymmetrical hypergroup (JPH).

A direct consequence of this definition is that for a nonidentity element x, $S(x) \cup \{e\} = x\backslash e = e/x$, when x is attractive, and $S(x) = x\backslash e = e/x$, when x is non attractive.

Example 22. *Let K be a field and G a subgroup of its multiplicative group. In K we define a hypercomposition "$\dotplus$" as follows:*

$$x \dotplus y = \{z \in K \mid z = xp + yq, \ p, q \in G\}$$

Then $(K, \dotplus)$ is a join polysymmetrical hypergroup having the 0 of K as its neutral element. Since $x \dotplus 0 = \{xq, \ q \in G\}$, the neutral element 0 is neither scalar nor strong. The symmetric set of an element x of K is $S(x) = \{-xp \mid p \in G\}$.

Example 23. *Let $(A_i, \cdot)$, $i \in I$, be a family of fortified transposition hypergroups that consist only of attractive elements, and suppose that the hypergroups A_i, $i \in I$ have a common identity e. Then $T = \bigcup_{i \in I} A_i$ becomes a transposition polysymmetrical hypergroup under the hypercomposition:*

$a \odot b = ab$ *if a, b are elements of the same hypergroup A_i*
$a \odot b = \{a, e, b\}$ *if $a \in A_i \cdot \cdot \{e\}$, $b \in A_j \cdot \cdot \{e\}$ and $i \neq j$*

Observe that e is a strong identity in T. Moreover, if $a \in A_i$, then $S(a) = (T \cdot \cdot A_i) \cup \{a'\}$ where a' is the inverse of a in A_i.

Example 24. *Let Δ_i, $i \in I$ be a family of totally ordered sets that have a common minimum element e. The set $\Delta = \bigcup_{i \in I} \Delta_i$ with hypercomposition:*

$$xy = \begin{cases} [\min\{x,y\}, \max\{x,y\}] & \text{if } x, y \in \Delta_i, i \in I \\ [e, x] \cup [e, y] & \text{if } x \in \Delta_i, y \in \Delta_j \text{ and } i \neq j, i, j \in I \end{cases}$$

becomes a JPH with neutral element e.

Proposition 18. *If x is an attractive element of a transposition polysymmetrical hypergroup, then $S(x)$ consists of attractive elements.*

Proof. Let $e \in ex$. Then $x \in e \backslash e$. Moreover, $x \in e/x'$ for any $x' \in S(x)$. Thus, $e/x' \cap e \backslash e \neq \varnothing$, which, by the transposition axiom, gives $ee \cap ex' \neq \varnothing$, or $e \in ex'$. Hence x' is attractive. □

Proposition 19. *The result of the hypercomposition of two attractive elements in a transposition polysymmetrical hypergroup consists of attractive elements only, while the result of the hypercomposition of an attractive element with a non attractive element consists of non attractive elements.*

Proposition 20. *If the identity in a transposition polysymmetrical hypergroup is strong, then $\{x, y\} \subseteq xy$ and $x \in x/y$, $x \in y \backslash x$, for any two attractive elements x, y.*

The algebraic properties of transposition polysymmetrical hypergroups are studied in [145,146].

Definition 27. *A quasicanonical polysymmetrical hypergroup is a hypergroup H with a unique scalar identity e, which satisfies the axioms:*
i. *For every $x \in H$ there exists at least one element $x' \in H$, called symmetric of x, such that $e \in xx'$ and $e \in x'x$.*
ii. *If $z \in xy$, there exist $x', y' \in S(x)$ such that $x \in zy'$ and $y \in x'z$.*

A commutative quasicanonical polysymmetrical hypergroup is called a canonical polysymmetrical hypergroup.

Example 25. *Let H be a set that is totally ordered and symmetric around a center, denoted by $0 \in H$. Then H, equipped with the hypercomposition:*

$$x + y = \begin{cases} y, & \text{if } |x| < |y| \text{ for every } x, y \in H^- \cup \{0\} \text{ or } x, y \in \{0\} \cup H^+ \\ [x, y], & \text{if } x \in H^- \text{ and } y \in H^+ \end{cases}$$

becomes a canonical polysymmetrical hypergroup. Suppose now that $x, y, a, b \in H^+$ and $x < y < a < b$. Then $x/y = a/b = H^-$. Thus, $x/y \cap a/b \neq \varnothing$. However, $x + b = \{b\}$ and $y + a = \{a\}$. Therefore $(x + b) \cap (y + a) = \varnothing$, and so the transposition axiom is not valid.

The canonical polysymmetrical hypergroup was introduced by J. Mittas [181]. In addition, J. Mittas and Ch. Massouros, while studying the applications of hypergroups in the linear spaces, defined the generalized canonical polysymmetrical hypergroup [31]. Moreover, J. Mittas, in his paper [174], motivated by an observation about algebraically closed fields, discovered a special type of completely regular polysymmetrical hypergroup, which, later on, C. Yatras called *M-polysymmetrical hypergroup*. C. Yatras, in a series of papers, studied this hypergroup and its properties in detail [182–184]. J. Mittas also defined the *generalized M-polysymmetrical hypergroups* that were studied by himself and by Ch. Massouros [185,186].

Definition 28. *A M-polysymmetrical hypergroup H is a commutative hypergroup with an idempotent identity e that also satisfies the axioms:*

i. $x \in xe = ex$,
ii. *For every* $x \in H \cdot \cdot \{e\}$ *there exists at least one element* $x' \in H \cdot \cdot \{e\}$, *such that* $e \in xx'$, *and* $e \in x'x$, *(symmetric of x)*,
iii. *If* $z \in xy$ *and* $x' \in S(x)$, $y' \in S(y)$, $z' \in S(z)$, *then* $z' \in x'y'$.

Proposition 21. [182,183] *Every M-polysymmetrical hypergroup is a join hypergroup.*

In addition to the aforementioned hypergroups, other hypergroups have been defined and studied. Among them, are the complete hypergroups and the complete semihypergroups [17,102,107,112,187–189], the 1-hypergroups [102,107,190], the hypergroups of type U [191–198], the hypergroups of type C [199,200], and the cambiste hypergroup [28].

7. Subhypergroups and Cosets

Decompositions and partitions play an important role in the study of algebraic structures. Undoubtedly, this study is of particular interest in the theory of hypercompositional algebra. It has recently been addressed in various papers from different perspectives (e.g., [188,201–204]). Moreover, in [33] it is proved that general decomposition theorems that are valid in hypergroups give as corollaries well-known decomposition theorems in convex sets. A far-reaching concept of abstract group theory is the decomposition of a group into cosets by its subgroups. The hypergroup, though, being a more general structure than that of the group, has various types of subhypergroups. In contrast to groups, where any subgroup decomposes the group into cosets, in the hypergroups, not all the subhypergroups can define such a partition. This section presents the subhypergroups that can create a partition in the hypergroup and the relevant partitions.

Definition 29. *A nonempty subset K of H is a semi-subhypergroup when it is stable under the hypercomposition, i.e., it has the property* $xy \subseteq K$ *for all* $x, y \in K$. *K is a subhypergroup of H if it satisfies the reproductive axiom, i.e., if the equality* $xK = Kx = K$ *is valid for all* $x \in K$.

7.1. Closed and Ultra-Closed Subhypergroups

From the above Definition 29, it derives that when K is a subhypergroup and $a, b \in K$, the relations $a \in bx$ and $a \in yb$ always have solutions in K. If, for any two elements a and b in K, all the solutions of the relation $a \in yb$ lie inside K, then K is called *right closed* in H. Similarly, K is *left closed* when all the solutions of the relation $a \in bx$ lie in K. K is *closed* when it is both right and left closed. Note that the concepts subhypergroup and closed subhypergroup are self-dual. A direct consequence of the definition of the closed subhypergroup is the proposition:

Proposition 22. *The nonvoid intersection of two closed subhypergroups is a closed subhypergroup.*

The relevant property is not valid for every subhypergroup, since, although the nonvoid intersection of two subhypergroups is stable under the hypercomposition, the validity of the reproductive axiom fails. This was one of the reasons that led, from the very beginning of the hypergroup theory, to the consideration of more special types of subhypergroups, one of which is the above defined closed subhypergroup (e.g., see [5,13]). An equivalent definition of the closed hypergroup is the following one:

Definition 30. [205,206] *A subhypergroup K of H is called right closed if K is stable under the right division, i.e., if* $a/b \subseteq K$ *for all* $a, b \in K$. *K is called left closed if K is stable under the left division, i.e., if* $b \backslash a \subseteq K$, *for all* $a, b \in K$. *K is called closed when it is both right and left closed.*

Proposition 23. *If the hypercomposition in a hypergroup H is closed (containing), then H has no proper closed subhypergroups.*

Proof. According to Proposition 1, if a hypercomposition is closed, then $a/a = H$, for all $a \in H$. Consequently, the only closed subhypergroup of H is H itself. □

Proposition 24. *If K is a subset of a hypergroup H such that $a/b \subseteq K$ and $b\backslash a \subseteq K$, for all $a, b \in K$, then K is a closed subhypergroup of H.*

Proof. Initially, it will be proved that K is a hypergroup, i.e., that $aK = Ka = K$, for any a in K. Let $x \in K$. Then $a\backslash x \subseteq K$. Therefore $x \in aK$. Hence $K \subseteq aK$. For the reverse inclusion, now suppose that $y \in aK$. Then $K/y \subseteq K/aK$. So $K \cap (K/aK)y \neq \varnothing$. Thus, $y \in (K/aK)\backslash K$. Per mixed associativity, the equality $K/aK = (K/K)/a$ is valid. Thus:

$$(K/aK)\backslash K = ((K/K)/a)\backslash K \subseteq (K/a)\backslash K \subseteq (K/K)\backslash K \subseteq K\backslash K \subseteq K$$

Hence $y \in K$ and so $aK \subseteq K$. Therefore $aK = K$. The equality $Ka = K$ follows by duality. The rest comes from Definition 30. □

Proposition 25. *If K is a closed hypergroup of a hypergroup H, then:*

$$K = K/a = a/K = a\backslash K = K\backslash a$$

for all a in K.

Proposition 26. *If K is a subhypergroup of H, then:*

$$H \cdot\cdot K \subseteq (H \cdot\cdot K)x \text{ and } H \cdot\cdot K \subseteq x(H \cdot\cdot K)$$

for all $x \in K$.

Proof. Let $y \in H \cdot\cdot K$ and $y \notin (H \cdot\cdot K)x$. Per reproductive axiom, $y \in Hx$ and since $y \notin (H \cdot\cdot K)x$, y must be a member of Kx. Thus, $y \in Kx \subseteq KK = K$. This contradicts the assumption, and so $H \cdot\cdot K \subseteq (H \cdot\cdot K)x$. The second inclusion follows by duality. □

Proposition 27. *i. A subhypergroup K of H is right closed in H if and only if $(H \cdot\cdot K)x = H \cdot\cdot K$, for all $x \in K$.*
ii. A subhypergroup K of H is left closed in H if and only if $x(H \cdot\cdot K) = H \cdot\cdot K$, for all $x \in K$.
iii. A subhypergroup K of H is closed in H if and only if $x(H \cdot\cdot K) = (H \cdot\cdot K)x = H \cdot\cdot K$, for all $x \in K$.

Proof. (i) Let K be right closed in H. Suppose that $z \in H \cdot\cdot K$ and $zx \cap K \neq \varnothing$. Then there exists an element y in K such that $y \in zx$, or equivalently $z \in y/x$. Therefore $z \in K$, which is absurd. Hence $(H \cdot\cdot K)x \subseteq H \cdot\cdot K$. Next, because of Proposition 26, $H \cdot\cdot K \subseteq (H \cdot\cdot K)x$ and therefore $H \cdot\cdot K = (H \cdot\cdot K)x$. Conversely now: suppose that $(H \cdot\cdot K)x = H \cdot\cdot K$ for all $x \in K$. Then $(H \cdot\cdot K)x \cap K = \varnothing$ for all $x \in K$. Hence $t \notin rx$ and so $r \notin t/x$ for all $x, t \in K$ and $r \in H \cdot\cdot K$. Therefore $t/x \cap (H \cdot\cdot K) = \varnothing$ which implies that $t/x \subseteq K$. Thus K is right closed in H. (ii) follows by duality and (iii) is an obvious consequence of (i) and (ii). □

Corollary 11. *i. If K is a right closed subhypergroup in H, then $xK \cap K = \varnothing$, for all $x \in H \cdot\cdot K$.*
ii. If K is a left closed subhypergroup in H, then $Kx \cap K = \varnothing$, for all $x \in H \cdot\cdot K$.
iii. If K is a closed subhypergroup in H, then $xK \cap K = \varnothing$ and $Kx \cap K = \varnothing$, for all $x \in H \cdot\cdot K$.

Proposition 28. *Let A be a nonempty subset of a hypergroup H and suppose that*

$$A_0 = A \cup AA \cup A/A \cup A\backslash A$$
$$A_1 = A_0 \cup A_0 A_0 \cup A_0/A_0 \cup A_0\backslash A_0$$
$$\vdots$$
$$A_n = A_{n-1} \cup A_{n-1} A_{n-1} \cup A_{n-1}/A_{n-1} \cup A_{n-1}\backslash A_{n-1}$$

Then $\langle A \rangle = \bigcup_{n \geq 0} A_n$ *is the least closed subhypergroup of H, which contains A.*

If A is a singleton, then the above procedure constructs the monogene closed subhypergroup of the hypergroup H. It is easy to see that if the hypercomposition is open, then $\langle a \rangle = \{a\}$, while if it is closed, $\langle a \rangle = H$ for all $a \in H$. The notion of the monogene subhypergroups was introduced by J. Mittas in [177] for the case of the canonical hypergroups. In [67], there is a detailed study of the monogene symmetric subhypergroups of the fortified join hypergroups. These studies highlight the existence of two types of the order of an element: the *principal order*, which is an integer, and the *associated order*, which is a function. A subcategory of the monogene subhypergroups is the cyclic subhypergroups; that is, subhypergroups of the form:

$$\bigcup_{k \in \mathbb{N}} a^k,\ a \in H$$

It is easy to observe that in the case of the open hypercompositions, the cyclic subhypergroup, which is generated by an element a, is the element a itself. The study of cyclic subhypergroups has attracted the interest of many researchers [207–218]. A detailed study and thorough review of cyclic hypergroups is given in [219].

The closed subhypergroups are directly connected with the ultra-closed subhypergroups [17,220]. The properties of these subhypergroups, together with their ability to create cosets in special hypergroups, were studied in a long series of joint papers by M. De Salvo, D. Freni, D. Fasino, and G. Lo Faro [191–195]; D. Freni [196–198]; M. Gutan et al. [198–200]; and L. Haddad and Y. Sureau [221,222]. A new definition for the ultra-closed subhypergroups, with the use of the induced hypercompositions, is given below. After that, Theorem 24 proves that this definition is equivalent to Sureau's definition.

Definition 31. *A subhypergroup K of a hypergroup H is called right ultra-closed if it is closed and $a/a \subseteq K$ for each $a \in H$. K is called left ultra-closed if it is closed and $a\backslash a \subseteq K$ for each $a \in H$. If K is both right and left ultra-closed, then it is called ultra-closed.*

Theorem 23. *i. If K is right ultra-closed in H, then, either $a/b \subseteq K$, or $a/b \cap K = \varnothing$, for all $a, b \in H$. Moreover, if $a/b \subseteq K$, then $b/a \subseteq K$.*
ii. If K is left ultra-closed in H, then, either $b\backslash a \subseteq K$, or $b\backslash a \cap K = \varnothing$, for all $a, b \in H$. Moreover, if $b\backslash a \subseteq K$, then $a\backslash b \subseteq K$.

Proof. Suppose that $a/b \cap K \neq \varnothing, a, b \in H$. Then $a \in sb$, for some $s \in K$. Next, assume that $b/a \cap (H \cdot \cdot K) \neq \varnothing$. Then $b \in ta$, $t \in H \cdot \cdot K$. Thus $a \in s(ta) = (st)a$. Since K is closed, per Proposition 27, $st \subseteq H \cdot \cdot K$. So $a \in ra$, for some $r \in H \cdot \cdot K$. Therefore $a/a \cap (H \cdot \cdot K) \neq \varnothing$, which is absurd. Hence $b/a \subseteq K$. Now let x be an element in K such that $b \in xa$. If $a/b \cap (H \cdot \cdot K) \neq \varnothing$, there exists $y \in H \cdot \cdot K$ such that $a \in yb$. Therefore $b \in x(yb) = (xy)b$. Since K is closed, per Proposition 27, $xy \subseteq H \cdot \cdot K$. So, $b \in zb$ for some $z \in H \cdot \cdot K$. Therefore $b/b \cap (H \cdot \cdot K) \neq \varnothing$, which is absurd. Hence $a/b \subseteq K$. Duality gives (ii). □

Theorem 24. *Let H be a hypergroup and K a subhypergroup of H. Then:*
i. K is a right ultra-closed subhypergroup of H if and only if $Ka \cap (H \cdot \cdot K)a = \varnothing$, for all $a \in H$.
ii. K is a left ultra-closed subhypergroup of H if and only if $aK \cap a(H \cdot \cdot K) = \varnothing$, for all $a \in H$.

Proof. Suppose that K is a right ultra-closed subhypergroup of H. Then $a/a \subseteq K$ for all $a \in H$. Since K is right closed, then $(a/a)/s \subseteq K$ is valid, or equivalently, $a/(as) \subseteq K$ for all $s \in K$. Theorem 23 yields $(as)/a \subseteq K$ for all $s \in K$. If $Ka \cap (H \cdot \cdot K)a \neq \varnothing$, then there exist $s \in K$ and $r \in H \cdot \cdot K$, such that $sa \cap ra \neq \varnothing$, which implies that $r \in as/a$. But $(as)/a \subseteq K$, hence $r \in K$, which is absurd. Conversely now: let $Ka \cap (H \cdot \cdot K)a = \varnothing$ for all $a \in H$. If $a \in K$, then $K \cap (H \cdot \cdot K)a = \varnothing$. Therefore $s \notin ra$, for every $s \in K$ and $r \in H \cdot \cdot K$. Equivalently, $s/a \cap (H \cdot \cdot K) = \varnothing$, for all $s \in K$. Hence $s/a \subseteq K$ for all $s \in K$ and $a \in K$. So K is right closed. Next, suppose that $a/a \cap (H \cdot \cdot K) \neq \varnothing$ for some $a \in H$. Then $a \in (H \cdot \cdot K)a$, or $Ka \subseteq K(H \cdot \cdot K)a$. Since K is closed, per Proposition 27, $K(H \cdot \cdot K) \subseteq H \cdot \cdot K$ is valid. Thus $Ka \subseteq (H \cdot \cdot K)a$, which contradicts the assumption. Duality gives (ii). □

7.2. Invertible Subhypergroups and their Cosets

Definition 32. *A subhypergroup K of a hypergroup H is called right invertible if $a/b \cap K \neq \varnothing$ implies $b/a \cap K \neq \varnothing$, with $a, b \in H$, while it is called left invertible if $b \backslash a \cap K \neq \varnothing$ implies $a \backslash b \cap K \neq \varnothing$, with $a, b \in H$. K is invertible when it is both right and left invertible.*

Note that the concept of the invertible subhypergroup is self-dual. Theorem 4 in [134] gives an interesting example of an invertible subhypergroup in a join hypergroup of partial differential operators.

Theorem 25. *If K is invertible in H, then K is closed in H.*

Proof. Let $x \in K/K$. Then $K \approx xK$, thus $K \approx x \backslash K$. Since K is invertible, $K \approx x \backslash K$ implies $K \approx K \backslash x$ from which it follows that $x \in KK$. Since K is a subhypergroup, $KK = K$. Therefore $x \in K$. Consequently $K/K \subseteq K$. Since invertibility is self-dual, $K \backslash K \subseteq K$ is valid as well. □

The converse of Theorem 25 is not true. In [17] there are examples of closed hypergroups that are not invertible. On the other hand, the following proposition is a direct consequence of Theorem 23.

Proposition 29. *i. The right (left) ultra-closed subhypergroups of a hypergroup are right (left) invertible. ii. The ultra-closed subhypergroups of a hypergroup are invertible.*

Theorem 25 and Proposition 23 give the following result:

Proposition 30. *A hypergroup H has no proper invertible subhypergroups if its hypercomposition is closed (containing).*

Proposition 31. *A hypergroup H has no proper invertible subhypergroups if the hypercomposition is open.*

Proof. Suppose that K is an invertible subhypergroup of H and a is an element of H that does not belong to K. Because of the reproductive axiom, there exists $b \in H$ such that $a \in Kb$. Therefore $a/b \cap K \neq \varnothing$, which, per invertibility of K, implies that $b/a \cap K \neq \varnothing$. Hence $b \in Ka$, and so
$$a \in Kb \subseteq K(Ka) = (KK)a = Ka.$$
Thus, $a/a \cap K \neq \varnothing$. But, according to Proposition 1, $a/a = a$. Therefore $a \in K$, which contradicts the assumption. □

Proposition 32. *i. K is right invertible in H, if and only if the implication $a \in Kb \Rightarrow b \in Ka$ is valid for all $a, b \in H$.*
ii. K is left invertible in H if and only if the implication $b \in aK \Rightarrow a \in bK$ is valid for all $a, b \in H$.

Lemma 1. *i. If the implication $Ka \neq Kb \Rightarrow Ka \cap Kb = \varnothing$ is valid for all $a, b \in H$, then $c \in Kc$, for each $c \in H$.*
ii. If the implication $aK \neq bK \Rightarrow aK \cap bK = \varnothing$ is valid for all $a, b \in H$, then $c \in cK$, for each $c \in H$.

Proof. (i) There exists $x \in H$ such that $c \in Kx$. Hence $Kc \subseteq Kx$. Therefore $Kc \cap Kx \neq \varnothing$. So, $c \in Kx = Kc$. (ii) is the dual of (i). □

Theorem 26. *i. K is right invertible in H, if and only if the implication $Ka \neq Kb \Rightarrow Ka \cap Kb = \varnothing$ is valid for all $a, b \in H$.*
ii. K is left invertible in H, if and only if the implication $aK \neq bK \Rightarrow aK \cap bK = \varnothing$ is valid for all $a, b \in H$.

Proof. (i) Let K be right invertible in H. Assume that $Ka \neq Kb$ and that $c \in Ka \cap Kb$. Then $Kc \subseteq Ka$ and $Kc \subseteq Kb$. Moreover, $c \in Ka \cap Kb$ implies that $c/a \cap K \neq \varnothing$ and $c/b \cap K \neq \varnothing$. But K is invertible, consequently we have the sequence of the following equivalent statements:

$$a/c \cap K \neq \varnothing \text{ and } b/c \cap K \neq \varnothing; \quad a \in Kc \text{ and } b \in Kc; \quad Ka \subseteq Kc \text{ and } Kb \subseteq Kc$$

Therefore, $Ka = Kc$ and $Kb = Kc$. Thus, $Ka = Kb$, which contradicts the assumption. So, $Ka \cap Kb = \varnothing$. Conversely now: suppose that $Ka \neq Kb \Rightarrow Ka \cap Kb = \varnothing$ is valid for all $a, b \in H$ and moreover, assume that $a/b \cap K \neq \varnothing$. Then $a \in Kb$, consequently $Ka \subseteq Kb$. Since $Ka \cap Kb \neq \varnothing$, it derives that $Ka = Kb$. Per Lemma 1, $b \in Kb$. Therefore $b \in Ka$, which implies that $b/a \cap K \neq \varnothing$. (ii) is the dual of (i). □

Theorem 27. *i. If K is right invertible in H, then $H \nearrow K = \{Kx \mid x \in H\}$ is a partition in H.*
ii. If K is left invertible in H, then $H \swarrow K = \{xK \mid x \in H\}$ is a partition in H.
iii. If K is invertible in H, then $H \nwarrow K = \{KxK \mid x \in H\}$ is a partition in H.

Proof. The proofs of (i) and (ii) are similar to the proof of (iii). So, we prove (iii).
(iii) According to Lemma 1, if c is an arbitrary element in H, then $c \in Kc$ and $c \in cK$. $c \in Kc$ implies $cK \subseteq KcK$, and since $c \in cK$, it derives that $c \in KcK$. Now suppose that there exist $a, b \in H$ such that $KaK \cap KbK \neq \varnothing$. Let $c \in KaK \cap KbK$. $c \in KaK$ implies that there exists $y \in Ka$ such that $c \in yK$. However, due to the invertibility of K, it follows that $a \in Ky$ and $y \in cK$. Hence $a \in KcK$ and therefore $KaK \subseteq KcK$. Moreover, $c \in KaK$ implies that $KcK \subseteq KaK$. Consequently $KaK = KcK$. In the same way $KbK = KcK$, and therefore $KaK = KbK$. □

Theorem 28. *Let K be a right invertible subhypergroup of a hypergroup H and let $X = Kx, x \in H$. Then $X * Y = XY \nearrow K$ is a hypercomposition in $H \nearrow K = \{Kx \mid x \in H\}$ and $(H \nearrow K, *)$ is a hypergroup.*

Corollary 12. *If K is a subgroup of a group G, then $(G \nearrow K, *)$ is a hypergroup.*

Theorem 29. *Suppose that K is a subgroup of a group G, and $K \subseteq M \subseteq G$. Then $(M \nearrow K, *)$ is a subhypergroup of $(G \nearrow K, *)$ if and only if M is a subgroup of a group G.*

Proof. Suppose that $M \nearrow K$ is a subhypergroup of $G \nearrow K$. Then:

$$Kx * (M \nearrow K) = M \nearrow K = (M \nearrow K) * Kx, \text{ for all } Kx \in M \nearrow K$$

Therefore $KxM = M = MKx$, for all $x \in M$. Since $e \in K$, it follows that $xM = M$ and $Mx = M$ for all $x \in M$. Thus, M is a subgroup of G. Conversely now, let M be a subgroup of G and $x \in M$. Then:

$$Kx * (M \nearrow K) = KxM \nearrow K = KM \nearrow K = M \nearrow K$$

and
$$(M \nearrow K) * Kx = MKx \nearrow K = Mx \nearrow K = M \nearrow K.$$

Consequently, $(M \nearrow K, *)$ is a subhypergroup of $(G \nearrow K, *)$. □

7.3. Reflexive, Closed Subhypergroups and their Cosets

As shown above, when the hypercomposition is open or closed, there do not exist proper invertible subhypergroups. So, in such cases, the hypergroup cannot be decomposed into cosets with the use of the previous techniques, and different methods need to be developed in order to solve the decomposition problem. Such a method that uses a special type of closed subhypergroups is presented below, for the case of the transposition hypergroups.

Definition 33. *A subhypergroup N of a hypergroup H is called normal or invariant if $aN = Na$, for all $a \in H$.*

Proposition 33. *N is an invariant subhypergroup of a hypergroup H if and only if $N \backslash a = a/N$, for all $a \in H$.*

Proof. Suppose that N is invariant in H. Then we have the sequence of equivalent statements:
$$x \in N \backslash a;\ a \in Nx;\ a \in xN;\ x \in a/N$$

Therefore $N \backslash a = a/N$. Conversely now, suppose that $N \backslash a = a/N$. Then we have the following equivalent statements:
$$x \in aN;\ a \in x/N;\ a \in N \backslash x;\ x \in Na$$

Consequently, N is invariant. □

Definition 34. *A subhypergroup R of a hypergroup H is called reflexive if $a \backslash R = R/a$, for all $a \in H$.*

Obviously, all the subhypergroups of the commutative hypergroups are invariant and reflexive. The closed and reflexive subhypergroups have a special interest in transposition hypergroups because they can create cosets.

Theorem 30. *If L is a closed and reflexive subhypergroup of a transposition hypergroup T, then the sets L/x form a partition in T.*

Proof. Let $L/x \cap L/y \neq \varnothing$. Then $x \backslash L \cap L/y \neq \varnothing$, and therefore $xL \cap Ly \neq \varnothing$. Consequently, $x \in Ly/L$. Next, using the properties mentioned in Propositions 11 and 3 successively, we have:
$$L/x \subseteq L/(Ly/L) \subseteq LL/Ly = L/Ly = (L/y)/L = L \backslash (L/y) = (L \backslash L)/y = L/y$$

Hence $L/x \subseteq L/y$. By symmetry, $L/y \subseteq L/x$, and so $L/x = L/y$. □

According to the above theorem, for each $x \in T$ there exists a unique class that contains x. This unique class is denoted by L_x. The set of the classes modulo L is denoted by T_L. Note that $L \in T_L$. Indeed, since L is closed, $L/x = L$ for any $x \in L$.

Proposition 34. *If $xy \cap L \neq \emptyset$, then $L/x = L_y$ and $L/y = L_x$.*

Proof. $xy \cap L \neq \emptyset$ implies both $x \in L/y$ and $y \in x\backslash L = L/x$. Therefore $L_x = L/y$ and $L_y = L/x$. □

From the above proposition, it becomes evident that L_x and L/x are different classes in T_L. The following theorems reveal the algebraic form of the class L_x, $x \in T$.

Theorem 31. *If L_x is a class, then L/L_x is also a class modulo L and $L/L_x = L/y$ for some $y \in L_x$.*

Proof. Let $y \in L_x$. It suffices to prove that $L/L_x = L/y$, because this implies that L/L_x is a class modulo L. Since $y \in L_x$, it follows that $L/y \subseteq L/L_x$. For the proof of the reverse inclusion, let $z \in L/L_x$. Then $L \cap zL_x \neq \emptyset$, hence $L_x \cap z\backslash L = L_x \cap L/z \neq \emptyset$. By Theorem 30, $L_x = L/z$, and so $y \in L/z$. Then, $yz \cap L \neq \emptyset$ and $z \in y\backslash L = L/y$. Consequently, $L/L_x \subseteq L/y$, hence the theorem. □

Theorem 32. *If L_x is the class modulo L of an element x in T, then:*

$$L_x = Lx/L = L\backslash Lx = L/(L/x) = (x\backslash L)\backslash L.$$

Proof. Let $x \in T$. Since L is reflexive, $x\backslash L = L/x$. Applying the transposition axiom, we get $xL \approx Lx$, which implies that $x \in Lx/L$ and $x \in L\backslash xL$. Next, from Proposition 11 (ii), it follows that $L/(L/x) \subseteq Lx/L$. For the proof of the reverse inclusion, Corollary 2 and Proposition 11 (ii) sequentially give:

$$Lx/L \subseteq [L/(Lx/L)]\backslash L \subseteq (LL/Lx)\backslash L = (L/Lx)\backslash L$$

Since L is reflexive, the equality $(L/Lx)\backslash L = L/(L/Lx)$ holds. Therefore, Proposition 3 and Proposition 11 (ii) give:

$$L/(L/Lx) = L/[(L/x)/L] \subseteq LL/(L/x) = L/(L/x)$$

Consequently, $L/(L/x) = Lx/L$. Duality yields the rest. □

Corollary 13. *If T is a quasicanonical hypergroup, then $L_x = xL$.*

Proof. Per Proposition 16, $x/L = xL^{-1}$, thus:

$$L_x = (x/L)/L = \left(xL^{-1}\right)L^{-1} = (xL)L = x(LL) = xL.\ \square$$

Proposition 35. *If L is a reflexive closed subhypergroup of a transposition hypergroup T, then:*

$$(L/x)(L/y) \subseteq L/xy,\ x,y \in T.$$

Proof. The successive application of Proposition 11 (iii) and mixed associativity gives:

$$(L/x)(L/y) = (x\backslash L)(L/y) \subseteq (x\backslash LL)/y = (x\backslash L)/y = (L/x)/y = L/xy.\ \square$$

A direct consequence of the above proposition is that the partition which is defined by L is regular. Therefore, a hypercomposition "·" is defined in T_L by

$$(L/x) \cdot (L/y) = \{L/z \mid z \in xy\}.$$

Next, it is apparent that L is the neutral element of the above hypercomposition and that the transposition is valid. Hence the theorem:

Theorem 33. *If L is a reflexive closed subhypergroup of a transposition hypergroup T, then $(T_L, \cdot, L)$ is a quasicanonical hypergroup.*

Example 26. *A three-dimensional Euclidean space becomes a hypergroup under the hypercomposition defined in Example 4. It is easy to verify that this is a commutative transposition hypergroup of idempotent elements; that is, a join space. The notation E_J^3 is used for this hypergroup. A line L of the Euclidean space is a reflexive closed subhypergroup of E_J^3. The cosets that L defines in E_J^3 are the halfplanes L/x, which are drawn in the following Figure 4. $\left(E_J^3\right)_L$ is a quasicanonical hypergroup, L is its neutral element, and the symmetric of any element L/x is the element $L/(L/x)$.*

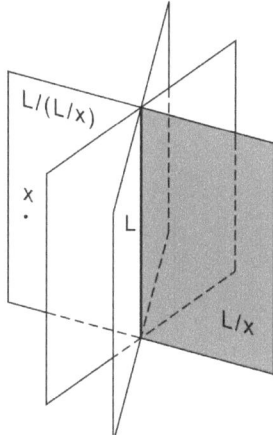

Figure 4. Example 26.

7.4. Symmetric Subhypergroups and their Cosets

As mentioned above, when the hypercomposition in a hypergroup is closed (containing), then there do not exist closed subhypergroups. In this case, other types of subhypergroups must be used for the decomposition of the hypergroup. An example of such a hypergroup is the transposition hypergroup with idempotent identity, which consists of attractive elements only. As shown in the sixth section of this paper, the transposition hypergroups can have one (fortified transposition hypergroups) or more (transposition polysymmetrical hypergroups) symmetric elements for each one of their elements. Obviously, a transposition polysymmetrical hypergroup becomes a fortified transposition hypergroup when the set $S(x)$ of the symmetric elements of any one of its elements x is a singleton, and moreover, when its identity is strong. The quite general case of the decomposition of the transposition polysymmetrical hypergroups with strong identity into cosets is pesented below. Since the result of the hypercomposition between two attractive elements contains these two elements, there do not exist proper closed subhypergroups. There exist, though, subhypergroups that contain the symmetric element of each one of their elements. These subhypergroups are the ones that decompose the transposition polysymmetrical hypergroups into cosets. In the following, P signifies a transposition polysymmetrical hypergroup with strong identity.

Definition 35. *A subhypergroup M of a transposition polysymmetrical hypergroup is called symmetric if $S(x) \subseteq M$, for all x in M.*

Proposition 36. *Let x, y be two elements in P such that $e \notin xS(y)$, then $xS(y) = x/y \cup S(y)$ and $S(y)x = y\backslash x \cup S(y)$.*

Proof. $xS(y) \subseteq x(e/y) \subseteq xe/y = \{x,e\}/y = x/y \cup e/y = x/y \cup \{e\} \cup S(y)$. Since e is not in $xS(y)$, it follows that $xS(y) \subseteq x/y \cup S(y)$. Moreover, since $e \in yS(y)$, it follows that $y \in e/S(y)$. Therefore $x/y \subseteq x/(e/S(y))$. Hence:

$$x/y \cup S(y) \subseteq x/(e/S(y)) \cup S(y) \subseteq xS(y)/e \cup S(y) = xS(y) \cup S(y) = xS(y)$$

Thus, $xS(y) = x/y \cup S(y)$. Dually, $S(y)x = y\backslash x \cup S(y)$. □

Proposition 37. *Let M be a symmetric subhypergroup of P. If $x \notin M$, then:*
i. $x/M \cap M = \emptyset$ and $M\backslash x \cap M = \emptyset$,
ii. $xM = x/M \cup M$ and $Mx = M\backslash x \cup M$,
iii. $M/x = M\,S(x)$ and $x\backslash M = S(x)M$,
iv. $(x/M)M = xM$ and $M(M\backslash x) = Mx$.

Proof. The two statements in each one of (i)–(iv) are dual, and therefore it is sufficient to prove only one of them.
(i) Let $x \notin M$ and $y \in M$ such that $x/y \cap M \neq \emptyset$. Then, $x \in My = M$, which contradicts the assumption for x. Thus, $x/M \cap M = \emptyset$.
(ii) Since M is symmetric, $S(M) = M$. Thus, $e \notin xM$, since $x \notin M$. So, via Proposition 36:

$$xM = xS(M) = x/M \cup S(M) = x/M \cup M.$$

(iii) According to Proposition 36, the equality $M\,S(x) = M/x \cup S(x)$ holds, therefore

$$M/x = M/x \cup e/x = M/x \cup S(x) = M\,S(x).$$

(iv) Since $x \in x/M$, it follows that $xM \subseteq (x/M)M$. In addition, because of (ii), $x/M \subseteq xM$, thus,

$$(x/M)M \subseteq (xM)M = x(MM) = xM. \quad \square$$

Theorem 34. *Let M be a symmetric subhypergroup of P. If $x, y \notin M$, then:*
i. $x/M \approx y/M$ implies $x/M = y/M$,
ii. $M\backslash x \approx M\backslash y$ implies $M\backslash x = M\backslash y$,
iii. $M\backslash(x/M) \approx M\backslash(y/M)$ implies $M\backslash(x/M) = M\backslash(y/M)$,
iv. $(M\backslash x)/M \approx (M\backslash y)/M$ implies $(M\backslash x)/M = (M\backslash y)/M$.

Proof. (i) $x/M \cap y/M \neq \emptyset$ implies that $x \in (y/M)M$. Since $y \notin M$, from (iv) and (ii) of Proposition 37, it follows that $(y/M)M = yM \subseteq y/M \cup M$. Thus, $x \in y/M \cup M$. Since $x \notin M$, it follows that $x \in y/M$. Thus, $x/M \subseteq (y/M)/M = y/(MM) = y/M$. By symmetry, $y/M \subseteq x/M$. Hence $x/M = y/M$.
(ii) is the dual of (i).
(iii) Per Corollary 1, mixed associativity, and Propositions 11 and 37 (iv), we have:

$$M\backslash(x/M) \approx M\backslash(y/M) \Rightarrow (M\backslash x)/M \approx M\backslash(y/M) \Rightarrow M\backslash x \approx [M\backslash(y/M)]M \Rightarrow$$

$$\Rightarrow M\backslash x \approx M\backslash[(y/M)M] \Rightarrow M\backslash x \approx M\backslash yM \Rightarrow x \in yM \Rightarrow y \in x/M \Rightarrow$$

$$\Rightarrow y/M \subseteq (x/M)/M \Rightarrow y/M \subseteq x/(MM) \Rightarrow y/M \subseteq x/M \Rightarrow M\backslash(y/M) \subseteq M\backslash(x/M).$$

By symmetry, $M\backslash(x/M) \subseteq M\backslash(y/M)$, thus the equality is valid.
Finally, (iv) is true because $(M\backslash x)/M = M\backslash(x/M)$ and $(M\backslash y)/M = M\backslash(y/M)$. □

If $x \in P$ and M is a nonempty symmetric subhypergroup of P, then $x_{\underset{M}{\leftarrow}}$ (i.e., the *left coset* of M determined by x) and dually, $x_{\underset{M}{\rightarrow}}$ (i.e., the *right coset* of M determined by x) are given by:

$$x_{\underset{M}{\leftarrow}} = \begin{cases} M, & \text{if } x \in M \\ x/M, & \text{if } x \notin M \end{cases} \quad \text{and} \quad x_{\underset{M}{\rightarrow}} = \begin{cases} M, & \text{if } x \in M \\ M \backslash x, & \text{if } x \notin M \end{cases}$$

Since, per Corollary 1, the equality $(B \backslash A)/C = B \backslash (A/C)$ is valid in any hypergroup, the double coset of M determined by x can be defined by:

$$x_M = \begin{cases} M, & \text{if } x \in M \\ M \backslash (x/M) = (M \backslash x)/M, & \text{if } x \notin M \end{cases}$$

Per Theorem 34, the distinct left cosets and right cosets, as well as the double cosets, are disjoint. Thus, each one of the families:

$$P \swarrow M = \left\{ x_{\underset{M}{\leftarrow}} \mid x \in P \right\}, \ P \nearrow M = \left\{ x_{\underset{M}{\rightarrow}} \mid x \in P \right\} \text{ and } P \swarrow^{\nearrow} M = \{ x_M \mid x \in P \}$$

forms a partition of P. If M is normal, then it follows that $x_{\underset{M}{\leftarrow}} = x_{\underset{M}{\rightarrow}} = x_M$. Therefore:

$$P \swarrow M = P \nearrow M = P \swarrow^{\nearrow} M$$

Proposition 38. *Let M be a symmetric subhypergroup of P. Then:*

i. $x \in x_{\underset{M}{\leftarrow}}$, $x \in x_{\underset{M}{\rightarrow}}$ and $x \in x_M$,

ii. $x_{\underset{M}{\leftarrow}} \subseteq x_M$ and $x_{\underset{M}{\rightarrow}} \subseteq x_M$,

iii. $x_M = \left(x_{\underset{M}{\leftarrow}} \right)_{\underset{M}{\rightarrow}} = \left(x_{\underset{M}{\rightarrow}} \right)_{\underset{M}{\leftarrow}}$.

Proposition 39. *Let M be a symmetric subhypergroup of P. Then:*

$$xM = x/M \cup M \text{ and } Mx = M \backslash x \cup M.$$

Proof. The two equalities are dual. Per Proposition 36, the equality $xS(M) = x/M \cup S(M)$ is valid. But M is symmetric, so $S(M) = M$. Therefore, $xM = x/M \cup M$. □

Proposition 40. *Let M be a symmetric subhypergroup of P. Then:*

i. $x_{\underset{M}{\leftarrow}} M = xM = x_{\underset{M}{\leftarrow}} \cup M$,

ii. $M x_{\underset{M}{\rightarrow}} = Mx = x_{\underset{M}{\rightarrow}} \cup M$.

Proof. (i) If $x \in M$, the equality is true, because each one of its parts equals to M. If $x \notin M$, then the sequential application of Propositions 37 (iv) and 39 gives:

$$x_{\underset{M}{\leftarrow}} M = (x/M)M = xM = x/M \cup M = x_{\underset{M}{\leftarrow}} \cup M.$$

Duality gives (ii). □

Corollary 14. *Let Q be a nonvoid subset of P and M a symmetric subhypergroup of P. Then:*

$$Q_{\underset{M}{\leftarrow}} M = QM = Q_{\underset{M}{\leftarrow}} M \cup M \text{ and } M Q_{\underset{M}{\rightarrow}} = MQ = Q_{\underset{M}{\rightarrow}} \cup M$$

Proposition 41. *Let M be a symmetric subhypergroup of P. Then:*

$$M x_M = M x_{\underset{M}{\leftarrow}} = x_M \cup M = MxM = x_{\underset{M}{\rightarrow}} M = x_M M$$

Proof. Per Proposition 40 (i): $MxM = M\left(x_{\overleftarrow{M}} \cup M\right) = Mx_{\overleftarrow{M}} \cup MM = Mx_{\overleftarrow{M}} \cup M$, and since the hypercomposition is closed, $Mx_{\overleftarrow{M}} \cup M = Mx_{\overleftarrow{M}}$. Per duality: $M\tilde{x}M = x_{\overrightarrow{M}}M$. Next, per Propositions 38 (iii) and 40, we have:

$$Mx_M = M\left(x_{\overleftarrow{M}}\right)_{\overrightarrow{M}} = Mx_{\overleftarrow{M}} = \left(x_{\overleftarrow{M}}\right)_{\overrightarrow{M}} \cup M = x_M \cup M$$

Duality yields the rest. □

Proposition 42. *Let M be a symmetric subhypergroup of P. Then:*

$$(xy)_{\overleftarrow{M}} M \subseteq x_{\overleftarrow{M}} y_{\overleftarrow{M}} \cup M \text{ and } (xy)_{\overrightarrow{M}} M \subseteq x_{\overrightarrow{M}} y_{\overrightarrow{M}} \cup M$$

Proof. Per Corollary 14, $(xy)_{\overleftarrow{M}} \subseteq (xy)_{\overleftarrow{M}} M = (xy)M$. But $x \in x/M$, thus we have: $(xy)M \subseteq (x/M)yM = x_{\overleftarrow{M}}(yM)$. Now, per Proposition 40:

$$x_{\overleftarrow{M}}(yM) = x_{\overleftarrow{M}}\left(y_{\overrightarrow{M}} \cup M\right) = x_{\overleftarrow{M}} y_{\overrightarrow{M}} \cup x_{\overleftarrow{M}} M = x_{\overleftarrow{M}} y_{\overrightarrow{M}} \cup \left(x_{\overleftarrow{M}} \cup M\right) = \left(x_{\overleftarrow{M}} y_{\overrightarrow{M}} \cup x_{\overleftarrow{M}}\right) \cup M$$

Since the hypercomposition is closed, $x_{\overleftarrow{M}} y_{\overrightarrow{M}} \cup x_{\overleftarrow{M}} = x_{\overleftarrow{M}} y_{\overrightarrow{M}}$ is valid, and therefore:

$$\left(x_{\overleftarrow{M}} y_{\overrightarrow{M}} \cup x_{\overleftarrow{M}}\right) \cup M = x_{\overleftarrow{M}} y_{\overrightarrow{M}} \cup M.$$

Duality gives the second inclusion. □

Corollary 15. *If X, Y are nonvoid subsets of P and M is a symmetric subhypergroup of P, then:*

$$(XY)_{\overleftarrow{M}} M \subseteq X_{\overleftarrow{M}} Y_{\overleftarrow{M}} \cup M \text{ and } (XY)_{\overrightarrow{M}} M \subseteq X_{\overrightarrow{M}} Y_{\overrightarrow{M}} \cup M$$

Proposition 43. *Let M be a symmetric subhypergroup of P. Then:*

$$(xy)_M = x_M y_M \cup M$$

Proof. Per Proposition 38 (iii) and Corollary 15:

$$(xy)_M = \left((xy)_{\overleftarrow{M}}\right)_{\overrightarrow{M}} \subseteq \left(x_{\overleftarrow{M}} y_{\overleftarrow{M}} \cup M\right)_{\overrightarrow{M}} = \left(x_{\overleftarrow{M}} y_{\overleftarrow{M}}\right)_{\overrightarrow{M}} \cup M_{\overrightarrow{M}} \subseteq$$
$$\subseteq \left(x_{\overleftarrow{M}}\right)_{\overrightarrow{M}} \left(x_{\overleftarrow{M}}\right)_{\overrightarrow{M}} \cup M = x_M y_M \cup M \quad \square$$

Corollary 16. *If X, Y are nonvoid subsets of P and M is a symmetric subhypergroup of P, then:*

$$(XY)_M = X_M Y_M \cup M$$

Corollary 17. *If X, Y are nonvoid subsets of P and M is a symmetric subhypergroup of P, then:*
i. $X_M Y_M \cap M \neq \emptyset$ implies $(X_M Y_M)_M = X_M Y_M \cup M$,
ii. $X_M Y_M \cap M = \emptyset$ implies $(X_M Y_M)_M = X_M Y_M$.

A hypercomposition that derives from P's hypercomposition can be defined in each one of the families $P \swarrow M$, $P \nearrow M$ and $P \measuredangle M$. Thus in $P \measuredangle M$, this hypercomposition is:

$$x_M \cdot y_M = \{z_M \mid z \in x_M y_M\}.$$

If the induced hypercompositions of · are denoted by ∕ and ∖, then:

$$x_M \mathbin{/} y_M = \{z_M \mid x_M \in z_M \cdot y_M\} = \{z_M \mid x \in z_M y_M\}$$

and

$$y_M \mathbin{\backslash} x_M = \{z_M \mid x_M \in y_M \cdot z_M\} = \{z_M \mid x \in y_M z_M\}.$$

It is obvious that $x_M \mathbin{/} y_M \neq \varnothing$ and $y_M \mathbin{\backslash} x_M \neq \varnothing$. Therefore, according to Theorem 14, the following proposition is valid:

Proposition 44. *The reproductive axiom is valid in* $\left(P\slash M, \cdot\right)$.

In the families $P \slash M$ and $P \nearrow M$, the associativity may fail. However, it is valid in $P\slash M$, as per the next Proposition:

Proposition 45. *The associative axiom is valid in* $\left(P\slash M, \cdot\right)$.

Proof. It must be proved that

$$(x_M \cdot y_M) \cdot z_M = x_M \cdot (y_M \cdot z_M).$$

This is true if and only if $((x_M y_M)_M z_M)_M = (x_M (y_M z_M)_M)_M$. So, if $x_M y_M \cap M = \varnothing$, then Corollary 17 (ii) yields $(x_M y_M)_M = x_M y_M$ and the above equality is obvious. If $x_M y_M \cap M \neq \varnothing$, then Corollary 17 (i) yields $(x_M y_M)_M = x_M y_M \cup M$. Hence:

$$(x_M y_M) z_M = (x_M y_M \cup M) z_M = x_M y_M z_M \cup M z_M$$
$$= x_M y_M z_M \cup z_M \cup M = x_M y_M z_M \cup M$$

Since $x_M y_M \cap M \neq \varnothing$ and $x_M y_M \subseteq x_M y_M z_M$, it follows that $M \subseteq (x_M y_M z_M)_M$ is valid. Therefore:

$$((x_M y_M)_M z_M)_M = (x_M y_M z_M \cup M)_M = (x_M y_M z_M)_M \cup M = (x_M y_M z_M)_M.$$

Duality yields $(x_M y_M z_M)_M = (x_M (y_M z_M)_M)_M$, and so the associativity is valid. □

Proposition 46. *The transposition axiom is valid in* $\left(P\slash M, \cdot\right)$.

Proof. Suppose that $y_M \mathbin{\backslash} x_M \cap z_M \mathbin{/} w_M \neq \varnothing$. Then:

$$\{p_M \mid x_M \in y_M \cdot p_M\} \cap \{q_M \mid z_M \in q_M \cdot w_M\} \neq \varnothing \Leftrightarrow$$
$$\Leftrightarrow \{p_M \mid x \in y_M p_M\} \cap \{q_M \mid z \in q_M w_M\} \neq \varnothing.$$

Therefore $y'\backslash x \cap z/w' \neq \varnothing$ for some $y' \in y_M$ and $w' \in w_M$. Thus, $xw' \cap y'z \neq \varnothing$, and so

$$x_M \cdot w_M \cap z_M \cdot y_M \neq \varnothing. \ \Box$$

Propositions 44, 45, and 46 give the theorem:

Theorem 35. *If M is a symmetric subhypergroup of P, then* $\left(P\slash M, \cdot\right)$ *is a transposition hypergroup.*

A consequence of Proposition 41 is that $M \cdot x_M = x_M \cdot M = \{x_M, M\}$ for every x_M in $P\slash M$. Hence:

Proposition 47. *M is a strong identity in the hypergroup $P\slash M$.*

Proposition 48. $P\swarrow M$ consists only of attractive elements.

Since $\{x_M, y_M\} \subseteq x_M \cdot y_M$ for all $x_M, y_M \in P\swarrow M$, the following is true:

Proposition 49. The hypercomposition in $\left(P\swarrow M, \cdot\right)$ is closed.

Proposition 50. If $y \in S(x)$, then $M \in x_M \cdot y_M$, for all $x_M \in P\swarrow M$.

Propositions 47 and 50 give as a consequence the following theorem:

Theorem 36. If M is a symmetric subhypergroup of P, then $\left(P\swarrow M, \cdot\right)$ is a transposition polysymmetrical hypergroup.

Corollary 18. If M is a symmetric subhypergroup of a fortified transposition hypergroup of attractive elements T, then $\left(T\swarrow M, \cdot\right)$ is a fortified transposition hypergroup, M is its strong identity, and each one of its elements is attractive.

7.5. The Cosets in Quasicanonical Hypergroups

It is apparent that the subgroups of a group are symmetric, closed, and invertible. The same is true in the case of the quasicanonical hypergroups. Indeed, per Proposition 16, $x/y = xy^{-1}$ and $y\backslash x = y^{-1}x$, therefore a subhypergroup U of a quasicanonical hypergroup Q is symmetric if and only if it is closed. Moreover, since the reversibility is valid in the quasicanonical hypergroups, if U is a symmetric subhypergroup of a quasicanonical hypergroup Q, then the implications $a \in rb \Rightarrow b \in r^{-1}a$ and $a \in br \Rightarrow b \in ar^{-1}$ hold for every $a, b \in Q$, $r \in U$. Therefore, because of Proposition 32, U is invertible. Hence the left and right cosets of a subhypergroup U of a quasicanonical hypergroup Q are of the form aU and Ua, $a \in Q$, respectively, and create partitions in Q. The quotient hypergroup $Q \swarrow U$ defined by the left cosets and the quotient hypergroup $Q \nearrow U$ defined by the right cosets are transposition hypergroups, but not necessarily quasicanonical ones. For example, although U is a right scalar identity for the left cosets, since $aU \cdot U = aU$, it need not be a left scalar identity as well, since $aU \approx U \cdot aU$.

Double cosets have the form UaU and also create a partition in Q. It is easy to observe that $(UaU)^{-1} = Ua^{-1}U$ and that U is a bilateral scalar identity. Hence:

Theorem 37. If U is a symmetric subhypergroup of a quasicanonical hypergroup Q, then the quotient hypergroup of the double cosets $Q\swarrow U$ is a quasicanonical hypergroup with scalar identity U.

Corollary 19. If U is a normal symmetric subhypergroup of a quasicanonical hypergroup Q, then the left, right, and double cosets coincide and the quotient hypergroup $Q\swarrow U$ is a quasicanonical hypergroup with U as its scalar identity.

8. Conclusions and Open Problems

This paper presents the structural relation between the hypergroup and the group. As it has been proved, these two algebraic structures satisfy the exact same axioms and their only difference appears in the law of synthesis of their elements. The law of synthesis in the hypergroups is so general that it allowed this structure's enrichment with further axioms which created a significant number of special hypergroups. Many of these special hypergroups are presented in this paper, along with examples, their fundamental properties, and the different types of their subhypergroups and applications. Among them, the transposition hypergroup has important applications in geometry and computer science [67]. This hypergroup satisfies the following axiom:

$$b\backslash a \cap c/d \neq \varnothing \text{ implies } ad \cap bc \neq \varnothing, \text{ for all } a,b,c,d \in H,$$

which was named *transposition axiom* [24]. If this axiom's implication is reversed, the following new axiom is created:

$$ad \cap bc \neq \varnothing \text{ implies } b\backslash a \cap c/d \neq \varnothing, \text{ for all } a,b,c,d \in H.$$

We will call this axiom *rev-transposition axiom* (i.e., *reverse transposition axiom*) and the relevant hypergroup *rev-transposition hypergroup* (i.e., *reverse transposition hypergroup*) or *rev-join hypergroup* (i.e., *reverse join hypergroup*) in the commutative case. The relation between these two hypergroups is very interesting. For example, the following proposition is valid in the rev-transposition hypergroups:

Proposition 51. *The following are true in any rev-transposition hypergroup:*

i. $ab/c \subseteq a(b/c)$ and $c\backslash ba \subseteq (c\backslash b)a$,
ii. $ab/c \subseteq a/(c/b)$ and $c\backslash ba \subseteq (b\backslash c)\backslash a$,
iii. $(b\backslash ac)/d = b\backslash (ac/d) \subseteq (b\backslash a)(c/d)$,
iv. $(b\backslash ad)/c = b\backslash (ad/c) \subseteq (b\backslash a)/(c/d)$,
v. $(a\backslash bc)/d = a\backslash (bc/d) \subseteq (b\backslash a)\backslash (c/d)$.

The comparison of Proposition 51 with Proposition 11 reveals that the reversal of the implication in the transposition axiom reverses the inclusion relations in properties (i)–(v). The study of this hypergroup and its potential applications present an interesting open problem of the theory of the hypercompositional algebra.

However, the following property also applies:

$$b\backslash a \cap c/d \neq \varnothing \Leftrightarrow ad \cap bc \neq \varnothing, \text{ for all } a,b,c,d \in H$$

We will name this axiom *bilateral transposition axiom*. Under this axiom, the inclusion relations of Propositions 11 and 51 become equalities. Examples of hypergroups verifying the bilateral transposition axiom are the quasicanonical hypergroups, the canonical hypergroups, and of course, the groups and the abelian groups. So, the following question arises: *Do there exist other hypergroups satisfying the bilateral transposition axiom apart from the quasicanonical and the canonical ones?*

In addition, as it is shown in Section 7, many questions remain open in the decomposition of hypergroups. Addressing these quesions necessitates the introduction of new tools and techniques, as hypercompositional algebra is an off-the-map region of abstract algebra, which requires great caution in the application of processes for achieving correct results. It is very easy to be led to wrong conclusions when relying on methods and results of classical algebra. Such a case is indicated in remark 2 in [67].

Author Contributions: Both authors contributed equally to this work. All authors have read and agreed to the published version of the manuscript.

Funding: This research received no external funding. The APC was funded by the MPDI journal *Mathematics*.

Institutional Review Board Statement: Not applicable.

Informed Consent Statement: Not applicable.

Data Availability Statement: Not applicable.

Acknowledgments: We wish to extend our sincere thanks to I. Cristea, S. Hoskova-Mayerova, and M. Novák for having urged us to write this paper during the round tables of the 1st Symposium on Hypercompositional Algebra—New Developments and Applications, Thessaloniki, Greece 2017, ICNAAM 2017.

Conflicts of Interest: The authors declare no conflict of interest.

References

1. Marty, F. Sur une Généralisation de la Notion de Groupe. Huitième Congrès des Mathématiciens Scand. Stockholm. 1934; 45–49.
2. Marty, F. Rôle de la notion de hypergroupe dans l' étude de groupes non abéliens. *C. R. Acad. Sci. (Paris)* **1935**, *201*, 636–638.
3. Marty, F. Sur les groupes et hypergroupes attachés à une fraction rationnelle. *Ann. L' Ecole Norm.* **1936**, *3*, 83–123.
4. Krasner, M. Sur la primitivité des corps B-adiques. *Mathematica* **1937**, *13*, 72–191.
5. Krasner, M. La loi de Jordan—Holder dans les hypergroupes et les suites generatrices des corps de nombres P—adiqes, (I). *Duke Math. J.* **1940**, *6*, 120–140, (II) *Duke Math. J.* **1940**, *7*, 121–135. [CrossRef]
6. Krasner, M. La caractérisation des hypergroupes de classes et le problème de Schreier dans ces hypergroupes. *C. R. Acad. Sci. (Paris)* **1941**, *212*, 948–950.
7. Krasner, M. Hypergroupes moduliformes et extramoduliformes. *Acad. Sci. (Paris)* **1944**, *219*, 473–476.
8. Krasner, M.; Kuntzmann, J. Remarques sur les hypergroupes. *C.R. Acad. Sci. (Paris)* **1947**, *224*, 525–527.
9. Kuntzmann, J. Opérations multiformes. Hypergroupes. *C. R. Acad. Sci. (Paris)* **1937**, *204*, 1787–1788.
10. Kuntzmann, J. Homomorphie entre systémes multiformes. *C. R. Acad. Sci. (Paris)* **1937**, *205*, 208–210.
11. Wall, H.S. Hypergroups. *Am. J. Math.* **1937**, *59*, 77–98. [CrossRef]
12. Ore, O. Structures and group theory, I. *Duke Math. J.* **1937**, *3*, 149–174. [CrossRef]
13. Dresher, M.; Ore, O. Theory of multigroups. *Am. J. Math.* **1938**, *60*, 705–733. [CrossRef]
14. Eaton, E.J.; Ore, O. Remarks on multigroups. *Am. J. Math.* **1940**, *62*, 67–71. [CrossRef]
15. Eaton, E.J. Associative Multiplicative Systems. *Am. J. Math.* **1940**, *62*, 222–232. [CrossRef]
16. Griffiths, L.W. On hypergroups, multigroups, and product systems. *Am. J. Math.* **1938**, *60*, 345–354. [CrossRef]
17. Corsini, P. *Prolegomena of Hypergroup Theory*; Aviani Editore: Tricesimo Udine, Italy, 1993.
18. Bourbaki, N. *Éléments de Mathématique, Algèbre*; Hermann: Paris, France, 1971.
19. Prenowitz, W. Projective Geometries as multigroups. *Am. J. Math.* **1943**, *65*, 235–256. [CrossRef]
20. Prenowitz, W. Descriptive Geometries as multigroups. *Trans. Am. Math. Soc.* **1946**, *59*, 333–380. [CrossRef]
21. Prenowitz, W. Spherical Geometries and mutigroups. *Can. J. Math.* **1950**, *2*, 100–119. [CrossRef]
22. Prenowitz, W. A Contemporary Approach to Classical Geometry. *Am. Math. Month.* **1961**, *68*, 1–67. [CrossRef]
23. Prenowitz, W.; Jantosciak, J. Geometries and Join Spaces. *J. Reine Angew. Math.* **1972**, *257*, 100–128.
24. Prenowitz, W.; Jantosciak, J. *Join Geometries. A Theory of Convex Sets and Linear Geometry*; Springer: Berlin/Heidelberg, Germany, 1979.
25. Jantosciak, J. Classical geometries as hypergroups. In Proceedings of the Atti del Convegno su Ipergruppi altre Structure Multivoche et loro Applicazioni, Udine, Italy, 15–18 October 1985; pp. 93–104.
26. Jantosciak, J. A brif survey of the theory of join spaces. In Proceedings of the 5th Intern. Congress on Algebraic Hyperstructures and Applications, Iasi, Romania, 4–10 July 1993; Hadronic Press: Palm Harbor, FL, USA, 1994; pp. 109–122.
27. Barlotti, A.; Strambach, K. Multigroups and the foundations of Geometry. *Rend. Circ. Mat. Palermo* **1991**, *XL*, 5–68. [CrossRef]
28. Freni, D. Sur les hypergroupes cambistes. *Rend. Ist. Lomb.* **1985**, *119*, 175–186.
29. Freni, D. Sur la théorie de la dimension dans les hypergroupes. *Acta Univ. Carol. Math. Phys.* **1986**, *27*, 67–80.
30. Massouros, C.G. Hypergroups and convexity. *Riv. Mat. Pura Appl.* **1989**, *4*, 7–26.
31. Mittas, J.; Massouros, C.G. Hypergroups defined from linear spaces. *Bull. Greek Math. Soc.* **1989**, *30*, 63–78.
32. Massouros, C.G. Hypergroups and Geometry. *Mem. Acad. Romana Math. Spec. Issue* **1996**, *XIX*, 185–191.
33. Massouros, C.G. On connections between vector spaces and hypercompositional structures. *Ital. J. Pure Appl. Math.* **2015**, *34*, 133–150.
34. Dramalidis, A. On geometrical hyperstructures of finite order. *Ratio Math.* **2011**, *21*, 43–58.
35. Massouros, G.G.; Mittas, J. Languages—Automata and hypercompositional structures. In Proceedings of the 5th Intern. Congress on Algebraic Hyperstructures and Applications, Xanthi, Greece, 27–30 June 1990; World Scientific: Singapore, 1991; pp. 137–147.
36. Massouros, G.G. Automata, Languages and Hypercompositional Structures. Ph.D. Thesis, National Technical University of Athens, Athens, Greece, 1993.
37. Massouros, G.G. Automata and hypermoduloids. In Proceedings of the 5th Intern. Congress on Algebraic Hyperstructures and Applications, Iasi, Romania, 4–10 July 1993; Hadronic Press: Palm Harbor, FA, USA, 1994; pp. 251–265.
38. Massouros, G.G. An automaton during its operation. In Proceedings of the 5th Internation Congress on Algebraic Hyperstructures and Applications, Iasi, Romania, 4–10 July 1993; Hadronic Press: Palm Harbor, FA, USA, 1994; pp. 267–276.
39. Massouros, G.G. On the attached hypergroups of the order of an automaton. *J. Discret. Math. Sci. Cryptogr.* **2003**, *6*, 207–215. [CrossRef]
40. Massouros, G.G. Hypercompositional structures in the theory of languages and automata. *An. Sti. Univ. Al. I. Cuza Iasi Sect. Inform.* **1994**, *III*, 65–73.
41. Massouros, G.G. Hypercompositional structures from the computer theory. *Ratio Math.* **1999**, *13*, 37–42.

42. Massouros, C.G.; Massouros, G.G. Hypergroups associated with graphs and automata. *AIP Conf. Proc.* **2009**, *1168*, 164–167. [CrossRef]
43. Massouros, C.G. On path hypercompositions in graphs and automata. *MATEC Web Conf.* **2016**, *41*, 5003. [CrossRef]
44. Chvalina, J.; Chvalinova, L. State hypergroups of Automata. *Acta Math. Inform. Univ. Ostrav.* **1996**, *4*, 105–120.
45. Chvalina, J.; Křehlík, S.; Novák, M. Cartesian composition and the problem of generalizing the MAC condition to quasi-multiautomata. *An. St. Univ. Ovidius Constanta* **2016**, *24*, 79–100. [CrossRef]
46. Chvalina, J.; Novák, M.; Křehlík, S. Hyperstructure generalizations of quasi-automata induced by modelling functions and signal processing. *AIP Conf. Proc.* **2019**, *2116*, 310006. [CrossRef]
47. Novák, M.; Křehlík, S.; Stanek, D. n-ary Cartesian composition of automata. *Soft Comput.* **2019**, *24*, 1837–1849. [CrossRef]
48. Novák, M. Some remarks on constructions of strongly connected multiautomata with the input semihypergroup being a centralizer of certain transformation operators. *J. Appl. Math.* **2008**, *I*, 65–72.
49. Chvalina, J.; Novák, M.; Smetana, B. Staněk, D. Sequences of Groups, Hypergroups and Automata of Linear Ordinary Differential Operators. *Mathematics* **2021**, *9*, 319. [CrossRef]
50. Křehlík, S. n-Ary Cartesian Composition of Multiautomata with Internal Link for Autonomous Control of Lane Shifting. *Mathematics* **2020**, *8*, 835. [CrossRef]
51. Chorani, M.; Zahedi, M.M. Some hypergroups induced by tree automata. *Aust. J. Basic Appl. Sci.* **2012**, *6*, 680–692.
52. Chorani, M. State hyperstructures of tree automata based on lattice-valued logic. *RAIRO Theor. Inf. Appl.* **2018**, *52*, 23–42.
53. Krasner, M. Approximation des Corps Valués Complets de Caractéristique p≠0 par Ceux de Caractéristique 0, Colloque d' Algèbre Supérieure (Bruxelles, Decembre 1956), Centre Belge de Recherches Mathématiques, Établissements Ceuterick, Louvain, Librairie Gauthier-Villars, Paris. 1957; 129–206.
54. Viro, O. On basic concepts of tropical geometry. *Proc. Steklov Inst. Math.* **2011**, *273*, 252–282. [CrossRef]
55. Viro, O. Hyperfields for Tropical Geometry I. Hyperfields and dequantization. *arxiv* **2010**, arXiv:1006.3034.
56. Baker, M.; Bowler, N. Matroids over hyperfields. *arxiv* **2017**, arXiv:1601.01204.
57. Jun, J. Geometry of hyperfields. *arxiv* **2017**, arXiv:1707.09348.
58. Lorscheid, O. Tropical geometry over the tropical hyperfield. *arxiv* **2019**, arXiv:1907.01037.
59. Connes, A.; Consani, C. The hyperring of adèle classes. *J. Number Theory* **2011**, *131*, 159–194. [CrossRef]
60. Connes, A.; Consani, C. From monoids to hyperstructures: In search of an absolute arithmetic. *arXiv* **2010**, arXiv:1006.4810.
61. Massouros, C.G. Free and cyclic hypermodules. *Ann. Mat. Pura Appl.* **1988**, *150*, 153–166. [CrossRef]
62. Mittas, J. Espaces vectoriels sur un hypercorps. Introduction des hyperspaces affines et Euclidiens. *Math. Balk.* **1975**, *5*, 199–211.
63. Vahedi, V.; Jafarpour, M.; Aghabozorgi, H.; Cristea, I. Extension of elliptic curves on Krasner hyperfields. *Comm. Algebra* **2019**, *47*, 4806–4823. [CrossRef]
64. Vahedi, V.; Jafarpour, M.; Cristea, I. Hyperhomographies on Krasner Hyperfields. *Symmetry* **2019**, *11*, 1442. [CrossRef]
65. Vahedi, V.; Jafarpour, M.; Hoskova-Mayerova, S.; Aghabozorgi, H.; Leoreanu-Fotea, V.; Bekesiene, S. Derived Hyperstructures from Hyperconics. *Mathematics* **2020**, *8*, 429. [CrossRef]
66. Freni, D. Strongly Transitive Geometric Spaces: Applications to Hypergroups and Semigroups Theory. *Commun. Algebra* **2004**, *32*, 969–988. [CrossRef]
67. Massouros, G.G.; Massouros, C.G. Hypercompositional algebra, computer science and geometry. *Mathematics* **2020**, *8*, 1338. [CrossRef]
68. Corsini, P.; Leoreanu, V. *Applications of Hyperstructures Theory*; Kluwer Academic Publishers: Dordrecht, The Netherlands, 2003.
69. Hoskova-Mayerova, S.; Maturo, A. Decision-making process using hyperstructures and fuzzy structures in social sciences. In *Soft Computing Applications for Group Decision Making and Consensus Modeling*; Studies in Fuzziness and Soft Computing 357; Collan, M., Kacprzyk, J., Eds.; Springer International Publishing: Cham, Switzerland, 2018; pp. 103–111.
70. Hoskova-Mayerova, S.; Maturo, A. Algebraic hyperstructures and social relations. *Ital. J. Pure Appl. Math.* **2018**, *39*, 701–709.
71. Hoskova-Mayerova, S.; Maturo, A. Fuzzy Sets and Algebraic Hyperoperations to Model Interpersonal Relations. In *Recent Trends in Social Systems: Quantitative Theories and Quantitative Models*; Maturo, A., Hoskova-Mayerova, S., Soitu, D.T., Kacprzyk, J., Eds.; Studies in Systems, Decision and Control 66; Springer International Publishing: Cham, Switzerland, 2017; pp. 211–221.
72. Hoskova-Mayerova, S.; Maturo, A. An analysis of social relations and social group behaviors with fuzzy sets and hyperstructures. *Int. J. Algebraic Hyperstruct. Appl.* **2015**, *2*, 91–99.
73. Hoskova-Mayerova, S.; Maturo, A. Hyperstructures in social sciences. *AWER Procedia Inf. Technol. Comput. Sci.* **2013**, *3*, 547–552.
74. Golzio, A.C. Non-Deterministic Matrices: Theory and Applications to Algebraic Semantics. Ph.D. Thesis, IFCH, University of Campinas, Campinas, Brazil, 2017.
75. Golzio, A.C. A brief historical Survey on hyperstructures in Algebra and Logic. *S. Am. J. Log.* **2018**, *4*, 91–119.
76. Massouros, C.G.; Massouros, G.G. On open and closed hypercompositions. *AIP Conf. Proc.* **2017**, *1978*, 340002-1–340002-4. [CrossRef]
77. Massouros, C.G.; Dramalidis, A. Transposition Hv-groups. *ARS Comb.* **2012**, *106*, 143–160.
78. Novák, M.; Křehlík, Š. EL-hyperstructures revisited. *Soft Comput.* **2018**, *22*, 7269–7280. [CrossRef]
79. Tsitouras, C.G.; Massouros, C.G. On enumeration of hypergroups of order 3. *Comput. Math. Appl.* **2010**, *59*, 519–523. [CrossRef]
80. Vougiouklis, T. *Hyperstructures and Their Representations*; Hadronic Press: Palm Harbor, FL, USA, 1994.
81. Corsini, P. Binary relations and hypergroupoids. *Ital. J. Pure Appl. Math.* **2000**, *7*, 11–18.
82. Corsini, P. On the hypergroups associated with a binary relation. *Multi. Val. Log.* **2000**, *5*, 407–419.

83. Corsini, P.; Leoreanu, V. Hypergroups and binary relations. *Algebra Univers.* **2000**, *43*, 321–330. [CrossRef]
84. Davvaz, B.; Leoreanu-Fotea, V. Binary relations on ternary semihypergroups. *Comm. Algebra* **2010**, *38*, 3621–3636. [CrossRef]
85. Rosenberg, I.G. Hypergroups and join spaces determined by relations. *Ital. J. Pure Appl. Math.* **1998**, *4*, 93–101.
86. Cristea, I. Several aspects on the hypergroups associated with n-ary relations. *An. Stiint. Univ. Ovidius Constanta Ser. Mat.* **2009**, *17*, 99–110.
87. Cristea, I.; Stefanescu, M. Binary relations and reduced hypergroups. *Discret. Math.* **2008**, *308*, 3537–3544. [CrossRef]
88. Kankaras, M.; Cristea, I. Fuzzy reduced hypergroups. *Mathematics* **2020**, *8*, 263. [CrossRef]
89. Cristea, I.; Stefanescu, M.; Angheluta, C. About the fundamental relations defined on the hypergroupoids associated with binary relations. *Electron. J. Combin.* **2011**, *32*, 72–81. [CrossRef]
90. Cristea, I.; Jafarpour, M.; Mousavi, S.S.; Soleymani, A. Enumeration of Rosenberg hypergroups. *Comput. Math. Appl.* **2011**, *32*, 72–81. [CrossRef]
91. DeSalvo, M.; LoFaro, G. Hypergroups and binary relations. *Multi. Val. Log.* **2002**, *8*, 645–657.
92. DeSalvo, M.; LoFaro, G. A new class of hypergroupoids associated to binary relations. *J. Mult. Val. Log. Soft Comput.* **2003**, *9*, 361–375.
93. Pelea, C.; Purdea, I. Direct limits of direct systems of hypergroupoids associated to binary relations, in Advances in Mathematics of Uncertainty. In Proceedings of the Symposium "Mathematics of Uncertainty", ECIT 2006, Lasi, Romania, 20–23 September 2006; pp. 31–42.
94. Breaz, S.; Pelea, C.; Purdea, I. Products of hypergroupoids associated to binary relations. *Carpathian J. Math.* **2009**, *25*, 23–36.
95. Massouros, C.G.; Tsitouras, C.G. Enumeration of hypercompositional structures defined by binary relations. *Ital. J. Pure Appl. Math.* **2011**, *28*, 43–54.
96. Tsitouras, C.G.; Massouros, C.G. Enumeration of Rosenberg type hypercompositional structures defined by binary relations. *Eur. J. Comb.* **2012**, *33*, 1777–1786. [CrossRef]
97. Corsini, P.; Tofan, I. On fuzzy hypergroups. *Pure Math. Appl.* **1997**, *8*, 29–37.
98. Tofan, I.; Volf, A.C. On some connections between hyperstructures and fuzzy sets. *Ital. J. Pure Appl. Math.* **2000**, *7*, 63–68.
99. Tofan, I.; Volf, A.C. Algebraic aspects of information organization. In *Systematic Organization of Information in Fuzzy Systems*; Melo-Pinto, P., Teodorescu, H.N., Fukuda, T., Eds.; 184 of NATO Science Series, III: Computer and Systems Sciences; IOS Press: Amsterdam, The Netherlands, 2003; pp. 71–87.
100. Corsini, P. Join Spaces, Power Sets, Fuzzy Sets. In Proceedings of the 5th International Congress on Algebraic Hyperstructures and Applications, Iasi, Romania, 4–10 July 1993; Hadronic Press: Palm Harbor, FL, USA, 1994; pp. 45–52.
101. Corsini, P.; Leoreanu, V. Fuzzy sets and join spaces associated with rough sets. *Rend. Circ. Mat. Palermo* **2002**, *51*, 527–536. [CrossRef]
102. Corsini, P.; Cristea, I. Fuzzy sets and non complete 1-hypergroups. *An. Stalele Univ. Ovidius Constanta* **2005**, *13*, 27–54.
103. Corsini, P.; Leoreanu-Fotea, V. On the grade of a sequence of fuzzy sets and join spaces determined by a hypergraph. *Southeast Asian Bull. Math.* **2010**, *34*, 231–242.
104. Leoreanu, V. Direct limit and inverse limit of join spaces associated with fuzzy sets. *Pure Math. Appl.* **2000**, *11*, 509–516.
105. Leoreanu-Fotea, V. Fuzzy join n-ary spaces and fuzzy canonical n-ary hypergroups. *Fuzzy Set. Syst.* **2010**, *161*, 3166–3173. [CrossRef]
106. Leoreanu, V. About hyperstructures associated with fuzzy sets of type 2. *Ital. J. Pure Appl. Math.* **2005**, *17*, 127–136.
107. Cristea, I. Complete hypergroups, 1-hypergroups and fuzzy sets. *An. St. Univ. Ovidius Constanta* **2000**, *10*, 25–38.
108. Cristea, I. A property of the connection between fuzzy sets and hypergroupoids. *Ital. J. Pure Appl. Math.* **2007**, *21*, 73–82.
109. Stefanescu, M.; Cristea, I. On the fuzzy grade of hypergroups. *Fuzzy Set. Syst.* **2008**, *159*, 1097–1106. [CrossRef]
110. Cristea, I.; Davvaz, B. Atanassov's intuitionistic fuzzy grade hypergroups. *Inf. Sci.* **2010**, *180*, 1506–1517. [CrossRef]
111. Davvaz, B.; Sadrabadi, E.H.; Cristea, I. Atanassov's intuitionistic fuzzy grade of ips hypergroups of order less than or equal to 6. *Iran. J. Fuzzy Syst.* **2012**, *9*, 71–97.
112. Angheluta, C.; Cristea, I. On Atanassov's intuitionistic fuzzy grade of complete hypergroups. *J. Mult. Val. Log. Soft Comput.* **2013**, *20*, 55–74.
113. Cristea, I. Hyperstructures and fuzzy sets endowed with two membership functions. *Fuzzy Sets Syst.* **2009**, *160*, 1114–1124. [CrossRef]
114. Cristea, I.; Hoskova, S. Fuzzy Pseudotopological Hypergroupoids. *Iran. J. Fuzzy Syst.* **2009**, *6*, 11–19.
115. Kehagias, A. An example of L-fuzzy Join Space. *Rend. Circ. Mat. Palermo* **2002**, *51*, 503–526. [CrossRef]
116. Kehagias, A. Lattice-fuzzy Meet and Join Hyperoperations. In Proceedings of the 8th Intern. Congress on Algebraic Hyperstructures and Applications, Samothraki, Greece, 1–9 September 2002; Spanidis Press: Xanthi, Greece, 2003; pp. 171–182.
117. Ameri, R.; Zahedi, M.M. Hypergroup and join space induced by a fuzzy subset. *Pure Math. Appl.* **1997**, *8*, 155–168.
118. Ameri, R. Fuzzy transposition hypergroups. *Ital. J. Pure Appl. Math.* **2005**, *18*, 147–154.
119. Zahedi, M.M.; Bolurian, M.; Hasankhani, A. On polygroups and fuzzy sub-polygroups. *J. Fuzzy Math.* **1995**, *3*, 1–15.
120. Zahedi, M.M.; Hasankhani, A. F-Polygroups. *Int. J. Fuzzy Math.* **1996**, *4*, 533–548.
121. Zahedi, M.M.; Hasankhani, A. F-Polygroups (II). *Inf. Sci.* **1996**, *89*, 225–243. [CrossRef]
122. Hasankhani, A.; Zahedi, M.M. Fuzzy sub-F-polygroups. *Ratio Math.* **1997**, *12*, 35–44.
123. Chowdhury, G. Fuzzy transposition hypergroups. *Iran. J. Fuzzy Syst.* **2009**, *6*, 37–52.
124. Davvaz, B.; Cristea, I. *Fuzzy Algebraic Hyperstructures*; Springer: Berlin/Heidelberg, Germany, 2015.

125. Massouros, C.G.; Massouros, G.G. On certain fundamental properties of hypergroups and fuzzy hypergroups—Mimic fuzzy hypergroups. *Intern. J. Risk Theory* **2012**, *2*, 71–82.
126. Massouros, C.G.; Massouros, G.G. On 2-element Fuzzy and Mimic Fuzzy Hypergroups. *AIP Conf. Proc.* **2012**, *1479*, 2213–2216. [CrossRef]
127. Hoskova-Mayerova, S.; Al Tahan, M.; Davvaz, B. Fuzzy multi-hypergroups. *Mathematics* **2020**, *8*, 244. [CrossRef]
128. Jantosciak, J. Transposition hypergroups, Noncommutative Join Spaces. *J. Algebra* **1997**, *187*, 97–119. [CrossRef]
129. Corsini, P. Graphs and Join Spaces. *J. Comb. Inf. Syst. Sci.* **1991**, *16*, 313–318.
130. Corsini, P. Binary relations, interval structures and join spaces. *J. Appl. Math. Comput.* **2002**, *10*, 209–216. [CrossRef]
131. Leoreanu, V. On the heart of join spaces and of regular hypergroups. *Riv. Mat. Pura Appl.* **1995**, *17*, 133–142.
132. Leoreanu-Fotea, V.; Rosenberg, I.G. Join Spaces Determined by Lattices. *J. Mult. Val. Log. Soft Comput.* **2010**, *16*, 7–16.
133. Hoskova, S. Abelization of join spaces of affine transformations of ordered field with proximity. *Appl. Gen. Topol.* **2005**, *6*, 57–65. [CrossRef]
134. Chvalina, J.; Hoskova, S. Modelling of join spaces with proximities by first-order linear partial differential operators. *Ital. J. Pure Appl. Math.* **2007**, *21*, 177–190.
135. Hoskova, S.; Chvalina, J.; Rackova, P. Transposition hypergroups of Fredholm integral operators and related hyperstructures (Part I). *J. Basic Sci.* **2006**, *3*, 11–17.
136. Hoskova, S.; Chvalina, J.; Rackova, P. Transposition hypergroups of Fredholm integral operators and related hyperstructures (Part II). *J. Basic Sci.* **2008**, *4*, 55–60.
137. Chvalina, J.; Hoskova-Mayerova, S. On Certain Proximities and Preorderings on the Transposition Hypergroups of Linear First-Order Partial Differential Operators. *An. Stalele Univ. Ovidius Constanta Ser. Mat.* **2014**, *22*, 85–103. [CrossRef]
138. Chvalina, J.; Chvalinova, L. Transposition hypergroups formed by transformation operators on rings of differentiable functions. *Ital. J. Pure Appl. Math.* **2004**, *15*, 93–106.
139. Massouros, C.G. Canonical and Join Hypergroups. *An. Sti. Univ. Al. I. Cuza Iasi* **1996**, *42*, 175–186.
140. Massouros, C.G. Normal homomorphisms of fortified join hypergroups. In Proceedings of the 5th Intern. Congress on Algebraic Hyperstructures and Applications, Iasi, Rommania, 4–10 July 1993; Hadronic Press: Palm Harbor, FL, USA, 1994; pp. 133–142.
141. Jantosciak, J.; Massouros, C.G. Strong Identities and fortification in Transposition hypergroups. *J. Discret. Math. Sci. Cryptogr.* **2003**, *6*, 169–193. [CrossRef]
142. Massouros, C.G. Isomorphism theorems in fortified transposition hypergroups. *AIP Conf. Proc.* **2013**, *1558*, 2059–2062. [CrossRef]
143. Massouros, C.G. On the enumeration of rigid hypercompositional structures. *AIP Conf. Proc.* **2015**, *1648*, 740005-1–740005-6. [CrossRef]
144. Massouros, G.G.; Massouros, C.G.; Mittas, J. Fortified join hypergroups. *Ann. Math. Blaise Pascal* **1996**, *3*, 155–169. [CrossRef]
145. Massouros, G.G.; Zafiropoulos, F.A.; Massouros, C.G. Transposition polysymmetrical hypergroups. In Proceedings of the 8th Intern. Congress on Algebraic Hyperstructures and Applications, Samothraki, Greece, 1–9 September 2002; Spanidis Press: Xanthi, Greece, 2003; pp. 191–202.
146. Massouros, C.G.; Massouros, G.G. Transposition polysymmetrical hypergroups with strong identity. *J. Basic Sci.* **2008**, *4*, 85–93.
147. Massouros, C.G.; Massouros, G.G. The transposition axiom in hypercompositional structures. *Ratio Math.* **2011**, *21*, 75–90.
148. Massouros, C.G.; Massouros, G.G. On subhypergroups of fortified transposition hypergroups. *AIP Conf. Proc.* **2013**, *1558*, 2055–2058. [CrossRef]
149. Massouros, C.G.; Massouros, G.G. On the algebraic structure of transposition hypergroups with idempotent identity. *Iran. J. Math. Sci. Inform.* **2013**, *8*, 57–74.
150. Massouros, C.G.; Massouros, G.G. Transposition hypergroups with idempotent identity. *Int. J. Algebr. Hyperstruct. Appl.* **2014**, *1*, 15–27.
151. Massouros, G.G. Fortified join hypergroups and join hyperrings. *An. Sti. Univ. Al. I. Cuza Iasi Sect. I Mat.* **1995**, *XLI*, 37–44.
152. Massouros, G.G. The subhypergroups of the fortified join hypergroup. *Ital. J. Pure Appl. Math.* **1997**, *2*, 51–63.
153. Nieminen, J. Join Space Graphs. *J. Geom.* **1988**, *33*, 99–103. [CrossRef]
154. Nieminen, J. Chordal Graphs and Join Spaces. *J. Geom.* **1989**, *34*, 146–151. [CrossRef]
155. Ameri, R. Topological transposition hypergroups. *Ital. J. Pure Appl. Math.* **2003**, *13*, 181–186.
156. Massouros, C.G.; Massouros, G.G. Identities in multivalued algebraic structures. *AIP Conf. Proc.* **2010**, *1281*, 2065–2068. [CrossRef]
157. Bonansinga, P. Sugli ipergruppi quasicanonici. *Atti Soc. Pelor. Sc. Fis. Mat. Nat.* **1981**, *27*, 9–17.
158. Massouros, C.G. Quasicanonical hypergroups. In Proceedings of the 4th Internation Congress, on Algebraic Hyperstructures and Applications, Xanthi, Greece, 27–30 June 1990; World Scientific: Singapore, 1991; pp. 129–136.
159. Ioulidis, S. Polygroups et certaines de leurs propriétiés. *Bull. Greek Math. Soc.* **1981**, *22*, 95–104.
160. Comer, S. Polygroups derived from cogroups. *J. Algebra* **1984**, *89*, 397–405. [CrossRef]
161. Davvaz, B. *Polygroup Theory and Related Systems*; World Scientific: Singapore, 2013.
162. Harrison, D.K. Double coset and orbit spaces. *Pacific, J. Math.* **1979**, *80*, 451–491. [CrossRef]
163. Mittas, J. Sur une classe d'hypergroupes commutatifs. *C. R. Acad. Sci. Paris* **1969**, *269*, 485–488.
164. Mittas, J. Hypergroupes values et hypergroupes fortement canoniques. *Proc. Acad. Athens* **1969**, *44*, 304–312.
165. Krasner, M. A class of hyperrings and hyperfields. *Int. J. Math. Math. Sci.* **1983**, *6*, 307–312. [CrossRef]
166. Massouros, C.G. Methods of constructing hyperfields. *Intern. J. Math. Math. Sci.* **1985**, *8*, 725–728. [CrossRef]

167. Massouros, C.G. On the theory of hyperrings and hyperfields. *Algebra Log.* **1985**, *24*, 728–742. [CrossRef]
168. Massouros, C.G. A class of hyperfields and a problem in the theory of fields. *Math. Montisnigri* **1993**, *1*, 73–84.
169. Massouros, C.G. A Field Theory Problem Relating to Questions in Hyperfield Theory. *AIP Conf. Proc.* **2011**, *1389*, 1852–1855. [CrossRef]
170. Nakassis, A. Recent results in hyperring and hyperfield theory. *Intern. J. Math. Math. Sci.* **1988**, *11*, 209–220. [CrossRef]
171. Mittas, J. Sur les hyperanneaux et les hypercorps. *Math. Balk.* **1973**, *3*, 368–382.
172. Mittas, J. Hypergroupes canoniques hypervalues. *C. R. Acad. Sci. Paris* **1970**, *271*, 4–7.
173. Mittas, J. Les hypervaluations strictes des hypergroupes canoniques. *C. R. Acad. Sci. Paris* **1970**, *271*, 69–72.
174. Mittas, J. Hypergroupes et hyperanneaux polysymetriques. *C. R. Acad. Sc. Paris* **1970**, *271*, 920–923.
175. Mittas, J. Contributions a la théorie des hypergroupes, hyperanneaux, et les hypercorps hypervalues. *C. R. Acad. Sc. Paris* **1971**, *272*, 3–6.
176. Mittas, J. Hypergroupes canoniques values et hypervalues. *Math. Balk* **1971**, *1*, 181–185.
177. Mittas, J. Hypergroupes canoniques. *Math. Balk.* **1972**, *2*, 165–179.
178. Mittas, J. Contributions a la théorie des structures ordonnées et des structures valuées. *Proc. Acad. Athens* **1973**, *48*, 319–331.
179. Mittas, J. Hypergroupes canoniques values et hypervalues. Hypergroupes fortement et superieurement canoniques. *Bull. Greek Math. Soc.* **1982**, *23*, 55–88.
180. Mittas, J. Certaines remarques sur les hypergroupes canoniques hypervaluables et fortement canoniques. *Riv. Mat. Pura Appl.* **1991**, *9*, 61–67.
181. Mittas, J. Hypergroupes polysymetriques canoniques. In Proceedings of the Atti del Convegno su Ipergruppi, Altre Strutture Multivoche e loro Applicazioni, Udine, Italy, 15–18 October 1985; pp. 1–25.
182. Yatras, C. M-polysymmetrical hypergroups. *Riv. Mat. Pura Appl.* **1992**, *11*, 81–92.
183. Yatras, C. Subhypergroups of M-polysymmetrical hypergroups. In Proceedings of the 5th Intern. Congress on Algebraic Hyperstructures and Applications, Iasi, Rommania, 4–10 July 1993; Hadronic Press: Palm Harbor, FA, USA, 1994; pp. 123–132.
184. Yatras, C. Types of Polysymmetrical Hypergroups. *AIP Conf. Proc.* **2018**, *1978*, 340004-1–340004-5. [CrossRef]
185. Mittas, J. Generalized M-Polysymmetric Hypergroups. In Proceedings of the 10th Internation Congress on Algebraic Hyperstructures and Applications, Brno, Czech Republic, 3–9 September 2008; pp. 303–309.
186. Massouros, C.G.; Mittas, J. On the theory of generalized M-polysymmetric hypergroups. In Proceedings of the 10th Intern. Congress, on Algebraic Hyperstructures and Applications, Brno, Czech Republic, 3–9 September 2008; pp. 217–228.
187. Massouros, G.G. On the Hypergroup Theory. *Fuzzy Syst. A.I. Rep. Lett. Acad. Romana* **1995**, *IV*, 13–25.
188. Sonea, A.C.; Cristea, I. The Class Equation and the Commutativity Degree for Complete Hypergroups. *Mathematics* **2020**, *8*, 2253. [CrossRef]
189. De Salvo, M.; Lo Faro, G. On the n*-complete hypergroups. *Discret. Math.* **1999**, *208/209*, 177–188. [CrossRef]
190. De Salvo, M.; Fasino, D.; Freni, D.; Lo Faro, G. 1-hypergroups of small size. *Mathematics* **2021**, *9*, 108. [CrossRef]
191. De Salvo, M.; Freni, D.; Lo Faro, G. A new family of hypergroups and hypergroups of type U on the right of size five. *Far East J. Math. Sci.* **2007**, *26*, 393–418.
192. De Salvo, M.; Freni, D.; Lo Faro, G. On the hypergroups of type U on the right of size five, with scalar identity. *J. Mult. Valued Log. Soft Comput.* **2011**, *17*, 425–441.
193. De Salvo, M.; Fasino, D.; Freni, D.; Lo Faro, G. On strongly conjugable extensions of hypergroups of type U with scalar identity. *Filomat* **2013**, *27*, 977–994.
194. Fasino, D.; Freni, D. Existence of proper semihypergroups of type U on the right. *Discret. Math.* **2007**, *307*, 2826–2836. [CrossRef]
195. Fasino, D.; Freni, D. Minimal order semihypergroups of type U on the right. *Mediterr. J. Math.* **2008**, *5*, 295–314. [CrossRef]
196. Freni, D. Structure des hypergroupes quotients et des hypergroupes de type U. *Ann. Sci. Univ. Clermont-Ferrand II Math.* **1984**, *22*, 51–77.
197. Freni, D. Minimal order semihypergroups of type U on the right, II. *J. Algebra* **2011**, *340*, 77–89. [CrossRef]
198. Freni, D.; Gutan, M. Sur les hypergroupes de type U. *Mathematica (Cluj)* **1994**, *36*, 25–32.
199. Gutan, M. Sur une classe d'hypergroupes de type C. *Ann. Math. Blaise Pascal* **1994**, *1*, 1–19. [CrossRef]
200. Gutan, M.; Sureau, Y. Hypergroupes de type C 'a petites partitions. *Riv. Mat. Pura Appl.* **1995**, *16*, 13–38.
201. Heidari, D.; Cristea, I. Breakable semihypergroups. *Symmetry* **2019**, *11*, 100. [CrossRef]
202. Heidari, D.; Cristea, I. On Factorizable semihypergroups. *Mathematics* **2020**, *8*, 1064. [CrossRef]
203. Bordbar, H.; Cristea, I. Regular Parameter Elements and Regular Local Hyperrings. *Mathematics* **2021**, *9*, 243. [CrossRef]
204. Massouros, C.G. Separation and Relevant Properties in Hypergroups. *AIP Conf. Proc.* **2016**, *1738*, 480051-1–480051-4. [CrossRef]
205. Massouros, C.G. On the semi-subhypergroups of a hypergroup. *Int. J. Math. Math. Sci.* **1991**, *14*, 293–304. [CrossRef]
206. Massouros, C.G. Some properties of certain subhypergroups. *Ratio Math.* **2013**, *25*, 67–76.
207. Freni, D. Ipergruppi ciclici e torsione negli ipergruppi. *Matematica (Catania)* **1980**, *35*, 270–286.
208. De Salvo, M.; Freni, D. Sugli ipergruppi ciclici e completi. *Matematiche (Catania)* **1980**, *35*, 211–226.
209. De Salvo, M.; Freni, D. Semi-ipergruppi e ipergruppi ciclici. *Atti Sem. Mat. Fis. Univ. Modena* **1981**, *30*, 44–59.
210. DeSalvo, M.; Freni, D.; LoFaro, G. Hypercyclic subhypergroups of finite fully simple semihypergroups. *J. Mult. Valued Log. Soft Comput.* **2017**, *29*, 595–617.
211. Vougiouklis, T. Cyclicity in a special class of hypergroups. *Acta Univ. Carolin. Math. Phys.* **1981**, *22*, 3–6.

212. Leoreanu, V. About the simplifiable cyclic semihypergroups. *Ital. J. Pure Appl. Math.* **2000**, *7*, 69–76.
213. Al Tahan, M.; Davvaz, B. On a special single-power cyclic hypergroup and its automorphisms. *Discret. Math. Algorithms Appl.* **2016**, *7*, 1650059. [CrossRef]
214. Al Tahan, M.; Davvaz, B. Commutative single power cyclic hypergroups of order three and period two. *Discret. Math. Algorithms Appl.* **2017**, *9*, 1750070. [CrossRef]
215. Al Tahan, M.; Davvaz, B. On some properties of single power cyclic hypergroups and regular relations. *J. Algebra Appl.* **2017**, *16*, 1750214. [CrossRef]
216. Al Tahan, M.; Davvaz, B. The cyclic hypergroup associated with Sn its automorphism group and its fuzzy grade. *Discret. Math. Algorithms Appl.* **2018**, *10*, 1850070. [CrossRef]
217. Al Tahan, M.; Davvaz, B. Hypermatrix representations of single power cyclic hypergroups. *Ital. J. Pure Appl. Math.* **2017**, *38*, 679–696.
218. Gu, Z. On cyclic hypergroups. *J. Algebra Appl.* **2019**, *18*, 1950213. [CrossRef]
219. Novak, M.; Krehlik, S.; Cristea, I. Cyclicity in EL-hypergroups. *Symmetry* **2018**, *10*, 611. [CrossRef]
220. Sureau, Y. Sous-hypergroupe engendre par deux sous-hypergroupes et sous-hypergroupe ultra-clos d'un hypergroupe. *C. R. Acad. Sc. Paris* **1977**, *284*, 983–984.
221. Haddad, L.; Sureau, Y. Les cogroupes et les D-hypergroupes. *J. Algebra* **1988**, *118*, 468–476. [CrossRef]
222. Haddad, L.; Sureau, Y. Les cogroupes et la construction de Utumi. *Pac. J. Math.* **1990**, *145*, 17–58. [CrossRef]

Article

Hypercompositional Algebra, Computer Science and Geometry

Gerasimos Massouros [1,*] **and Christos Massouros** [2]

[1] School of Social Sciences, Hellenic Open University, Aristotelous 18, GR 26335 Patra, Greece
[2] Core Department, Euripus Campus, National and Kapodistrian University of Athens, GR 34400 Euboia, Greece; ChrMas@uoa.gr or Ch.Massouros@gmail.com
* Correspondence: germasouros@gmail.com

Received: 21 May 2020; Accepted: 14 July 2020; Published: 11 August 2020

Abstract: The various branches of Mathematics are not separated between themselves. On the contrary, they interact and extend into each other's sometimes seemingly different and unrelated areas and help them advance. In this sense, the Hypercompositional Algebra's path has crossed, among others, with the paths of the theory of Formal Languages, Automata and Geometry. This paper presents the course of development from the hypergroup, as it was initially defined in 1934 by F. Marty to the hypergroups which are endowed with more axioms and allow the proof of Theorems and Propositions that generalize Kleen's Theorem, determine the order and the grade of the states of an automaton, minimize it and describe its operation. The same hypergroups lie underneath Geometry and they produce results which give as Corollaries well known named Theorems in Geometry, like Helly's Theorem, Kakutani's Lemma, Stone's Theorem, Radon's Theorem, Caratheodory's Theorem and Steinitz's Theorem. This paper also highlights the close relationship between the hyperfields and the hypermodules to geometries, like projective geometries and spherical geometries.

Keywords: hypergroup; hyperfield; formal languages; automata; convex set; vector space; geometry

---

1. Introduction

This paper is written in the context of the special issue "Hypercompositional Algebra and Applications" in "*Mathematics*" and it aims to shed light on two areas where the Hypercompositional Algebra has expanded and has interacted with them: Computer Science and Geometry.

Hypercompositional Algebra is a branch of Abstract Algebra which appeared in the 1930s via the introduction of the hypergroup.

It is interesting that the group and the hypergroup are two algebraic structures which satisfy exactly the same axioms, i.e., the associativity and the reproductivity, but they differ in the law of synthesis. In the first one, the law of synthesis is a composition, while in the second one it is a hypercomposition. This difference makes the hypergroup a much more general algebraic structure than the group, and for this reason the hypergroup has been gradually enriched with further axioms, which are either more powerful or less powerful, leading thus to a significant number of special hypergroups. Among them, there exist hypergroups that were proved to be very useful for the study of Formal Languages and Automata, as well as convexity in Euclidian vector spaces. Furthermore, based on these hypergroups, there derived other hypercompositional structures, which are equally as useful in the study of Geometries (spherical, projective, tropical, etc.) and Computer Science.

A *binary operation* (or *composition*) " · " on a non-void set E is a rule which assigns a unique element of E to each element of $E \times E$. The notation $a_i \cdot a_j = a_k$, where a_i, a_j, a_k are elements of E, indicates that a_k is the result of the operation " · " performed on the operands a_i and a_j. When no confusion arises, the operation symbol "·" may be omitted, and we write $a_i a_j = a_k$. If the binary operation is *associative*, that is, if it satisfies the equality

$$a(bc) = (ab)c \text{ for all } a, b, c \in E$$

the pair $(E, \cdot)$ is called *semigroup*. An element $e \in E$ is an *identity* of $(E, \cdot)$ if for all $a \in E$, $ea = ae = a$. A triplet $(E, \cdot, e)$ is a *monoid* if $(E, \cdot)$ is a semigroup and e is its identity. If no ambiguity arises, we can denote a semigroup $(E, \cdot)$ or a monoid $(E, \cdot, e)$ simply by E. A binary operation on E is *reproductive* if it satisfies:

$$aE = Ea = E \text{ for all } a \in E$$

Definition 1. *(Definition of the group) The pair $(G, \cdot)$ is called group if G is a non-void set and $\cdot$ is an associative and reproductive binary operation on G.*

This definition does not appear in group theory books, but it is equivalent to the one mentioned in them. We introduce it here as we consider it to be the most appropriate in order to demonstrate the close relationship between the group and the hypergroup. Thus, the following two properties of the group derive from the above definition:

Property 1. *In a group G there is an element e, called the identity element, such that $ae = ea = a$ for all a in G.*

Property 2. *For each element a of a group G there exists an element a' of G, called the inverse of a, such that $a \cdot a' = a' \cdot a = e$.*

The proof of the above properties can be found in [1,2]. If no ambiguity arises, we may abbreviate $(G, \cdot)$ to G.

Example 1. *(The free monoid) Often, a finite non-empty set A is referred to as an alphabet. The elements of $A^{\{1,\ldots,l\}}$, i.e., the functions from $\{1, \ldots, l\}$ to A, are called strings (or words) of length l. When $l = 0$, then we have $A^\varnothing$ which is equal to $\{\varnothing\}$. The empty set $\varnothing$ is the only string of length 0 over A. This string is called the empty string and it is denoted by λ. If x is a string of length l over A and $x(i) = a_i$, $i = 1, \ldots, l$, we write $x = (a_1, a_2, \ldots, a_l)$. The set*

$$\text{strings}(A) = A^* = \bigcup_{n=0}^{\infty} A^{\{1,\ldots,n\}}$$

becomes a monoid if we define

$$(a_1, a_2, \ldots, a_l)(b_1, b_2, \ldots b_k) = (a_1, a_2, \ldots, a_l, b_1, b_2, \ldots b_k)$$

The identity element of this binary operation, which is called string concatenation, is λ. The strings of length 1 generate the monoid, since every element $(a_1, \ldots, a_l)$ is a finite product $(a_1) \ldots (a_l)$ of strings of length 1. The function g from A to $A^{\{1\}}$ which is defined by

$$g(a) = (a), \ a \in A$$

is a bijection. Thus we may identify a string (a), of length 1 over A, with its only element a. This means that we can regard the sets $A^{\{1\}}$ and A as identical and consequently we may regard the elements of A^* as words $a_1 a_2 \ldots a_l$ in the alphabet A. It is obvious that for all $x \in A^*$ there is exactly one natural number n and exactly one sequence of elements $a_1, a_2, \ldots, a_n$ of A, such that $x = a_1 a_2 \ldots a_n$. A^* is called the free monoid on the set (or alphabet) A.

A generalization of the binary operation is the *binary hyperoperation* or *hypercomposition* "·" on a non-void set E, which is a rule that assigns to each element of $E \times E$ a unique element of the power set $\mathcal{P}(E)$ of E. Therefore, if a_i, a_j are elements of E, then $a_i \cdot a_j \subseteq E$. When there is no likelihood of confusion, the symbol "·" can be omitted and we write $a_i a_j \subseteq E$. If A, B are subsets of E, then $A \cdot B$ signifies the union $\bigcup_{(a,b) \in A \times B} a \cdot b$. In particular if $A = \varnothing$ or $B = \varnothing$, then $AB = \varnothing$. In both cases, aA and Aa have the same meaning as $\{a\}A$ and $A\{a\}$ respectively. Generally, the singleton $\{a\}$ is identified with its member a.

Definition 2. *(Definition of the hypergroup) The pair $(H, \cdot)$ is called hypergroup, if H is a non-void set and $\cdot$ is an associative and reproductive binary hypercomposition on H.*

The following Proposition derives from the definition of the hypergroup:

Proposition 1. *In a hypergroup the result of the hypercomposition of any two elements is non-void.*

The proof of Proposition 1 can be found in [3,4]. If no ambiguity arises, we may abbreviate $(H, \cdot)$ to H. Two significant types of hypercompositions are the closed and the open ones. A hypercomposition is called *closed* [5] (or *containing* [6], or *extensive* [7]) if the two participating elements always belong to the result of the hypercomposition, while it is called *open* if the result of the hypercomposition of any two elements different from each other does not contain the two participating elements.

The notion of the hypergroup was introduced in 1934 by F. Marty, who used it in order to study problems in non-commutative algebra, such as cosets determined by non-invariant subgroups [8–10]. From Definitions 1 and 2 it is evident that both groups and hypergroups satisfy the same axioms and their only difference is that the law of synthesis of two elements is a composition in groups, while it is a hypercomposition in hypergroups. This difference makes the hypergroups much more general algebraic structures than the groups, to the extent that properties similar to the previous 1 and 2 generally cannot be proved for the hypergroups. Furthermore, in the hypergroups there exist different types of identities [11–13]. In general, an element $e \in H$ is an identity if $a \in ae \cap ea$ for all $a \in H$. An identity is called *scalar* if $a = ae = ea$ for all $a \in H$, while it is called *strong* if $a \in ae = ea \subseteq \{e, a\}$ for all $a \in H$. Obviously, if the hypergroup has an identity, then the hypercomposition cannot be open.

Besides, in groups, both equations $a = xb$ and $a = bx$ have a unique solution, while, in the hypergroups, the analogous relations $a \in xb$ and $a \in bx$ do not have unique solutions. Thus F. Marty in [8] defined the two induced hypercompositions (right and left division) that derive from the hypergroup's hypercomposition:

$$\frac{a}{\mid b} = \{x \in H \mid a \in xb\} \text{ and } \frac{a}{b \mid} = \{x \in H \mid a \in bx\}.$$

If H is a group, then $\frac{a}{\mid b} = ab^{-1}$ and $\frac{a}{b \mid} = b^{-1}a$. It is obvious that if "·" is commutative, then the right and the left division coincide. For the sake of notational simplicity, $a \,/\, b$ or $a : b$ is used to denote the right division, or right hyperfraction, or just the division in the commutative hypergroups and $b \setminus a$ or $a..b$ is used to denote the left division, or left hyperfraction [14,15]. Using the induced hypercomposition we can create an axiom equivalent to the reproductive axiom, regarding which, the following Proposition is valid [3,4]:

Proposition 2. *In a hypergroup H, the non-empty result of the induced hypercompositions is equivalent to the reproductive axiom.*

W. Prenowitz enriched a commutative hypergroup of idempotent elements with one more axiom, in order to use it in the study of Geometry [16–21]. More precisely, in a commutative hypergroup H, all the elements of which satisfy the properties $aa = a$ and $a\,/\,a = a$, he introduced the *transposition axiom*:

$$a\,/\,b \cap c\,/\,d \neq \varnothing \text{ implies } ad \cap bc \neq \varnothing \text{ for all } a,b,c,d \in H.$$

He named this new hypergroup *join space*. For the sake of terminology unification, a commutative hypergroup which satisfies the transposition axiom is called *join hypergroup*. Prenowitz was followed by J. Jantosciak [20–23], V. W. Bryant, R. J. Webster [24], D. Freni, [25,26], J. Mittas, C. G. Massouros [2,27–29], A. Dramalidis [30,31], etc. In the course of his research, J. Jantosciak generalized the transposition axiom in an arbitrary hypergroup as follows:

$$b \setminus a \cap c\,/\,d \neq \varnothing \text{ implies } ad \cap bc \neq \varnothing \text{ for all } a,b,c,d \in H$$

and he named this hypergroup *transposition hypergroup* [15]. These algebraic structures attracted the interest of a big number of researchers, among whom there are, J. Jantosciak [15,22,23], I. Cristea [32–35], P. Corsini, [35–42], V. Leoreanu-Fortea [40–46], S. Hoskova-Mayerova, [47–51], J. Chvalina [48–52], P. Rackova [49,50], C. G. Massouros [3,6,12,13,53–62], G. G. Massouros [3,12,13,53–62], R. Ameri [63–65], M. M. Zahedi [63], I. Rosenberg [66], etc.

Furthermore, it has been proved that these hypergroups also comprise a useful tool in the study of Languages and Automata [67–71] and a constructive origin for the development of other, new, hypercompositional structures [53,57,58,72,73].

Definition 3. *A transposition hypergroup, which has a scalar identity e, is called quasicanonical hypergroup [74,75] or polygroup [76–78].*

In the quasicanonical hypergroups, there exist properties analogous to 1 and 2 which are valid in the groups:

Proposition 3. [15,53] *If $(Q, \cdot, e)$ is a quasicanonical hypergroup, then:*

(i) *for each $a \in Q$ there exists one and only one $a' \in Q$ such that $e \in aa' = a'a$*
(ii) *$c \in ab \Rightarrow a \in cb' \Rightarrow b \in a'c$*

The inverse is also true:

Proposition 4. [15,53] *If a hypergroup Q has a scalar identity e and*

(i) *for each $a \in Q$ there exists one and only one $a' \in Q$ such that $e \in aa' = a'a$*
(ii) *$c \in ab \Rightarrow a \in cb' \Rightarrow b \in a'c$*

then, the transposition axiom is valid in Q.

A commutative quasicanonical hypergroup is called *canonical hypergroup*. This hypergroup was first used by M. Krasner [79] but it owes its name to J. Mittas [80,81].

A non-empty subset K of a hypergroup H is called *semi-subhypergroup* when it is stable under the hypercomposition, i.e., $xy \subseteq K$ for all $x,y \in K$. K is a *subhypergroup* of H if it satisfies the reproductive axiom, i.e., if the equality $xK = Kx = K$ is valid for all $x \in K$. Since the structure of the hypergroup is much more complicated than that of the group, there are various kinds of subhypergroups. A subhypergroup K of H is called *closed from the right* (in H), (resp. *from the left*) if, for every element x in the complement K^c of K, it holds that $(xK) \cap K = \varnothing$ (resp.$(Kx) \cap K = \varnothing$). K is called *closed* if it is closed both, from the right and from the left [82–84]. It has been proved [1] that a subhypergroup is closed if and only if it is stable under the induced hypercompositions, i.e.,

$$a / b \subseteq K \text{ and } b \setminus a \subseteq K \text{ for all } a, b \in K.$$

A subhypergroup K of a hypergroup is *invertible* if $a / b \cap K \neq \varnothing$ implies $b / a \cap K \neq \varnothing$, and $b \setminus a \cap K \neq \varnothing$ implies $a \setminus b \cap K \neq \varnothing$. From this definition it derives that every invertible subhypergroup is also closed, but the opposite is not valid.

Proposition 5. [1,14] *If a subset K of a hypergroup H is stable under the induced hypercompositions, then K is a subhypergroup of H.*

Proposition 6. [1,14] *If K is a closed subhypergroup of a hypergroup H and $a \in K$, then:*

$$a / K = K / a = aK = K = Ka = K \setminus a = a \setminus K.$$

It has been proved [14,60,62] that the set of the semi-subhypergroups (resp. the set of the closed subhypergroups) which contains a non-void subset E is a complete lattice. Hence, the minimum (in the sense of inclusion) semi-subhypergroup of a hypergroup H, which contains a given non-empty subset E of H, can be assigned to E. This semi-subhypergroup is denoted by $[E]$ and it is called the generated by E semi-subhypergroup of H. Similarly, $\langle E \rangle$ is the generated by E closed subhypergroup of H. For notational simplicity, if $E = \{a_1, \ldots, a_n\}$, then $[E] = [a_1, \ldots, a_n]$ and $\langle E \rangle = \langle a_1, \ldots, a_n \rangle$ are used instead.

Duality. Two statements of the theory of hypergroups are *dual statements* (see [15,53]), if each one of them results from the other by interchanging the order of the hypercomposition "·", that is, interchanging any hyperproduct ab with ba. Observe that the reproductive and the associative axioms are self-dual. Moreover, observe that the induced hypercompositions / and \ have dual definitions; hence, they must be interchanged during the construction of a dual statement. So, the transposition axiom is self-dual as well. Therefore, the following principle of duality holds for the theory of hypergroups:

Given a Theorem, the dual statement, which results from the interchange of the order of the hypercomposition (and the necessary interchange of / and \), is also a Theorem.

Special notation: In the following pages, apart from the typical algebraic notations, we are using Krasner's notation for the complement and difference. So, we denote with $A..B$ the set of elements that are in the set A, but not in the set B.

2. Formal Languages, Automata Theory and Hypercompositional Structures

Mathematically, a language whose words are written with letters from an alphabet Σ, is defined as a subset of the free monoid Σ^* generated by Σ. The above definition of the language is fairly general and it includes all the written natural languages as well as the artificial ones. In general, a language is defined in two ways: It is either presented as an exhaustive list of all its valid words, i.e., through a dictionary, or it is presented as a set of rules defining the acceptable words. Obviously the first method can only be used when the language is finite. All the natural languages, such as English or Greek are finite and they have their own dictionaries. Artificial languages, on the other hand, may be infinite, and they can only be defined by the second way.

In the artificial languages, precision and no guesswork are required, especially when computers are concerned. The regular expressions, which are very precise language-defining symbols, were created and developed as a language-defining symbolism. The languages that are associated with these expressions are called *regular languages*.

The regular expressions were introduced by Kleene [85] who also proved that they are equivalent in expressive power to finite automata. McNaughton and Yamada gave their own proof to this [86], while Brzozowski [87,88] and Brzozowski and McClusky [89] further developed the theory of regular expressions. In regular languages the expression $x + y$ where x and y are strings of characters from an alphabet Σ means "either x or y". Therefore $x + y = \{x, y\}$. In this way the monoid Σ^* is enriched with a hypercomposition. This hypercomposition is named *B-hypercomposition* [67–69].

Proposition 7. [67,68] *A non-void set equipped with the B-hypecomposition is a join hypergroup.*

A hypergroup equipped with the B-hypercomposition is called *B-hypergroup* [67–69]. Moreover, the empty set of words and its properties in the theory of the regular languages leads to the following extension: Let $0 \notin \Sigma^*$. In the set $\overline{\Sigma^*} = \Sigma^* \cup \{0\}$ a hypercomposition, called *dilated B-hypercomposition*, is defined as follows:
$$x + y = \{x, y\} \quad \text{if } x, y \in \overline{\Sigma^*} \text{ and } x \neq y$$
$$x + x = \{x, 0\} \quad \text{for all } x \in \overline{\Sigma^*}$$

The associativity and the commutativity of the dilated B-hypercomposition derive without difficulty. Moreover, the transposition axiom is verified, since

$$x / y = y \setminus x = \begin{cases} H, & \text{if } x = y \\ \{x, y\}, & \text{if } x \neq y \text{ and } x = e \\ x, & \text{if } x \neq y \text{ and } x \neq e \end{cases}$$

This join hypergoup is called *dilated B-hypergroup* and it has led to the definition of a new class of hypergroups, the class of the *fortified transposition hypergroups* and *fortified join hypergroups*.

An *automaton* $\mathcal{A}$ is a collection of five objects $(\Sigma, S, \delta, s_0, F)$, where Σ is the alphabet of input letters (a finite nonempty set of symbols), S is a finite nonvoid set of states, s_0 is an element of S indicating the start (or initial) state, F is a (possibly empty) subset of S representing the set of the final (or accepting) states and δ is the state transition function with domain $S \times \Sigma$ and range S, in the case of a deterministic automaton (DFA), or $\mathcal{P}(S)$, the powerset of S, in the case of a nondeterministic automaton (NDFA). Σ^* denotes the set of words (or strings) formed by the letters of Σ –closure of Σ– and $\lambda \in \Sigma^*$ signifies the empty word. Given a DFA $\mathcal{A}$, the extended state transition function for $\mathcal{A}$, denoted δ^*, is a function with domain $S \times \Sigma^*$ and range S defined recursively as follows:

i. $\delta^*(s, a) = \delta(s, a)$ for all s in S and a in Σ
ii. $\delta^*(s, \lambda) = s$ for all s in S
iii. $\delta^*(s, ax) = \delta^*(\delta(s, a), x)$ for all s in S, x in Σ^* and a in Σ.

In [67–71] it is shown that the set of the states of an automaton, equipped with different hypercompositions, can be endowed with the structure of the hypergroup. The hypergroups that have derived in this way are named *attached hypergroups to the automaton*. To date, various types of attached hypergroups have been developed to represent the structure and operation of the automata with the use of the hypercompositional algebra tools. Between them are:

i.   the attached hypergroups of the order, and
ii.   the attached hypergroups of the grade.

Those two types of hypergroups were also used for the minimisation of the automata. In addition, in [69] another hypergroup, derived from a different definition of the hypercomposition, was attached to the set of the states of the automaton. Due to its definition, this hypergroup was named the *attached hypergroup of the paths* and it led to a new proof of Kleene's Theorem. Furthermore, in [70], the *attached hypergroup of the operation* was defined in automata. One of its applications is that this hypergroup can indicate all the states on which an automaton can be found after the t-clock pulse. For the purpose of defining the attached hypergroup of the operation, the notions of the Prefix and the Suffix of a word needed to be introduced. Let x be a word in Σ^*, then:

$$Prefix(x) = \{y \in \Sigma^* \mid yz = x \text{ for some } z \in \Sigma^*\} \text{ and } Suffix(x) = \{z \in \Sigma^* \mid yz = x \text{ for some } y \in \Sigma^*\}$$

Let s be an element of S. Then:

$$I_s = \{x \in \Sigma^* \mid \delta^*(s_0, x) = s\} \text{ and } P_s = \{\{s_i \in S \mid s_i = \delta^*(s_0, y), y \in Prefix(x), x \in I_s\}\}$$

Obviously, the states s_0 and s are in P_s.

Lemma 1. *If $r \in P_s$, then $P_r \subseteq P_s$.*

Proof. $P_r = \{s_i \in S \mid s_i = \delta^*(s_0, y), y \in Prefix(v), v \in I_r\}$ and since $r \in P_s$, it holds that $\delta^*(r, z) = s$, for some z in $Suffix(x)$, $x \in I_s$. Thus $\delta^*(s_i, y_i) = s$, $y_i \in Suffix$ and so the Lemma. □

With the use of the above notions, more hypercompositional structures can be attached on the set of the states of the automaton.

Proposition 8. *The set S of the states of an automaton equipped with the hypercomposition*

$$s + q = P_s \bigcup P_q \text{ for all } s, q \in S$$

becomes a join hypergroup.

Proof. Initially, notice that $s + S = \bigcup_{q \in S}(P_s \cup P_q) = S$. Hence, the reproductive axiom is valid. Next, the definition of the hypercomposition yields the equality:

$$s + (q + r) = s + (P_q \cup P_r) = P_s \cup (\bigcup_{u \in P_q \cup P_r} P_u)$$

Per Lemma 1, the right-hand side of the above equality is equal to $P_s \cup (P_q \cup P_r)$ which, however, is equal to $(P_s \cup P_q) \cup P_r$. Using again Lemma 1, we get the equality $(P_s \cup P_q) \cup P_r = (\bigcup_{v \in P_s \cup P_q} P_v) \cup P_r$

Thus:

$$(\bigcup_{v \in P_s \cup P_q} P_v) \cup P_r = (P_s \cup P_q) + r = (s + q) + r$$

and so, the associativity is valid. Next, observe that the hypercomposition is commutative and therefore:

$$s / q = q \backslash s = \begin{cases} S, & \text{if } s \in P_q \\ \{r \in S \mid P_s \subseteq P_r\}, & \text{if } s \notin P_q \end{cases}$$

Suppose that $s / q \cap p / r \neq \emptyset$. Then, $(s + r) \cap (q + p) = (P_s \cup P_r) \cap (P_q \cup P_p)$ which is non-empty, since it contains s_0. Hence the transposition axiom is valid and so the Proposition. □

Proposition 9. *The set S of the states of an automaton equipped with the hypercomposition*

$$s + q = P_s \cap P_q \text{ for all } s, q \in S$$

becomes a join semihypergroup.

Proof. Since $s_o \in P_r$, for all $r \in S$, the result of the hypercomposition is always non-void. On the other hand

$$s / q = q \setminus s = \begin{cases} S, & \text{if } s \in P_q \\ \varnothing, & \text{if } s \notin P_q \end{cases}$$

hence, since s / q, with s, q in S, is not always nonvoid, the reproductive axiom is not valid. The associativity can be verified in the same way as in the previous Proposition. Finally if $s / q \cap p / r \neq \varnothing$, then $s / q \cap p / r = S$ and so the intersection $(s + r) \cap (q + p)$ which is equal to $(P_s \cap P_r) \cap (P_q \cap P_p)$ is non-empty, since it contains s_o. □

G. G. Massouros [68–73], G. G. Massouros and J. D. Mittas [67] and after them J. Chvalina [90–92], L. Chvalinova [90], M. Novak [91–94], S. Křehlík [91–93], M. M. Zahedi [95], M. Ghorani [95,96] etc, studied automata using algebraic hypercompositional structures.

Formal Languages and Automata theory are very close to Graph theory. P. Corsini [97,98], M. Gionfriddo [99], Nieminen [100,101], I. Rosenberg [66], M. De Salvo and G. Lo Faro [102–104], I. Cristea et al. [105–108], C. Massouros and G. Massouros [109,110], C. Massouros and C. Tsitouras [111,112] and others studied hypergroups associated with graphs. In the following we will present how to attach a join hypergroup to a graph. In general, a *graph* is a set of points called *vertices* connected by lines, which are called *edges*. A *path* in a graph is a sequence of no repeated vertices $v_1, v_2, \ldots, v_n$, such that $\overline{v_1 v_2}, \overline{v_2 v_3}, \ldots, \overline{v_{n-1} v_n}$ are edges in the graph. A graph is said to be *connected* if every pair of its vertices is connected by a path. A *tree* is a connected graph with no cycles. Let $\mathcal{T}$ be a tree. In the set V of its vertices a hypercompostion "·" can be introduced as follows: for each two vertices x, y in V, $x \cdot y$ is the set of all vertices which belong to the path that connects vertex x with vertex y. Obviously this hypercomposition is a closed hypercomposition, i.e., x, y are contained in $x \cdot y$ for every x, y in V.

Proposition 10. *If V is the set of the vertices of a tree $\mathcal{T}$, then $(V, \cdot)$ is a join hypergroup.*

Proof. Since $\{x, y\} \subseteq xy$, it derives that $xV = V$ for each x in V and therefore the reproductive axiom is valid. Moreover, since $\mathcal{T}$ is an undirected graph, the hypecomposition is commutative. Next, let x, y, z be three arbitrary vertices of $\mathcal{T}$. If any of these three vertices, e.g., z, belongs to the path that the other two define, then $(xy)z = x(yz) = xy$. If x, y, z do not belong to the same path, then there exists only one vertex v in xy such that $vz \cap xy = \{v\}$. Indeed if there existed a second vertex w such that $wz \cap xy = \{w\}$, then the tree $\mathcal{T}$ would have a cycle, which is absurd. So $(xy)z = xy \cup vz$ and $x(yz) = xv \cup yz$. Since $xy \cup vz = xv \cup yz$, it derives that $(xy)z = x(yz)$. Now, for the transposition axiom, suppose that x, y, z, w are vertices of $\mathcal{T}$ such that $x / y \cap z / w \neq \varnothing$. If x, y, z, w are in the same path, then considering all their possible arrangements in their path, it derives that $xw \cap yz \neq \varnothing$. Next, suppose that the four vertices do not belong to the same path. Thus, suppose that z does not belong to the path defined by y, w. Then, $z \notin yw$. Consider zy and zw. As indicated above, since there are no cycles in $\mathcal{T}$, there exists only one vertex v in xy such that $zy = yv \cup vz$ and $zw = wv \cup vz$. Now, we distinguish between the cases:

(i) if x, y, w do not belong to the same path, then for the same reasons as above there exists only one s in xy such that $xy = ys \cup sx$ and $sw = ws \cup sx$. Since $x / y \cap z / w \neq \varnothing$, there exists r in V such that $x \in ry$ and $z \in rw$. Thus, since $\mathcal{T}$ contains no cycles, and in order for srv not to form a cycle, s and v must coincide. Hence, $v \in xw \cap yz$ and therefore $xw \cap yz \neq \varnothing$.

(ii) if x belongs to the same path with y and w, then:

   (ii$_a$) if $x \in yw$, then $yw = yx \cup xw$ and $xw \subseteq x / y$. Hence, $v = x$, $x \in xw \cap yz$ and therefore $xw \cap yz \neq \varnothing$.

   (ii$_b$) if $x \notin yw$, then $x / y \cap z / w = \varnothing$. □

A *spanning tree* of a connected graph is a tree whose vertex set is the same as the vertex set of the graph, and whose edge set is a subset of the edge set of the graph. Any connected graph has at least one spanning tree and there exist algorithms, which find such trees. Hence, any connected graph can be endowed with the join hypergroup structure through its spanning trees. Moreover, since a connected graph may have more than one spanning trees, more than one join hypergroups can be associated to a graph. On the other hand, in any connected or not connected graph, a hypergroup can be attached according to the following Proposition:

Proposition 11. *The set V of the vertices of a graph, is equipped with the structure of the hypergroup, if the result of the hypercomposition of two vertices v_i and v_j is the set of the vertices which appear in all the possible paths that connect v_i to v_j, or the set $\{v_i, v_j\}$, if there do not exist any connecting paths from vertex v_i to vertex v_j.*

2.1. Fortified Transposition Hypergroups

Definition 4. *A fortified transposition hypergroup (FTH) is a transposition hypergroup H with a unique strong identity e, which satisfies the axiom:*

for every $x \in H..\{e\}$ there exists one and only one element $y \in H..\{e\}$, such that $e \in xy$ and $e \in yx$.

y is denoted by x^{-1} and it is called inverse or symmetric of x. When the hypercomposition is written additively, the strong identity is denoted by 0, the unique element y is called opposite or negative instead of inverse and the notation $-x$ is used. If the hypercomposition is commutative, the hypergroup is called fortified join hypergroup (FJH).

It has been proved that every FTH consists of two types of elements, the *canonical (c-elements)* and the *attractive (a-elements)* [53,57]. An element x is called canonical if $ex = xe$ is the singleton $\{x\}$, while it is called attractive if $ex = xe = \{e, x\}$. We denote with A the set of the a-elements and with C the set of the c-elements. By convention $e \in A$.

Proposition 12. [53,57]

(i) if x is a non-identity attractive element, then $e / x = e \cdot x^{-1} = \{x^{-1}, e\} = x^{-1} \cdot e = x \setminus e$
(ii) if x is a canonical element, then $e / x = x^{-1} = x \setminus e$
(iii) if x, y are attractive elements and $x \neq y$, then $x \cdot y^{-1} = x / y \cup \{y^{-1}\}$ and $y^{-1} \cdot x = y \setminus x \cup \{y^{-1}\}$
(iv) if x, y are canonical elements, then $x \cdot y^{-1} = x / y$ and $y^{-1} \cdot x = y \setminus x$.

Theorem 1. [57]

(i) the result of the hypercomposition of two a-elements is a subset of A and it always contains these two elements.
(ii) the result of the hypercomposition of two non-symmetric c-elements consists of c-elements,
(iii) the result of the hypercomposition of two symmetric c-elements contains all the a-elements.
(iv) the result of the hypercomposition of an a-element with a c-element is the c-element.

Theorem 2. [53,57] *If H is a FTH, then the set A of the attractive elements is the minimum (in the sense of inclusion) closed subhypergroup of H.*

The proof of the above Theorems as well as other properties of the theory of the FTHs and FJHs can be found in [53,55–57,61]. The next two Propositions refer to the reversibility in FTHs.

Lemma 2. *If $w \in x \cdot y$, then $w \cdot x^{-1} \cap e \cdot y \neq \emptyset$ and $w \cdot y^{-1} \cap e \cdot x \neq \emptyset$*

Proof. $w \in x \cdot y$ implies $x \in w / y$ and $y \in x \setminus w$. Moreover $x \in x^{-1} \setminus e$ and $y \in e / y^{-1}$. Consequently, $w / y \cap x^{-1} \setminus e \neq \varnothing$ and $x \setminus w \cap y \in e / y^{-1} \neq \varnothing$. Next, the transposition axiom gives the Lemma. □

Proposition 13. *If $w \in x \cdot y$, and if any one of x, y is a canonical element, then*

$$x \in w \cdot y^{-1} \text{ and } y \in x^{-1} \cdot w$$

Proof. We distinguish between two cases:

(i) If $x, y \in C$, then $e \cdot x = x$ and $e \cdot y = y$. Next Lemma 2 applies and yields the Proposition.
(ii) Suppose that $x \in A$ and $y \in C$. Then, according to Theorem 1(iv), $x \cdot y = y$; thus, $w = y$. Via Theorem 1(iii), $A \subseteq y \cdot y^{-1}$; thus, $x \in y \cdot y^{-1}$. Per Theorem 2, $x^{-1} \in A$, consequently $y = x^{-1}y$.

Hence the Proposition is proved. □

Proposition 14. *Suppose that x, y are attractive elements and $w \in x \cdot y$.*

(i) *if $w = y = e$, then $e \in xe$ implies that $e \in ex$, while $x \notin ee$*
(ii) *if $w = x \neq y$, then $x \in x \cdot y$ implies that $x \in xy^{-1}$, while, generally, $y \notin xx^{-1}$*
(iii) *in any other case $w \in x \cdot y$ implies $x \in w \cdot y^{-1}$ and $y \in x^{-1} \cdot w$*

Sketch of Proof. Cases (i) and (ii) are direct consequences of the Theorem 1, while case (iii) derives from the application of Lemma 2. □

The property which is described in Proposition 13 is called *reversibility* and because of Proposition 14, this property holds partially in the case of TPH.

Another distinction between the elements of the FTH stems from the fact that the equality $\left(xx^{-1}\right)^{-1} = xx^{-1}$ (or $-(x - x) = x - x$ in the additive case) is not always valid. The elements that satisfy the above equality are called *normal*, while the others are called *abnormal* [53,57].

Example 2. *Let H be a totally ordered set, dense and symmetric around a center denoted by $0 \in H$. With regards to this center the partition $H = H^- \cup \{0\} \cup H^+$ can be defined, according to which, for every $x \in H^-$ and $y \in H^+$ it is $x < 0 < y$ and $x \leq y \Rightarrow -y \leq -x$ for every $x, y \in H$, where $-x$ is the symmetric of x with regards to 0. Then H, equipped with the hypercomposition:*

$$x + y = \{x, y\}, \quad \text{if } y \neq -x$$

and

$$x + (-x) = [0, |x|] \cup \{-|x|\}$$

becomes a FJH in which $x - x \neq -(x - x)$, for every $x \neq 0$.

Proposition 15. *The canonical elements of a FTH are normal.*

Proof. Let x be a canonical element. Because of Theorem 1, $A \subseteq xx^{-1}$ while, according to Theorem 2, $A^{-1} = A$. Thus $A^{-1} \subseteq xx^{-1}$ and therefore $A \subseteq \left(xx^{-1}\right)^{-1}$. Suppose that z is a canonical element in xx^{-1}. Per Proposition 13, $x \in zx$. So $xx^{-1} \subseteq z(xx^{-1})$. Hence, we have the sequence of implications:

$$e \in z\left(xx^{-1}\right); \ z^{-1} \in xx^{-1}; \ z \in \left(xx^{-1}\right)^{-1}$$

So, $xx^{-1} \subseteq (xx^{-1})^{-1}$. Furthermore $xx^{-1} \subseteq (xx^{-1})^{-1}$ implies that $(xx^{-1})^{-1} \subseteq \left[(xx^{-1})^{-1}\right]^{-1} = xx^{-1}$ and therefore the Proposition holds. □

An important Theorem that is valid for TFH [53] is the following structure Theorem:

Theorem 3. *A transposition hypergroup H containing a strong identity e is isomorphic to the expansion of the quasicanonical hypergroup $C \cup \{e\}$ by the transposition hypergroup A of all attractive elements through the identity e.*

The special properties of the FTH give different types of subhypergroups. There exist subhypergroups of a FTH that do not contain the symmetric of each one of their elements, while there exist others that do. This leads to the definition of the symmetric subhypergroups. A subhypergroup K of a FTH is *symmetric*, if $x \in K$ implies $x^{-1} \in K$. It is known that the intersection of two subhypergroups is not always a subhypergroup. In the case of the symmetric subhypergroups though, the intersection of two such subhypergroups is always a symmetric subhypergoup [57,62]. Therefore, the set of the symmetric subhypergroups of a FTH consist a complete lattice. It is proved that the lattice of the closed subhypergroups of a FTH is a sublattice of the lattice of the symmetric subhypergroups of the FTH [57,62]. An analytic and detailed study of these subhypergroups is provided in the papers [1,60,62]. Here, we will present the study of the *monogene symmetric subhypergroups*, i.e., symmetric subhypergroups generated by a single element. So, let H be a FJH, let x be an arbitrary element of H and let $M(x)$ be the monogene symmetric subhypergroup which is generated by this element. Then:

$$x^n = \begin{cases} x \cdot x \ldots \cdot x & (n \text{ times}) & \text{if } n > 0 \\ e & & \text{if } n = 0 \\ x^{-1} \cdot x^{-1} \ldots \cdot x^{-1} & (-n \text{ times}) & \text{if } n < 0 \end{cases} \quad (1)$$

and:

$$x^m \cdot x^n = \begin{cases} x^{m+n} & \text{if } mn > 0 \\ x^{m+n} \cdot (x \cdot x^{-1})^{\min\{|m|,|n|\}} & \text{if } mn < 0 \end{cases} \quad (2)$$

From the above, it derives that:

$$x^{m+n} \subseteq x^m \cdot x^n$$

Theorem 4. *If x is an arbitrary element of a FJH, then the monogene symmetric subhypergroup which is generated by this element is:*

$$M(x) = \bigcup_{(m,n) \in \mathbb{Z} \times \mathbb{N}_0} x^m \cdot (x \cdot x^{-1})^n$$

Proof. The symmetric subhypergroup of a normal FTH which is generated from a non-empty set X consists of the unions of all the finite products of the elements that are contained in the union $X^{-1} \cup X$ [62]; thus, from (1) we have:

$$M(x) = \bigcup_{(k,l) \in \mathbb{N}^2} x^k \cdot (x^{-1})^l = \bigcup_{(k,l) \in \mathbb{N}^2} x^k \cdot x^{-l}$$

According to (2), it is $x^k \cdot x^{-l} = x^{k-l} \cdot (x \cdot x^{-1})^{\min\{k,l\}}$. But $k - l \in \mathbb{Z}$; therefore, the Theorem is established. □

From the above Theorem, Proposition 13 and Theorem 1, we have the following Corollary:

Corollary 1. *Every monogene symmetric subhypergroup $M(x)$ with generator a canonical element x is closed, it contains all the attractive elements and also*

$$M(x) = \bigcup_{(m,n) \in \mathbb{Z}^2} x^m \cdot \left(x \cdot x^{-1}\right)^n$$

Remark 1.

(i) Since $e \in x \cdot x^{-1}$ the inclusion $x^m \cdot \left(x \cdot x^{-1}\right)^n \subseteq x^m \cdot \left(x \cdot x^{-1}\right)^q$ is valid for $n < q$.
(ii) for $x = e$, it is $M(e) = \{e\}$.

Let us define now a symbol $\omega(x)$ (which can even be the $+\infty$), and name it *order of x* and simultaneously *order of the monogene subhypergroup $M(x)$*. Two cases can appear such that one revokes the other:

I. For any $(m,n) \in \mathbb{Z} \times \mathbb{N}_0$, with $m \neq 0$, we have:

$$e \notin x^m \cdot \left(x \cdot x^{-1}\right)^n$$

Then we define the order of x and of $M(x)$ to be the infinity and we write $\omega(x) = +\infty$.

Proposition 16. *If $\omega(x) = +\infty$, then x is a canonical element.*

Proof. Suppose that x belongs to the set A of the attractive elements. Then, per Theorem 1.(i), $x^m \subseteq A$ and $x \in x^m$. Consequently:

$$e \in x \cdot e \subseteq x^m \cdot e \subseteq x^m \cdot \left(x \cdot x^{-1}\right)^n$$

This contradicts our assumption and therefore x is a canonical element. □

The previous Proposition and Theorem 1 result to the following Corollary:

Corollary 2. *If $\omega(x) = +\infty$, then x^m does not contain attractive elements for every $m \in \mathbb{Z}^*$.*

Proposition 17. *If $\omega(x) = +\infty$, then*

$$x^{m+n} \cap x^n = \varnothing, \text{ if } m > 0 \text{ and } x^{n-m} \cap x^n = \varnothing, \text{ if } m < 0$$

for any $(m,n) \in \mathbb{Z} \times \mathbb{N}_0$, with $m \neq 0$.

Proof. From $e \notin x^m \cdot \left(x \cdot x^{-1}\right)^n$ it derives that $x^m \cap \left(x \cdot x^{-1}\right)^{-n} = \varnothing$. According to Proposition 16, x is a canonical element and therefore, because of Proposition 15, x is normal; thus, $\left(x \cdot x^{-1}\right)^{-n} = \left(x \cdot x^{-1}\right)^n$. Therefore $x^m \cap \left(x \cdot x^{-1}\right)^n = \varnothing$ or $x^m \cap (x^n \cdot x^{-n}) = \varnothing$. So, the Proposition follows from the reversibility. □

Proposition 18. $\omega(x) = +\infty$, *if and only if*

(i) $x^m \subseteq C$, *for every $m \in \mathbb{Z}^*$*
(ii) $x^{m_1} \cap x^{m_2} = \varnothing$, *for every $m_1, m_2 \in \mathbb{Z}$ with $m_1 \neq m_2$.*

Proof. If $\omega(x) = +\infty$, per Corollary 2, x^m does not contain attractive elements for every $m \in \mathbb{Z}^*$. Moreover, if $n \neq 0$, then (ii) derives from Proposition 17. If $n = 0$, then $e \notin x^m$ and assuming that $m = m_1 + m_2$ with $m_1 m_2 > 0$, we successively have:

$$e \notin x^m;\ e \notin x^{m_1 + m_2};\ e \notin x^{m_1} x^{m_2};\ x^{-m_1} \cap x^{m_2} = \varnothing$$

Conversely now. If for every $m \in \mathbb{Z}^*$, the intersection $x^m \cap A$ is void and if for every $m_1, m_2 \in \mathbb{Z}$ with $m_1 \neq m_2$, the intersection $x^{m_1} \cap x^{m_2}$ is also void, then $e \notin x^{m_1} x^{-m_2}$ and therefore:

$$e \notin x^{m_1 - m_2}\ if\ m_1 m_2 < 0$$

and

$$e \notin x^{m_1 - m_2} (xx^{-1})^{\min\{|m_1|, |m_2|\}}\ if\ m_1 m_2 > 0$$

Thus,

$$e \notin x^m \cdot (x \cdot x^{-1})^n\ \text{for every } (m, n) \in \mathbb{Z} \times \mathbb{N}_0$$

So, the Proposition holds. □

II. There exist $(m, n) \in \mathbb{Z} \times \mathbb{N}_0$ with $m \neq 0$ such that:

$$e \in x^m \cdot (x \cdot x^{-1})^n$$

Proposition 19. *Let p be the minimum positive integer for which there exists $s \in \mathbb{N}_0$ such that $e \in x^p \cdot (x \cdot x^{-1})^s$. Then for a given $m \in \mathbb{Z}^*$ there exist $n \in \mathbb{N}$ such that $e \in x^m \cdot (x \cdot x^{-1})^n$ if and only if m is divided by p.*

Proof. Let $m = kp, k \in \mathbb{Z}$. From $e \in x^p \cdot (x \cdot x^{-1})^r$ it derives that

$$e \in x^{kp} \cdot (x \cdot x^{-1})^{kr} = x^m \cdot (x \cdot x^{-1})^n$$

Therefore, the Proposition.

Conversely now. If x is an a-element, then $e \in x \cdot (x \cdot x^{-1})^n$ for every $n \in \mathbb{N}$, so $p = 1$, and thus the Proposition. Next, if x is a c-element, and $e \in x^m \cdot (x \cdot x^{-1})^n$ with $m = kp + r, k \in \mathbb{Z}, 0 < r < p$. Then:

$$e \in x^m \cdot (x \cdot x^{-1})^n = x^{kp+r} \cdot (x \cdot x^{-1})^n \subseteq x^{kp} \cdot x^r \cdot (x \cdot x^{-1})^n$$

According to our hypothesis $e \in x^p \cdot (x \cdot x^{-1})^s$. Moreover, per Theorem 1, the sum of two non-opposite c-elements does not contain any a-elements. Consequently, there do not exist a-elements in x^p, and so

$$x^{-p} \subseteq x^{-p} \cdot x^p \cdot (x \cdot x^{-1})^n = (x^{-1} \cdot x)^p \cdot (x \cdot x^{-1})^n = (x \cdot x^{-1})^{p+n}$$

Thus

$$x^{kp} \subseteq (x \cdot x^{-1})^{-k(p+n)} = (x \cdot x^{-1})^{|k|(p+n)}$$

and therefore

$$e \in (x \cdot x^{-1})^{|k|(p+n)} \cdot x^r \cdot (x \cdot x^{-1})^n = x^r \cdot (x \cdot x^{-1})^{|k|(p+n)+n}$$

This contradicts the supposition, according to which p is the minimum element with the property $e \in x^p \cdot (x \cdot x^{-1})^s$. Thus $r = 0$, and so $m = kp$. □

For $m = kp$, $k \in \mathbb{Z}$, let q_k be the minimum non-negative integer for which $e \in x^{kp} \cdot (x \cdot x^{-1})^{q_k}$. Thus a function $q : \mathbb{Z} \to \mathbb{N}_0$ is defined which corresponds each k in $\mathbb{Z}$ to the non-negative integer q_k.

Definition 5. *The pair $\omega(x) = (p, q)$ is called order of x and of $M(x)$. The number p is called principal order of x and of $M(x)$, while the function q is called associated order of x and of $M(x)$.*

Consequently, according to the above definition, if x is an attractive element, then $e \in x \cdot (x \cdot x^{-1})$ and therefore $\omega(x) = (1, q)$ with $q(k) = 1$ for every $k \in \mathbb{Z}^*$. Moreover, if x is a self-inverse canonical element, then $e \in x^2 \cdot (x \cdot x^{-1})^0$, if $x \notin x \cdot x^{-1}$ and $e \in x \cdot (x \cdot x^{-1})$, if $x \in x \cdot x^{-1}$ and thus $\omega(x) = (2, q)$ with $q(k) = 0$ in the first case and $\omega(x) = (1, q)$ with $q(k) = 1$ in the second case (for every $k \in \mathbb{Z}$).

Moreover, we remark that the order of e is $\omega(e) = (1, q)$, with $q(k) = 0$, for every $k \in \mathbb{Z}$, and e is the only element which has this property. Yet, it is possible that there exist non-identity elements $x \in H$ with prime order 1, and this happens if and only if there exists an integer n such that $x^{-1} \in (x \cdot x^{-1})^n$, as for example when x is a self-inverse canonical element.

2.2. The Hyperringoid

Let Σ^* be the set of strings over an alphabet Σ. Then:

Proposition 20. *String concatenation is distributive over the B-hypecomposition.*

Proof. Let $a, b, c \in \Sigma^*$. Then, $a(b + c) = a\{b, c\} = \{ab, ac\} = ab + ac$. □

Via the thorough verification of the distributive axiom in all the different cases and taking into consideration that 0 is a bilaterally absorbing element with respect to the string concatenation on the set $\overline{\Sigma^*}$, it can also be proved that:

Proposition 21. *String concatenation is distributive over the dilated B-hypecomposition.*

Consequently, Σ^* and $\overline{\Sigma^*}$ are algebraic structures equipped with a composition and a hypercomposition which are related with the distributive law.

Definition 6. *A hyperringoid is a non-empty set Y equipped with an operation "·" and a hyperoperation "+" such that:*

i. *$(Y, +)$ is a hypergroup*
ii. *$(Y, \cdot)$ is a semigroup*
iii. *the operation "·" distributes on both sides over the hyperoperation "+ ".*

If the hypergroup $(Y, +)$ has extra properties, which make it a special hypergroup, it gives birth to corresponding special hyperringoids. So, if $(Y, +)$ is a join hypergroup, then the hyperringoid is called *join*. A distinct join hyperringoid is the *B-hyperringoid*, in which the hypergroup is a B-hypergroup. A *fortified join hyperringoid* or *join hyperring* is a hyperringoid whose additive part is a fortified join hypergroup and whose zero element is bilaterally absorbing with respect to the multiplication. A special join hyperring is the *join B-hyperring*, in which the hypergroup is a dilated B-hypergroup. If the additive part of a fortified join hyperringoid becomes a canonical hypergroup, then it is called *hyperring*. The hyperrigoid was introduced in 1990 [67] as the trigger for the study of languages and automata with the use of tools from hypercompositional algebra. An extensive study of the fundamental properties of hyperringoids can be found in [61,113–116].

Example 3. *Let $(R, +, \cdot)$ be a ring. If in R we define the hypercomposition:*

$$a \oplus b = \{a, b, a + b\}, \text{ for all } a, b \in R$$

then $(R, \oplus, \cdot)$ is a join hyperring.

Example 4. *Let $\leq$ be a linear order (also called a total order or chain) on Y, i.e., a binary reflexive and transitive relation such that for all $y, y' \in Y$, $y \neq y'$ either $y \leq y'$ or $y' \leq y$ is valid. For $y, y' \in Y$, $y < y'$, the set $\{z \in Y \mid y \leq z \leq y'\}$ is denoted by $[y, y']$ and the set $\{z \in Y \mid y < z < y'\}$ is denoted by $]y, y'[$. The order is dense if no $]y, y'[$ is void. Suppose that $(Y, \cdot, \leq)$ is a totally ordered group, i.e., $(Y, \cdot)$ is a group such that for all $y \leq y'$ and $x \in Y$, it holds that $x \cdot y \leq x \cdot y'$ and $y \cdot x \leq y' \cdot x$. If the order is dense, then the set Y can be equipped with the hypercomposition:*

$$x + y = \begin{cases} x & \text{if } x = y \\ ]\min\{x, y\}, \max\{x, y\}[& \text{if } x \neq y \end{cases}$$

and the triplet $(Y, +, \cdot)$ becomes a join hyperringoid. Indeed, since the equalities

$$x + y =]\min\{x, y\}, \max\{x, y\}[= y + x$$

and

$$(x + y) + z =]\min\{x, y, z\}, \max\{x, y, z\}[= x + (y + z)$$

are valid for every $x, y, z \in Y$, the hypercomposition is commutative and associative. Moreover,

$$x / y = y \setminus x = \begin{cases} x & \text{if } x = y \\ \{t \in Y : x < t\} & \text{if } y < x \\ \{t \in Y : t < x\} & \text{if } x < y \end{cases}$$

Thus, when the intersection $(x / y) \cap (z / w)$ is non-void, the intersection $(x + w) \cap (z + y)$ is also non-void. So, the transposition axion is valid. Therefore $(Y, +)$ is a join hypergroup. Moreover,

$$x \cdot (y + z) = \begin{cases} x \cdot y = x \cdot y + x \cdot z & \text{if } y = z \\ x \cdot]y, z[= x \cdot \bigcup_{y < t < z} \{t\} = \bigcup_{y < t < z} \{x \cdot t\} = x \cdot y + x \cdot z, & \text{if } y \neq z \end{cases}$$

It is worth mentioning that the hypercomposition:

$$x + y = [\min\{x, y\}, \max\{x, y\}], \text{ for all } x, y \in Y$$

endows $(Y, \cdot)$ with the join hyperringoid structure as well.

As per Proposition 20, the set of the words Σ^* over an alphabet Σ can be equipped with the structure of the B-hyperringoid. This hyperringoid has the property that each one of its elements, which are the words of the language, has a unique factorization into irreducible elements, which are the letters of the alphabet. So, this hyperringoid has a finite prime subset, that is a finite set of initial and irreducible elements, such that each one of its elements has a unique factorization with factors from this set. In this sense, this hyperringoid has a property similar to the one of the Gauss' rings. Moreover, because of Proposition 21, $\overline{\Sigma^*}$ can be equipped with the structure of the join B-hyperring which has the same property.

Definition 7. *A linguistic hyperringoid (resp. linguistic join hyperring) is a unitary B-hyperringoid (resp. join B-hyperring) which has a finite prime subset P and which is non-commutative for $|P| > 1$.*

It is obvious that every B-hyperringoid or join B-hyperring is not a linguistic one.

Proposition 22. *From every non-commutative free monoid with finite base, there derives a linguistic hyperringoid.*

Example 5. *Let $\{0,1\}^{2\times 2}$ express the set of 2×2 matrices, which consist of the elements 0,1, that is the following 16 matrices:*

$$A_1 = \begin{bmatrix} 1 & 0 \\ 0 & 0 \end{bmatrix}, A_2 = \begin{bmatrix} 0 & 1 \\ 0 & 0 \end{bmatrix}, A_3 = \begin{bmatrix} 0 & 0 \\ 1 & 0 \end{bmatrix}, A_4 = \begin{bmatrix} 0 & 0 \\ 0 & 1 \end{bmatrix}$$

$$B_1 = \begin{bmatrix} 1 & 1 \\ 0 & 0 \end{bmatrix}, B_2 = \begin{bmatrix} 0 & 1 \\ 0 & 1 \end{bmatrix}, B_3 = \begin{bmatrix} 0 & 0 \\ 1 & 1 \end{bmatrix}, B_4 = \begin{bmatrix} 1 & 0 \\ 1 & 0 \end{bmatrix}, B_5 = \begin{bmatrix} 1 & 0 \\ 0 & 1 \end{bmatrix}, B_6 = \begin{bmatrix} 0 & 1 \\ 1 & 0 \end{bmatrix}$$

$$C_1 = \begin{bmatrix} 0 & 1 \\ 1 & 1 \end{bmatrix}, C_2 = \begin{bmatrix} 1 & 0 \\ 1 & 1 \end{bmatrix}, C_3 = \begin{bmatrix} 1 & 1 \\ 1 & 0 \end{bmatrix}, C_4 = \begin{bmatrix} 1 & 1 \\ 0 & 1 \end{bmatrix}$$

$$D_1 = \begin{bmatrix} 1 & 1 \\ 1 & 1 \end{bmatrix}$$

$$E_1 = \begin{bmatrix} 0 & 0 \\ 0 & 0 \end{bmatrix}$$

Consider the set T of all 2×2 matrices deriving from products of the above matrices, except the zero matrix. T becomes a B-hyperringoid under B-hypercomposition and matrix multiplication. Observe that none of the matrices A_1, B_6, C_4 can be written as the product of any two matrices from the set T while all the matrices in T result from products of these three matrices. Therefore, T is a linguistic hyperringoid, whose prime subset is $\{A_1, B_6, C_4\}$. Furthermore, if T is enriched with the zero matrix, then it becomes a linguistic join hyperring.

M. Krasner was the first one who introduced and studied hypercompositional structures with an operation and a hyperoperation. The first structure of this kind was the hyperfield, an additive-multiplicative hypercompositional structure whose additive part is a canonical hypergroup and the multiplicative part a commutative group. The hyperfield was introduced by M. Krasner in [79] as the proper algebraic tool in order to define a certain approximation of complete valued fields by sequences of such fields. Later on, Krasner introduced the hyperring which is related to the hyperfield in the same way as the ring is related to the field [117]. Afterwards, J. Mittas introduced the *superring* and the *superfield*, in which both the addition and the multiplication are hypercompositions and more precisely, the additive part is a canonical hypergroup and the multiplicative part is a semi-hypergroup [118–120]. In the recent bibliography, a structure whose additive part is a hypergroup and the multiplicative part is a semi-group is also referred to with the term additive hyperring and similarly, the term multiplicative hyperring is used when the multiplicative part is a hypergroup.

Rings and Krasner's hyperrings have many common elementary algebraic properties, e.g., in both structures the following are true:

(i) $x(-y) = (-x)y = -xy$
(ii) $(-x)(-y) = xy$
(iii) $w(x-y) = wx - wy$, $(x-y)w = xw - yw$

In the hyperringoids though, these properties are not generally valid, as it can be seen in the following example:

Example 6. *Let S be a multiplicative semigroup having a bilaterally absorbing element 0. Consider the set:*

$$P = (\{0\} \times S) \cup (S \times \{0\})$$

With the use of the hypercomposition "+":

$$(x,0) + (y,0) = \{(x,0), (y,0)\}$$
$$(0,x) + (0,y) = \{(0,x), (0,y)\}$$
$$(x,0) + (0,y) = (0,y) + (x,0) = \{(x,0), (0,y)\} \text{ for } x \neq y$$
$$(x,0) + (0,x) = (0,x) + (x,0) = \{(x,0), (0,x), (0,0)\}$$

P becomes a fortified join hypergroup with neutral element $(0,0)$. If $(0,x)$ is denoted by $\bar{x}$ and $(0,0)$ by $\bar{0}$, then the opposite of $\bar{x}$ is $-\bar{x} = (x,0)$. Obviously this hypergroup has not c-elements. Now let us introduce in P a multiplication defined as follows:

$$(x_1, y_1) \cdot (x_2, y_2) = (x_1 x_2, y_1 y_2)$$

This multiplication makes $(P, +, \cdot)$ a join hyperring, in which

$$-(\bar{x}\,\bar{y}) = -[(0,x)(0,y)] = -(0,xy) = (xy,0) \neq \bar{0} \text{ while } \bar{x}(-\bar{y}) = (0,x)(y,0) = (0,0) = \bar{0}$$

Similarly, $(-\bar{x})\bar{y} = \bar{0} \neq -(\bar{x}\,\bar{y})$. Furthermore

$$(-\bar{x})(-\bar{y}) = (x,0)(y,0) = (xy,0) = -\overline{xy}$$

More examples of hyperringoids can be found in [61,113–116].

3. Hypercompositional Algebra and Geometry

It is very well known that there exists a close relation between Algebra and Geometry. So, as it should be expected, this relation also appears between Hypercompositional Algebra and Geometry. It is really of exceptional interest that the axioms of the hypergroup are directly related to certain Euclid's postulates [121]. Indeed, according to the first postulate of Euclid:

"Ἠιτήσθω ἀπό παντός σημείου ἐπί πᾶν σημεῖον εὐθεῖαν γραμμήν ἀγαγεῖν" [121]

(Let the following be postulated: to draw a straight line from any point to any point [122])

So, to any pair of points (a,b), the segment of the straight line ab can be mapped. This segment always exists and it is a nonempty set of points. In fact, it is a multivalued result of the composition of two elements. Thus, a hypercomposition has been defined in the set of the points. Next, according to the second postulate:

"Καὶ πεπερασμένην εὐθεῖαν κατά τό συνεχές ἐπ' εὐθείας ἐκβαλεῖν" [121]

(To produce a finite straight line continuously in a straight line [122])

The sets a/b and b/a are nonempty. Therefore, as per Proposition 2, the reproductive axiom is valid. Besides, it is easy to prove that the associativity holds in the set of the points. It is only necessary to keep in mind the definition of the equal figures given by Euclid in the "Common Notions":

"Τά τῷ αὐτῷ ἴσα καί ἀλλήλοις ἐστίν ἴσα" [121]

(Things which are equal to the same thing are also equal to one another [122])

So, the set of the points is a hypergroup. Moreover, through similar reasoning, it can be proved that any Euclidean space of dimension n can become a hypergroup. Indeed:

Proposition 23. *Let $(V, +)$ be a linear space over an ordered field $(F, +, \cdot)$. Then V, with the hypercomposition:*

$$xy = \{\kappa x + \lambda y \mid \kappa, \lambda \in F^*_+, \ \kappa + \lambda = 1\}$$

becomes a join hypergroup.

This hypergroup is called attached hypergroup. Properties of vector spaces can be found via the attached hypergroup. Thus, for example:

Proposition 24. *In a vector space V over an ordered field F, the elements a_i, $i = 1, \ldots, k$ are affinely dependent if and only if there exist distinct integers $s_1, \ldots, s_n, t_1, \ldots, t_m$ that belong to $\{1, \ldots, k\}$ such that:*

$$[a_{s_1}, \ldots, a_{s_n}] \cap [a_{t_1}, \ldots, a_{t_m}] \neq \varnothing$$

In fact, several hypergroups can be attached to a vector space [28]. The connection of hypercompositional structures with Geometry was initiated by W. Prenowitz [16–19]. The classical geometries, descriptive geometries, spherical geometries and projective geometries can be treated as certain kinds of hypergoups, all satisfying the transposition axiom. The hypercomposition plays the central role in this approach. It assigns the appropriate connection between any two distinct points. Thus, in Euclidian geometry, it gives the points of the segment; in spherical geometry, it gives the points of the minor arc of the great circle; in projective geometry it gives the point of the line. This development is dimension free and it is applicable to spaces of arbitrary dimension, finite or infinite.

3.1. Hypergroups and Convexity

Several geometric notions can be described with the use of the hypercomposition. One such notion is the convexity. It is known that a figure is called *convex*, if the segment joining any pair of its points lies entirely in it. As mentioned above, the set of the points of the plane, as well as the set of the points of any vector space V over an ordered field, becomes a hypergroup under the hypercomposition defined in Proposition 23. From this point of view, that is, with the use of the hypercomposition, a subset E of V is convex if $ab \subseteq E$, for all $a, b \in E$. However, a subset E of a hypergroup which has this property is a semi-subhypergroup [14]. Thus:

Proposition 25. *The convex subsets of a vector space V are the semi-subhypergroups of its attached hypergroup.*

Consequently, the properties of the convex sets of a vector space are simple applications of the properties of the semi-subhypergroups, or the subhypergroups of a hypergroup, and more precisely, the attached hypergroup. So, this approach, except from the fact that it leads to remarkable results, it also gives the opportunity to generalize the already known Theorems of the vector spaces in sets with fewer axioms than the ones of the vector spaces. Next, we will present some well-known named Theorems that arise as corollaries of more general Theorems which are valid in hypercompositional algebra.

In hypergroups the following Theorem holds [2,14]:

Theorem 5. *Let H be a hypergroup in which every set with cardinality greater than n has two disjoint subsets A, B such that $[A] \cap [B] \neq \varnothing$. If $(Y_i)_{i \in I}$ with card $I \geq n$ is a finite family of semi-subhypergroups of H, in which the intersection of every n elements is non-void, then all the sets Y_i have a non-void intersection.*

The combination of Propositions 24, 25 and Theorem 5 gives the corollary:

Corollary 3. *(Helly's Theorem). Let $(C_i)_{i \in I}$ be a finite family of convex sets in $\mathbb{R}^d$, with $d + 1 < $ card I. Then, if any $d + 1$ of the sets C_i have a non-empty intersection, all the sets C_i have a non-empty intersection.*

Next, the following Theorem stands for a join hypergroup:

Theorem 6. *Let A, B be two disjoint semi-subhypergroups in a join hypergroup and let x be an idempotent element not in the union $A \cup B$. Then $[A \cup \{x\}] \cap B = \varnothing$ or $[B \cup \{x\}] \cap A = \varnothing$*

The proof of Theorem 6 is found in [2,14] and it is repeated here for the purpose of demonstrating the techniques which are used for it.

Proof. Suppose that $[A \cup \{x\}] \cap B \neq \varnothing$ and $[B \cup \{x\}] \cap A \neq \varnothing$. Since x is idempotent the equalities $[A \cup \{x\}] = Ax$ and $[B \cup \{x\}] = Bx$ are valid. Thus, there exist $a \in A$ and $b \in B$, such that $ax \cap B \neq \varnothing$ and $bx \cap A \neq \varnothing$. Hence, $x \in B / a$ and $x \in A / b$. Thus, $B / a \cap A / b \neq \varnothing$. Next, the application of the transposition axiom, gives $Bb \cap Aa \neq \varnothing$. However, $Bb \subseteq B$ and $Aa \subseteq A$, since A, B are semi-subhypergroups. Therefore, $A \cap B \neq \varnothing$, which contradicts the Theorem's assumption. □

Corollary 4. *Let H be a join hypergroup endowed with an open hypercomposition. If A, B are two disjoint semi-subhypergroups of H and x is an element not in the union $A \cup B$, then:*

$$[A \cup \{x\}] \cap B = \varnothing \ \text{or} \ [B \cup \{x\}] \cap A = \varnothing.$$

The attached hypergroup of a vector space, which is defined in Proposition 23, is a join hypergroup whose hypercomposition is open, so Corollary 4 applies to it and we get the Kakutani's Lemma:

Corollary 5. *(Kakutani's Lemma). If A, B are disjoint convex sets in a vector space and x is a point not in their union, then either the convex envelope of $A \cup \{x\}$ and B or the convex envelope of $B \cup \{x\}$ and A are disjoint.*

Next in [2] it is proved that the following Theorem is valid:

Theorem 7. *Let H be a join hypergroup consisting of idempotent elements and suppose that A, B are two disjoint semi-subhypergroups in H. Then, there exist disjoint semi-subhypergroups M, N such that $A \subseteq M$, $B \subseteq N$ and $H = M \cup N$.*

A direct consequence of Theorem 7 is Stone's Theorem:

Corollary 6. *(Stone's Theorem). If A, B are disjoint convex sets in a vector space V, there exist disjoint convex sets M and N, such that $A \subseteq M$, $B \subseteq N$ and $V = M \cup N$.*

During his study of Geometry with hypercompositional structures, W. Prenowitz introduced the *exchange spaces* which are join spaces satisfying the axiom:

$$\text{if } c \in \langle a, b \rangle \text{ and } c \neq a, \text{ then } \langle a, b \rangle = \langle a, c \rangle$$

The above axiom enabled Prenowitz to develop a theory of linear independence and dimension of a type familiar to classical geometry. Moreover, a generalization of this theory has been achieved by Freni, who developed the notions of independence, dimension, etc., in a hypergroup H that satisfies only the axiom:

$$x \in \langle A \cup \{y\} \rangle, \ x \notin \langle A \rangle \implies y \in \langle A \cup \{x\} \rangle, \text{ for every } x, y \in H \text{ and } A \subseteq H.$$

Freni called these hypergroups *cambiste* [25,26]. A subset B of a hypergroup H is called *free* or *independent* if either $B = \varnothing$, or $x \notin \langle B - \{x\} \rangle$ for all $x \in B$, otherwise it is called *non-free* or *dependent*. B generates H, if $\langle B \rangle = H$, in which case B is a set of generators of H. A free set of generators is a *basis* of H. Freni proved that all the bases of a cambiste hypergroup have the same cardinality. The *dimension* of a cambiste hypergroup H is the cardinality of any basis of H. The dimension theory gives very interesting results in convexity hypergroups. A *convexity hypergroup* is a join hypergroup which satisfies the axioms:

i. the hypercomposition is open,
ii. $ab \cap ac \neq \varnothing$ implies $b = c$ or $b \in ac$ or $c \in ab$.

Prenowitz, defined this hypercompositional structure with equivalent axioms to the above, named it *convexity space* and used it, as did Bryant and Webster [24], for generalizing some of the theory of linear spaces. In [2] it is proved that every convexity hypergroup is a cambiste hypergroup. Moreover in [2] it is proved that the following Theorem stands for convexity hypergroups:

Theorem 8. *Every n+1 elements of a n-dimensional convexity hypergroup H are correlated.*

One can easily see that the attached hypergroup of a vector space is a convexity hypergroup and, moreover, if the dimension of the attached hypergroup H_V of a vector space V is n, then the dimension of V is $n-1$. Thus, we have the following corollary of Theorem 8:

Corollary 7. *(Radon's Theorem). Any set of d+2 points in $\mathbb{R}^d$ can be partitioned into two disjoint subsets, whose convex hulls intersect.*

Furthermore, the following Theorem is proved in [2]:

Theorem 9. *If x is an element of a n-dimensional convexity hypergroup H and $a_1, \ldots, a_n, a_{n+1}$ are n+1 elements of H such that $x \in a_1 \cdots a_n a_{n+1}$, then there exists a proper subset of these elements which contains x in their hyperproduct.*

A direct consequence of this Theorem is Caratheodory's Theorem:

Corollary 8. *(Caratheodory's Theorem). Any convex combination of points in $\mathbb{R}^d$ is a convex combination of at most d+1 of them.*

In addition, Theorems of the hypercompositional algebra are proved in [2], which give as corollaries generalizations and extensions of Caratheodory's Theorem.

An element a of a semi-subhypergroup S is called *interior* element of S if for each $x \in S$, $x \neq a$, there exists $y \in S$, $y \neq a$, such that $a \in xy$. In [2] it is proved that any interior element of a semi-subhypergroup S of a n-dimensional convexity hypergroup, is interior to a finitely generated semi-subhypergroup of S. More precisely, the following Theorem is valid [2]:

Theorem 10. *Let a be an interior element of a semi-subhypergroup S of a n-dimensional convexity hypergroup H. Then a is interior element of a semi-subhypergroup of S, which is generated by at most 2n elements.*

A corollary of this Theorem, when H is $\mathbb{R}^d$, is Steinitz's Theorem:

Corollary 9. *(Steinitz's Theorem). Any point interior to the convex hull of a set E in $\mathbb{R}^d$ is interior to the convex hull of a subset of E, containing 2d points at the most.*

D. Freni in [123] extended the use of the hypergroup in more general geometric structures, called geometric spaces. A *geometric space* is a pair (S, B) such that S is a non-empty set, whose elements are called points, and B is a non-empty family of subsets of S, whose elements are called blocks. Freni was followed by S. Mirvakili, S.M. Anvariyeh and B. Davvaz [124,125].

3.2. Hyperfields and Geometry

As it is mentioned in the previous Section 2.2, the hyperfield was introduced by M. Krasner in order to define a certain approximation of a complete valued field by a sequence of such fields [79]. The construction of this hyperfield, which was named by Krasner himself *residual hyperfield*, is also described in his paper [117].

Definition 8. *A hyperring is a hypercompositional structure $(H, +, \cdot)$, where H is a non-empty set, "$\cdot$" is an internal composition on H, and "+" is a hypercomposition on H. This structure satisfies the axioms:*

i. $(H, +)$ is a canonical hypergroup,
ii. $(H, \cdot)$ is a multiplicative semigroup in which the zero element 0 of the canonical hypergroup is a bilaterally absorbing element,
iii. the multiplication is distributive over the hypercomposition (hyperaddition), i.e.,

$$z(x+y) = zx + zy \text{ and } (x+y)z = xz + yz$$

for all $x, y, z \in H$.

If $H \backslash \{0\}$ is a multiplicative group then $(H, +, \cdot)$ is called hyperfield.

J. Mittas studied these hypercompositional structures in a series of papers [126–133]. Among the plenitude of examples which are found in these papers, we will mention the one which is presented in the first paragraph of [130].

Example 7. Let $(E, \cdot)$ be a totally ordered semigroup, having a minimum element 0, which is bilaterally absorbing with regards to the multiplication. The following hypercomposition is defined on E:

$$x + y = \begin{cases} \max\{x, y\} & \text{if } x \neq y \\ \{z \in E \mid z \leq x\} & \text{if } x = y \end{cases}$$

Then $(E, +, \cdot)$ is a hyperring. If $E \backslash \{0\}$ is a multiplicative group, then $(E, +, \cdot)$ is a hyperfield.

We referred to Mittas' example, because nowadays, this particular hyperfield is called *tropical hyperfield* (see, e.g., [134–138]) and it is proved to be an appropriate and effective algebraic tool for the study of tropical geometry.

M. Krasner worked on the occurrence frequency of such structures as the hyperrings and hyperfields and he generalized his previous construction of the residual hyperfields. He observed that, if R is a ring and G is a normal subgroup of R's multiplicative semigroup, then the multiplicative classes $\bar{x} = xG$, $x \in R$, form a partition of R and that the product of two such classes, as subsets of R, is a class modG as well, while their sum is a union of such classes. Next, he proved that the set $\bar{R} = R / G$ of these classes becomes a hyperring, if the product of $\bar{R}$'s two elements is defined to be their set-wise product and their sum to be the set of the classes contained in their set-wise sum [117]:

$$\bar{x} \cdot \bar{y} = xyG$$

and

$$\bar{x} + \bar{y} = \{zG \mid z \in xG + yG\}$$

He also proved that if R is a field, then R/G is a hyperfield. Krasner named these hypercompositional structures *quotient hyperring* and *quotient hyperfield*, respectively.

In the recent bibliography, there appear hyperfields with different and not always successful names, all of which belong to the class of the quotient hyperfields. For instance:

(a) starting from the papers [139,140] by A. Connes and C. Consani, there appeared many papers (e.g., [135–138]) which gave the name «Krasners' hyperfield» to the hyperfield which is constructed over the set $\{0, 1\}$ using the hypercomposition:

$$0 + 0 = 0, \ 0 + 1 = 1 + 0 = 1, \ 1 + 1 = \{0, 1\}$$

Oleg Viro, in his paper [135] is reasonably noticing that «To the best of my knowledge, K did not appear in Krasner's papers». Actually, this is a quotient hyperfield. Indeed, let F be a field and let F^* be its multiplicative subgroup. Then the quotient hyperfield $F / F^* = \{0, F^*\}$ is isomorphic to the hyperfield considered by A. Connes and C. Consani. Hence the two-element non-trivial hyperfield is isomorphic to a quotient hyperfield.

(b) Papers [139,140] by A. Connes and C. Consani, show the construction of the hyperfield, which is now called «*sign hyperfield*» in the recent bibliography, over the set {−1, 0, 1} with the following hypercomposition:

$$0+0 = 0,\ 0+1 = 1+0 = 1,\ 1+1 = 1,\ -1-1 = -1,\ 1-1 = -1+1 = \{-1,0,1\}$$

However, this hyperfield is a quotient hyperfield as well. Indeed, let F be an ordered field and let F^+ be its positive cone. Then the quotient hyperfield $F\ /\ F^+ = \{-F^+, 0, F^+\}$ is isomorphic to the hyperfield which is called sign hyperfield.

(c) The so called «*phase hyperfield*» (see e.g., [135,136]) in the recent bibliography, is just the quotient hyperfield $\mathbb{C}\ /\ \mathbb{R}^+$, where $\mathbb{C}$ is the field of complex numbers and $\mathbb{R}^+$ the set of the positive real numbers. The elements of this hyperfield are the rays of the complex field with origin the point (0,0). The sum of two elements $z\mathbb{R}^+$, $w\mathbb{R}^+$ of $\mathbb{C}\ /\ \mathbb{R}^+$ with $z\mathbb{R}^+ \neq w\mathbb{R}^+$ is the set $\{(zp + wq)\mathbb{R}^+\ |\ p,q \in \mathbb{R}^+\}$, which consists of all the interior rays $x\mathbb{R}^+$ of the convex angle which is created from these two elements, while the sum of two opposite elements gives the participating elements and the zero element. This hyperfield is presented in detail in [141].

Krasner, immediately realized that if all hyperrings could be isomorphically embedded into quotient hyperrings, then several conclusions of their theory could be deduced in a very straightforward manner, through the use of the ring theory. So, he raised the question whether all the hyperrings are isomorphic to subhyperrings of quotient hyperrings or not. He also raised a similar question regarding the hyperfields [117]. These questions were answered by C. Massouros [142–144] and then by A. Nakassis [145], via the following Theorems:

Theorem 11. [142,143] *Let* $(\Theta, \cdot)$ *be a multiplicative group. Let* $H = \Theta \cup \{0\}$, *where 0 is a multiplicatively absorbing element. If H is equipped with the hypercomposition:*

$$
\begin{aligned}
&x + 0 = 0 + x = x && \text{for all } x \in H, \\
&x + x = H..\{x\} && \text{for all } x \in \Theta, \\
&x + y = y + x = \{x, y\} && \text{for all } x, y \in \Theta \text{ with } x \neq y,
\end{aligned}
$$

then, $(H, +, \cdot)$ is a hyperfield, which does not belong to the class of quotient hyperfields when Θ is a periodic group.

Theorem 12. [144] *Let* $\overline{\Theta} = \Theta \otimes \{1, -1\}$ *be the direct product of the multiplicative groups Θ and $\{-1, 1\}$, where card $\Theta > 2$. Moreover, let $K = \overline{\Theta} \cup \{0\}$ be the union of $\overline{\Theta}$ with the multiplicatively absorbing element 0. If K is equipped with the hypercomposition:*

$$
\begin{aligned}
&w + 0 = 0 + w = w && \text{for all } w \in K, \\
&(x, i) + (x, i) = K..\{(x, i), (x, -i), 0\} && \text{for all } (x, i) \in \overline{\Theta}, \\
&(x, i) + (x, -i) = K..\{(x, i), (x, -i)\} && \text{for all } (x, i) \in \overline{\Theta}, \\
&(x, i) + (y, j) = \{(x, i),\ (x, -i),\ (y, j),\ (y, -j)\} && \text{for all } (x, i), (y, j) \in \overline{\Theta} \text{ with } (y, j) \neq (x, i),\ (x, -i),
\end{aligned}
$$

then, $(K, +, \cdot)$ is a hyperfield which does not belong to the class of quotient hyperfields when Θ is a periodic group.

Proposition 26. [145] *Let* $(T^*, \cdot)$ *be a multiplicative group of m, $m > 3$ elements. Let $T = T^* \cup \{0\}$, where 0 is a multiplicatively absorbing element. If T is equipped with the hypercomposition:*

$$
\begin{aligned}
&a + 0 = 0 + a = a && \text{for all } a \in T, \\
&a + a = \{0, a\} && \text{for all } a \in T^*, \\
&a + b = b + a = T..\{0, a, b\} && \text{for all } a, b \in T^* \text{ with } a \neq b,
\end{aligned}
$$

then, $(T, +, \cdot)$ is a hyperfield.

Theorem 13. [145] *If T^* is a finite multiplicative group of m, m > 3 elements and if the hyperfield T is isomorphic to a quotient hyperfield F / Q, then $Q \cup \{0\}$ is a field of m-1 elements while F is a field of $(m-1)^2$ elements.*

Clearly, we can choose the cardinality of T^* in such a way that T cannot be isomorphic to a quotient hyperfield. In [144,145] one can find non-quotient hyperrings as well.

Therefore, we know 4 different classes of hyperfields, so far: the class of the quotient hyperfields and the three ones which are constructed via the Theorems 11, 12 and 13.

The open and closed hypercompositions [5] in the hyperfields are of special interest. Regarding these, we have the following:

Proposition 27. *In a hyperfield K the sum $x + y$ of any two non-opposite elements $x, y \neq 0$ does not contain the participating elements if and only if, the difference $x - x$ equals to $\{-x, 0, x\}$, for every $x \neq 0$.*

Proposition 28. *In a hyperfield K the sum $x + y$ of any two non-opposite elements $x, y \neq 0$ contains these two elements if and only if, the difference $x - x$ equals to H, for every $x \neq 0$.*

For the proofs of the above Propositions 27 and 28, see [141]. With regard to Proposition 28, it is worth mentioning that there exist hyperfields in which, the sum $x + y$ contains only the two addends x, y, i.e. $x + y = \{x, y\}$, when $y \neq -x$ and $x, y \neq 0$ [141].

Theorem 14. [141] *Let $(K, +, \cdot)$ be a hyperfield. Let † be a hypercomposition on K, defined as follows:*

$$x \dagger y = (x + y) \cup \{x, y\} \quad \text{if } y \neq -x \text{ and } x, y \neq 0$$
$$x \dagger (-x) = K \quad \text{for all } x \in K \backslash \{0\}$$
$$x \dagger 0 = 0 \dagger x = x \quad \text{for all } x \in K$$

Then, $(K, \dagger, \cdot)$ is a hyperfield and moreover, if $(K, +, \cdot)$ is a quotient hyperfield, then $(K, \dagger, \cdot)$ is a quotient hyperfield as well.

Corollary 10. *If $(K, +, \cdot)$ is a field, then $(K, \dagger, \cdot)$ is a quotient hyperfield.*

The following problem in field theory is raised from the study of the isomorphism of the quotient hyperfields to the hyperfields which are constructed with the process given in Theorem 14:

when does a subgroup G of the multiplicative group of a field F have the ability to generate F via the subtraction of G from itself? [141,143]

A partial answer to this problem, which is available so far, regarding the finite fields is given with the following theorem:

Theorem 15. [146] *Let F be a finite field and G be a subgroup of its multiplicative group of index n and order m. Then, G-G = F, if and only if:*

$n = 2$ and $m > 2$,
$n = 3$ and $m > 5$,
$n = 4, -1 \in G$ and $m > 11$,
$n = 4, -1 \notin G$ and $m > 3$,
$n = 5, \text{char} F = 2$ and $m > 8$,
$n = 5, \text{char} F = 3$ and $m > 9$,
$n = 5, \text{char} F \neq 2, 3$ and $m > 23$.

Closely related to the hyperfield is the hypermodule and the vector space.

Definition 9. *A left hypermodule over a unitary hyperring P is a canonical hypergroup M with an external composition* $(a, m) \to am$, *from* $P \times M$ *to M satisfying the conditions:*

i. $a(m + n) = am + an$,
ii. $(a + b)m = am + bm$,
iii. $(ab)m = a(bm)$,
iv. $1m = m$ *and* $0m = 0$

for all $a, b \in P$ *and all* $m, n \in M$.

The *right hypermodule* is defined in a similar way. A hypermodule over a hyperfield is called *vector hyperspace*.

Suppose V and W are hypermodules over the hyperring P. The cartesian product $V \times W$ can become a hypercompositional structure over P, when the operation and the hyperoperation, for v, v_1, $v_2 \in V$, w, w_1, $w_2 \in W$, and $a \in P$, are defined componentwise, as follows:

$$(v_1, w_1) + (v_2, w_2) = \bigcup \{(v, w) \mid v \in v_1 + v_2, \ w \in w_1 + w_2\}$$

$$a(v, w) = (av, aw)$$

The resulting hypercompositional structure is called the *direct sum* of V and W.

Theorem 16. *The direct sum of the hypermodules is not a hypermodule.*

Proof. Let V and W be two hypermodules over a hyperring P. Then:

$$(a + b)(v, w) = \bigcup \{c(v, w) \mid c \in a + b\} = \bigcup \{(cv, cw) \mid c \in a + b\}$$

On the other hand:

$$a(v, w) + b(v, w) = (av, aw) + (bv, bw) = \bigcup \{(x, y) \mid x \in av + bv, \ y \in aw + bw\} =$$
$$= \bigcup \{(x, y) \mid x \in (a + b)v, \ y \in (a + b)w\} = \bigcup \{(sv, rw) \mid s, r \in a + b\}$$

Therefore:

$$(a + b)(v, w) \subseteq a(v, w) + b(v, w)$$

Consequently axiom (ii) is not valid. □

Remark 2. Errors in Published Papers. Unfortunately, there exist plenty of papers which incorrectly consider that the direct sum of hypermodules is a hypermodule. For instance, they mistakenly consider that if P is a hyperring or a hyperfield, then P^n is a hypermodule or a vector hyperspace over P respectively. Due to this error, a lot of, if not all the conclusions of certain papers are incorrect. We are not going to specifically refer to such papers, as we do not wish to add negative citations in our paper, but we refer positively to the paper by P. Ameri, M Eyvazi and S. Hoskova-Mayerova [120], where the authors have presented a counterexample which shows that the polynomials over a hyperring give a superring in the sense of Mittas [118,119] and not a hyperring, as it is mistakenly mentioned in a previously published paper which is referred there. This error can also be highlighted with the same method as the one in Theorem 16, since the polynomials over a hyperring P can be considered as the ordered sets $(a_0, a_1, \ldots)$ where a_i, i=0, 1, ... are their coefficients.

Following the above remark, we can naturally introduce the definition:

Definition 10. *A left weak hypermodule over a unitary hyperring P is a canonical hypergroup M with an external composition* $(a, m) \to am$, *from* $P \times M$ *to M satisfying the conditions (i), (iii), (iv) of the Definition 9 and, in place of (ii), the condition:*

ii'. $(a + b)m \subseteq am + bm$, *for all* $a, b \in P$ *and all* $m \in M$.

The quotient hypermodule over a quotient hyperring is constructed in [147], as follows:

Let M be a P–module, where P is a unitary ring, and let G be a subgroup of the multiplicative semigroup of P, which satisfies the condition $aG\, bG = abG$, for all $a, b \in P$. Note that this condition is equivalent to the normality of G only when $P..\{0\}$ is a group, which appears only in the case of division rings (see [144]). Next, we introduce in M the following equivalence relation:

$$x \sim y \Leftrightarrow x = ty, \ t \in G$$

After that, we equip $\overline{M}$ with the following hypercomposition, where $\overline{M}$ is the set of equivalence classes of M modulo $\sim$:

$$\overline{x} \dotplus \overline{y} = \left\{ \overline{w} \in \overline{M} \mid w = xp + yq, \ p, q \in G \right\}$$

i.e., $\overline{x} \dotplus \overline{y}$ consists of all the classes $\overline{w} \in \overline{M}$ which are contained in the set-wise sum of $\overline{x}, \overline{y}$. Then $(\overline{M}, \dotplus)$ becomes a canonical hypergroup. Let $\overline{P}$ be the quotient hyperring of P over G. We consider the external composition from $\overline{P} \times \overline{M}$ to $\overline{M}$ defined as follows:

$$\overline{a}\,\overline{x} = \overline{ax} \text{ for each } \overline{a} \in \overline{P}, \ \overline{x} \in \overline{M}.$$

This composition satisfies the axioms of the hypermodule and so $\overline{M}$ becomes a $\overline{P}$- hypermodule.

If M is a module over a division ring D, then, using the multiplicative group D^* of D we can construct the quotient hyperring $\overline{D} = D / D^* = \{0, D^*\}$ and the relevant quotient hypermodule $\overline{M}$. For any $\overline{a} \in \overline{M}$ it holds that $\overline{a} \dotplus \overline{a} = \{0, \overline{a}\}$. In [147] it is shown that this hypermodule is strongly related to the projective geometries. A. Connes and C. Consani, in [139,140] also prove that the projective geometries, in which the lines have at least four points, are exactly vector hyperspaces over the quotient hyperfield with two elements. Moreover, if V is a vector space over an ordered field F, then, using the positive cone F^+ of F we can construct the vector hyperspace $\overline{V}$ over the quotient hyperfield $\overline{F} = F / F^+ = \{F^-, 0, F^+\}$. In [147] it is shown that every Euclidean spherical geometry can be considered as a quotient vector hyperspace over the quotient hyperfield with three elements.

Modern algebraic geometry is based on abstract algebra which offers its techniques for the study of geometrical problems. In this sense, the hyperfields, were connected to the conic sections via a number of papers [148–150], where the definition of an elliptic curve over a field F was naturally extended to the definition of an elliptic hypercurve over a quotient Krasner hyperfield. The conclusions obtained in [148–150] were extended to cryptography as well.

4. Conclusions and Open Problems

In this paper we have initially presented the relationship between the groups and the hypergroups. It is interesting that the groups and the hypergroups are two algebraic structures which satisfy exactly the same axioms, i.e., the associativity and the reproductivity, but they differ in the law of synthesis. In the first ones, the law of synthesis is a composition, while in the second ones, it is a hypercomposition. This difference makes the hypergroups much more general algebraic structures than the groups, and for this reason, the hypergroups have been gradually enriched with further axioms, which are either more powerful or less powerful and they lead to a significant number of special hypergroups.

We have also presented the connection of the Hypercompositional Algebra to the Formal Languages and Automata theory as well as its close relationship to Geometry. It is very interesting that the transposition hypergroup, which appears in the Formal Languages, is the proper algebraic tool for the study of the convexity in Geometry. The study of this hypergroup has led to general Theorems which have as corollaries well known named Theorems in the vector spaces. Different types of transposition hypergroups, as for example the fortified transposition hypergroup, give birth to hypercompositional structures like the hyperringoid, the linguistic hyperringoid the join hyperring the algebraic structure of which is an area with a plentitude of hitherto open problems. Moreover, the hyperfield and the hypermodule describe fully and accurately the projective and the spherical geometries, while they are directly connected to other geometries as well. Moreover, the classification problem of hyperfields gives birth to the question:

when does a subgroup G of the multiplicative group of a field F have the ability to generate F via the subtraction of G from itself?

This question is answered for certain finite fields only and still remains to be answered in its entirety.

Author Contributions: Both authors contributed equally to this work. All authors have read and agreed to the published version of the manuscript.

Funding: This research received no external funding. The APC was funded by *Mathematics* MDPI

Conflicts of Interest: The authors declare no conflict of interest.

References

1. Massouros, C.G. Some properties of certain subhypergroups. *Ratio Math.* **2013**, *25*, 67–76.
2. Massouros, C.G. On connections between vector spaces and hypercompositional structures. *Ital. J. Pure Appl. Math.* **2015**, *34*, 133–150.
3. Massouros, C.G.; Massouros, G.G. The transposition axiom in hypercompositional structures. *Ratio Math.* **2011**, *21*, 75–90.
4. Massouros, C.G.; Massouros, G.G. On certain fundamental properties of hypergroups and fuzzy hypergroups—Mimic fuzzy hypergroups. *Internat. J. Risk Theory* **2012**, *2*, 71–82.
5. Massouros, C.G.; Massouros, G.G. On open and closed hypercompositions. *AIP Conf. Proc.* **2017**, *1978*, 340002. [CrossRef]
6. Massouros, C.G.; Dramalidis, A. Transposition Hv-groups. *Ars Comb.* **2012**, *106*, 143–160.
7. Novák, M.; Křehlík, Š. EL-hyperstructures revisited. *Soft Comput.* **2018**, *22*, 7269–7280. [CrossRef]
8. Marty, F. Sur une généralisation de la notion de groupe. Huitième Congrès des mathématiciens Scand. *Stockholm* **1934**, 45–49.
9. Marty, F. Rôle de la notion de hypergroupe dans l'étude de groupes non abéliens. *C.R. Acad. Sci.* **1935**, *201*, 636–638.
10. Marty, F. Sur les groupes et hypergroupes attachés à une fraction rationelle. *Ann. L'école Norm.* **1936**, *53*, 83–123. [CrossRef]
11. Massouros, C.G.; Massouros, G.G. Identities in multivalued algebraic structures. *AIP Conf. Proc.* **2010**, *1281*, 2065–2068. [CrossRef]
12. Massouros, C.G.; Massouros, G.G. On the algebraic structure of transposition hypergroups with idempotent identity. *Iran. J. Math. Sci. Inform.* **2013**, *8*, 57–74. [CrossRef]
13. Massouros, C.G.; Massouros, G.G. Transposition hypergroups with idempotent identity. *Int. J. Algebr. Hyperstruct. Appl.* **2014**, *1*, 15–27.
14. Massouros, C.G. On the semi-subhypergroups of a hypergroup. *Int. J. Math. Math. Sci.* **1985**, *8*, 725–728. [CrossRef]
15. Jantosciak, J. Transposition hypergroups, Noncommutative Join Spaces. *J. Algebra* **1997**, *187*, 97–119. [CrossRef]
16. Prenowitz, W. Projective Geometries as multigroups. *Am. J. Math.* **1943**, *65*, 235–256. [CrossRef]
17. Prenowitz, W. Descriptive Geometries as multigroups. *Trans. Am. Math. Soc.* **1946**, *59*, 333–380. [CrossRef]
18. Prenowitz, W. Spherical Geometries and mutigroups. *Can. J. Math.* **1950**, *2*, 100–119. [CrossRef]
19. Prenowitz, W. A Contemporary Approach to Classical Geometry. *Am. Math. Month* **1961**, *68*, 1–67. [CrossRef]
20. Prenowitz, W.; Jantosciak, J. Geometries and Join Spaces. *J. Reine Angew. Math.* **1972**, *257*, 100–128.
21. Prenowitz, W.; Jantosciak, J. *Join. Geometries. A Theory of Convex Sets and Linear Geometry*; Springer: Berlin/Heidelberg, Germany, 1979.

22. Jantosciak, J. Classical geometries as hypergroups. In Proceedings of the Atti Convegno Ipergruppi Altre Structure Multivoche e Loro Applicazioni, Udine, Italia, 15–18 October 1985; pp. 93–104.
23. Jantosciak, J. A brif survey of the theory of join spaces. In Proceedings of the 5th International Congress on AHA, Iasi, Rommania, 4–10 July 1993; Hadronic Press: Palm Harbor, FA, USA, 1994; pp. 109–122.
24. Bryant, V.W.; Webster, R.J. Generalizations of the Theorems of Radon, Helly and Caratheodory. *Monatsh. Math.* **1969**, *73*, 309–315. [CrossRef]
25. Freni, D. Sur les hypergroupes cambistes. *Rend. Ist. Lomb.* **1985**, *119*, 175–186.
26. Freni, D. Sur la théorie de la dimension dans les hypergroupes. *Acta Univ. Carol. Math. Phys.* **1986**, *27*, 67–80.
27. Massouros, C.G. Hypergroups and convexity. *Riv. Mat. Pura Appl.* **1989**, *4*, 7–26.
28. Mittas, J.; Massouros, C.G. Hypergroups defined from linear spaces. *Bull. Greek Math. Soc.* **1989**, *30*, 63–78.
29. Massouros, C.G. Hypergroups and Geometry, Mem. *Acad. Romana Math. Spec. Issue* **1996**, *19*, 185–191.
30. Dramalidis, A. On geometrical hyperstructures of finite order. *Ratio Math.* **2011**, *21*, 43–58.
31. Dramalidis, A. Some geometrical P-H_V-hyperstructures. In *New Frontiers in Hyperstructures*; Hadronic Press: Palm Harbor, FL, USA, 1996; pp. 93–102.
32. Cristea, I. Complete hypergroups, 1-hypergroups and fuzzy sets. *An. Stalele Univ. Ovidius Constanta* **2000**, *10*, 25–38.
33. Cristea, I. A property of the connection between fuzzy sets and hypergroupoids. *Ital. J. Pure Appl. Math.* **2007**, *21*, 73–82.
34. Stefanescu, M.; Cristea, I. On the fuzzy grade of hypergroups. *Fuzzy Sets Syst.* **2008**, *159*, 1097–1106. [CrossRef]
35. Corsini, P.; Cristea, I. Fuzzy sets and non complete 1-hypergroups. *An. Stalele Univ. Ovidius Constanta* **2005**, *13*, 27–54.
36. Corsini, P. Graphs and Join Spaces. *J. Comb. Inf. Syst. Sci.* **1991**, *16*, 313–318.
37. Corsini, P. *Prolegomena of Hypergroup Theory*; Aviani Editore: Udine, Italy, 1993.
38. Corsini, P. Join Spaces, Power Sets, Fuzzy Sets. In Proceedings of the 5th International Congress on Algebraic Hyperstructures and Applications, Iasi, Romania, 4–10 July 1993; Hadronic Press: Palm Harbor, FL, USA, 1994; pp. 45–52.
39. Corsini, P. Binary relations, interval structures and join spaces. *J. Appl. Math. Comput.* **2002**, *10*, 209–216. [CrossRef]
40. Corsini, P.; Leoreanu, V. Fuzzy sets and join spaces associated with rough sets. *Rend. Circ. Mat. Palermo* **2002**, *51*, 527–536. [CrossRef]
41. Corsini, P.; Leoreanu, V. *Applications of Hyperstructures Theory*; Kluwer Academic Publishers: Berlin, Germany, 2003.
42. Corsini, P.; Leoreanu-Fotea, V. On the grade of a sequence of fuzzy sets and join spaces determined by a hypergraph. *Southeast Asian Bull. Math.* **2010**, *34*, 231–242.
43. Leoreanu, V. On the heart of join spaces and of regular hypergroups. *Riv. Mat. Pura Appl.* **1995**, *17*, 133–142.
44. Leoreanu, V. Direct limit and inverse limit of join spaces associated with fuzzy sets. *PUMA* **2000**, *11*, 509–516.
45. Leoreanu-Fotea, V. Fuzzy join n-ary spaces and fuzzy canonical n-ary hypergroups. *Fuzzy Sets Syst.* **2010**, *161*, 3166–3173. [CrossRef]
46. Leoreanu-Fotea, V.; Rosenberg, I. Join Spaces Determined by Lattices. *J. Mult.-Valued Logic. Soft Comput.* **2010**, *16*, 7–16.
47. Hoskova, S. Abelization of join spaces of affine transformations of ordered field with proximity. *Appl. Gen. Topol.* **2005**, *6*, 57–65. [CrossRef]
48. Chvalina, J.; Hoskova, S. Modelling of join spaces with proximities by first-order linear partial differential operators. *Ital. J. Pure Appl. Math.* **2007**, *21*, 177–190.
49. Hoskova, S.; Chvalina, J.; Rackova, P. Transposition hypergroups of Fredholm integral operators and related hyperstructures (Part I). *J. Basic Sci.* **2006**, *3*, 11–17.
50. Hoskova, S.; Chvalina, J.; Rackova, P. Transposition hypergroups of Fredholm integral operators and related hyperstructures (Part II). *J. Basic Sci.* **2008**, *4*, 55–60.
51. Chvalina, J.; Hoskova-Mayerova, S. On Certain Proximities and Preorderings on the Transposition Hypergroups of Linear First-Order Partial Differential Operators. *An. Stalele Univ. Ovidius Constanta Ser. Mat.* **2014**, *22*, 85–103. [CrossRef]

52. Chvalina, J.; Chvalinova, L. Transposition hypergroups formed by transformation operators on rings of differentiable functions. *Ital. J. Pure Appl. Math.* **2004**, *15*, 93–106.
53. Jantosciak, J.; Massouros, C.G. Strong Identities and fortification in Transposition hypergroups. *J. Discret. Math. Sci. Cryptogr.* **2003**, *6*, 169–193. [CrossRef]
54. Massouros, C.G. Canonical and Join Hypergroups. *An. Sti. Univ. Al. I. Cuza Iasi* **1996**, *1*, 175–186.
55. Massouros, C.G. Normal homomorphisms of fortified join hypergroups, Algebraic Hyperstructures and Applications. In Proceedings of the 5th International Congress, Iasi, Romania, 4–10 July 1993; Hadronic Press: Palm Harbor, FL, USA, 1994; pp. 133–142.
56. Massouros, C.G. Isomorphism Theorems in fortified transposition hypergroups. *AIP Conf. Proc.* **2013**, *1558*, 2059–2062. [CrossRef]
57. Massouros, G.G.; Massouros, C.G.; Mittas, I.D. Fortified join hypergroups. *Ann. Math. Blaise Pascal* **1996**, *3*, 155–169. [CrossRef]
58. Massouros, G.G.; Zafiropoulos, F.A.; Massouros, C.G. Transposition polysymmetrical hypergroups. Algebraic Hyperstructures and Applications. In Proceedings of the 8th International Congress, Samothraki, Greece, 1–9 September 2002; Spanidis Press: Xanthi, Greece, 2003; pp. 91–202.
59. Massouros, C.G.; Massouros, G.G. Transposition polysymmetrical hypergroups with strong identity. *J. Basic Sci.* **2008**, *4*, 85–93.
60. Massouros, C.G.; Massouros, G.G. On subhypergroups of fortified transposition hypergroups. *AIP Conf. Proc.* **2013**, *1558*, 2055–2058. [CrossRef]
61. Massouros, G.G. Fortified join hypergroups and join hyperrings. *An. Sti. Univ. Al. I. Cuza, Iasi Sect. I Mat.* **1995**, *1*, 37–44.
62. Massouros, G.G. The subhypergroups of the fortified join hypergroup. *Italian J. Pure Appl. Math.* **1997**, *2*, 51–63.
63. Ameri, R.; Zahedi, M.M. Hypergroup and join space induced by a fuzzy subset. *Pure Math. Appl.* **1997**, *8*, 155–168.
64. Ameri, R. Topological transposition hypergroups. *Italian J. Pure Appl. Math.* **2003**, *13*, 181–186.
65. Ameri, R. Fuzzy transposition hypergroups. *Italian J. Pure Appl. Math.* **2005**, *18*, 147–154.
66. Rosenberg, I.G. Hypergroups and join spaces determined by relations. *Italian J. Pure Appl. Math.* **1998**, *4*, 93–101.
67. Massouros, G.G.; Mittas, J.D. Languages—Automata and hypercompositional structures, Algebraic Hyperstructures and Applications. In Proceedings of the 4th International Congress, Xanthi, Greece, 27–30 June 1990; World Scientific: Singapore, 1991; pp. 137–147.
68. Massouros, G.G. Automata, Languages and Hypercompositional Structures. Ph.D. Thesis, National Technical University of Athens, Athens, Greece, 1993.
69. Massouros, G.G. Automata and hypermoduloids, Algebraic Hyperstructures and Applications. In Proceedings of the 5th International Congress, Iasi, Romania, 4–10 July 1993; Hadronic Press: Palm Harbor, FL, USA, 1994; pp. 251–265.
70. Massouros, G.G. An automaton during its operation, Algebraic Hyperstructures and Applications. In Proceedings of the 5th International Congress, Iasi, Romania, 4–10 July 1993; Hadronic Press: Palm Harbor, FA, USA, 1994; pp. 267–276.
71. Massouros, G.G. On the attached hypergroups of the order of an automaton. *J. Discret. Math. Sci. Cryptogr.* **2003**, *6*, 207–215. [CrossRef]
72. Massouros, G.G. Hypercompositional structures in the theory of languages and automata. *An. Sti. Univ. Al. I. Cuza Iasi Sect. Inform.* **1994**, *3*, 65–73.
73. Massouros, G.G. Hypercompositional structures from the computer theory. *Ratio Math.* **1999**, *13*, 37–42.
74. Bonansinga, P. Sugli ipergruppi quasicanonici. *Atti Soc. Pelor. Sc. Fis. Mat. Nat.* **1981**, *27*, 9–17.
75. Massouros, C.G. Quasicanonical hypergroups. In Proceedings of the 4th Internat. Cong. on Algebraic Hyperstructures and Applications, Xanthi, Greece, 27–30 June 1990; World Scientific: Singapore, 1991; pp. 129–136.
76. Ioulidis, S. Polygroups et certaines de leurs propriétiés. *Bull. Greek Math. Soc.* **1981**, *22*, 95–104.
77. Comer, S. Polygroups derived from cogroups. *J. Algebra* **1984**, *89*, 397–405. [CrossRef]
78. Davvaz, B. *Polygroup Theory and Related Systems*; Publisher World Scientific: Singapore, 2013.

79. Krasner, M. *Approximation des corps valués complets de caractéristique p≠0 par ceux de caractéristique 0, Colloque d' Algèbre Supérieure (Bruxelles, Decembre 1956)*; Centre Belge de Recherches Mathématiques, Établissements Ceuterick, Louvain, Librairie Gauthier-Villars: Paris, France, 1957; pp. 129–206.
80. Mittas, J. Hypergroupes canoniques hypervalues, C.R. *Acad. Sci. Paris* **1970**, *271*, 4–7.
81. Mittas, J. Hypergroupes canoniques. *Math. Balk.* **1972**, *2*, 165–179.
82. Krasner, M. La loi de Jordan-Holder dans les hypergroupes et les suites generatrices des corps de nombres P-adiqes. *Duke Math. J.* **1940**, *6*, 120–140. [CrossRef]
83. Krasner, M. Cours d' Algebre superieure, Theories des valuation et de Galois. *Cours Fac. Sci. L' Univ. Paris* **1967**, *68*, 1–305.
84. Sureau, Y. Sous-hypergroupe engendre par deux sous-hypergroupes et sous-hypergroupe ultra-clos d'un hypergroupe, C.R. *Acad. Sc. Paris* **1977**, *284*, 983–984.
85. Kleene, S.C. Representation of Events in Nerve Nets and Finite Automata. In *Automata Studies*; Shannon, C.E., McCarthy, J., Eds.; Princeton University: Princeton, NJ, USA, 1956; pp. 3–42.
86. McNaughton, R.; Yamada, H. Regular Expressions and State Graphs for Automata. *IEEE Trans. Electron. Comput.* **1960**, *9*, 39–47. [CrossRef]
87. Brzozowski, J.A. A Survey of Regular Expressions and their Applications. *IEEE Trans. Electron. Comput.* **1962**, *11*, 324–335. [CrossRef]
88. Brzozowski, J.A. Derivatives of Regular Expressions. *J. ACM* **1964**, *11*, 481–494. [CrossRef]
89. Brzozowski, J.A.; McCluskey, E.J. Signal Flow Graph Techniques for Sequential Circuit State Diagrams. *IEEE Trans. Electron. Comput.* **1963**, *EC-12*, 67–76. [CrossRef]
90. Chvalina, J.; Chvalinova, L. State hypergroups of Automata. *Acta Math. Inform. Univ. Ostrav.* **1996**, *4*, 105–120.
91. Chvalina, J.; Křehlík, S.; Novák, M. Cartesian composition and the problem of generalizing the MAC condition to quasi-multiautomata. *An. Stalele Univ. Ovidius Constanta* **2016**, *24*, 79–100. [CrossRef]
92. Chvalina, J.; Novák, M.; Křehlík, S. Hyperstructure generalizations of quasi-automata induced by modelling functions and signal processing. *AIP Conf. Proc.* **2019**, *2116*, 310006. [CrossRef]
93. Novák, M.; Křehlík, S.; Stanek, D. n-ary Cartesian composition of automata. *Soft Comput.* **2019**, *24*, 1837–1849. [CrossRef]
94. Novák, M. Some remarks on constructions of strongly connected multiautomata with the input semihypergroup being a centralizer of certain transformation operators. *J. Appl. Math.* **2008**, *I*, 65–72.
95. Chorani, M.; Zahedi, M.M. Some hypergroups induced by tree automata. *Aust. J. Basic Appl. Sci.* **2012**, *6*, 680–692.
96. Chorani, M. State hyperstructures of tree automata based on lattice-valued logic. *RAIRO-Theor. Inf. Appl.* **2018**, *52*, 23–42.
97. Corsini, P. Binary relations and hypergroupoids. *Italian J. Pure Appl. Math.* **2000**, *7*, 11–18.
98. Corsini, P. Hypergraphs and hypergroups. *Algebra Universalis* **1996**, *35*, 548–555. [CrossRef]
99. Gionfriddo, M. *Hypergroups Associated with Multihomomorphisms between Generalized Graphs*; Corsini, P., Ed.; Convegno su Sistemi Binary e Loro Applicazioni: Taormina, Italy, 1978; pp. 161–174.
100. Nieminen, J. Join Space Graph. *J. Geom.* **1988**, *33*, 99–103. [CrossRef]
101. Nieminen, J. Chordal Graphs and Join Spaces. *J. Geom.* **1989**, *34*, 146–151. [CrossRef]
102. De Salvo, M.; Lo Faro, G. Hypergroups and binary relations. *Mult. Valued Log.* **2002**, *8*, 645–657.
103. De Salvo, M.; Lo Faro, G. A new class of hypergroupoids associated to binary relations. *J. Mult. Valued Logic. Soft Comput.* **2003**, *9*, 361–375.
104. De Salvo, M.; Freni, D.; Lo Faro, G. Fully simple semihypergroups. *J. Algebra* **2014**, *399*, 358–377. [CrossRef]
105. Cristea, I.; Jafarpour, M.; Mousavi, S.S.; Soleymani, A. Enumeration of Rosenberg hypergroups. *Comput. Math. Appl.* **2011**, *32*, 72–81. [CrossRef]
106. Cristea, I.; Stefanescu, M. Binary relations and reduced hypergroups. *Discrete Math.* **2008**, *308*, 3537–3544. [CrossRef]
107. Cristea, I.; Stefanescu, M.; Angheluta, C. About the fundamental relations defined on the hypergroupoids associated with binary relations. *Electron. J. Combin.* **2011**, *32*, 72–81. [CrossRef]
108. Aghabozorgi, H.; Jafarpour, M.; Dolatabadi, M.; Cristea, I. An algorithm to compute the number of Rosenberg hypergroups of order less than 7. *Ital. J. Pure Appl. Math.* **2019**, *42*, 262–270.

109. Massouros, C.G.; Massouros, G.G. Hypergroups associated with graphs and automata. *AIP Conf. Proc.* **2009**, *1168*, 164–167. [CrossRef]
110. Massouros, C.G. On path hypercompositions in graphs and automata. *MATEC Web Conf.* **2016**, *41*, 5003. [CrossRef]
111. Massouros, C.G.; Tsitouras, C.G. Enumeration of hypercompositional structures defined by binary relations. *Ital. J. Pure Appl. Math.* **2011**, *28*, 43–54.
112. Tsitouras, C.G.; Massouros, C.G. Enumeration of Rosenberg type hypercompositional structures defined by binary relations. *Eur. J. Comb.* **2012**, *33*, 1777–1786. [CrossRef]
113. Massouros, G.G. The hyperringoid. *Mult. Valued Logic.* **1998**, *3*, 217–234.
114. Massouros, G.G.; Massouros, C.G. Homomorphic relations on Hyperringoids and Join Hyperrings. *Ratio Mat.* **1999**, *13*, 61–70.
115. Massouros, G.G. Solving equations and systems in the environment of a hyperringoid, Algebraic Hyperstructures and Applications. In Proceedings of the 6th International Congress, Prague, Czech Republic, 1–9 September 1996; Democritus Univ. of Thrace Press: Komotini, Greece, 1997; pp. 103–113.
116. Massouros, C.G.; Massouros, G.G. On join hyperrings, Algebraic Hyperstructures and Applications. In Proceedings of the 10th International Congress, Brno, Czech Republic, 3–9 September 1996; pp. 203–215.
117. Krasner, M. A class of hyperrings and hyperfields. *Int. J. Math. Math. Sci.* **1983**, *6*, 307–312. [CrossRef]
118. Mittas, J. Sur certaines classes de structures hypercompositionnelles. *Proc. Acad. Athens* **1973**, *48*, 298–318.
119. Mittas, J. Sur les structures hypercompositionnelles, Algebraic Hyperstructures and Applications. In Proceedings of the 4th International Congress, Xanthi, Greece, 27–30 June 1990; World Scientific: Singapore, 1991; pp. 9–31.
120. Ameri, R.; Eyvazi, M.; Hoskova-Mayerova, S. Superring of Polynomials over a Hyperring. *Mathematics* **2019**, *7*, 902. [CrossRef]
121. Ευκλείδης (Euclid). Στοιχεία *(Elements) c. 300 BC*; Heiberg, I., Stamatis, E., Eds.; Ο.Ε.Δ.Β: Athens, Greece, 1975.
122. Heath, T.L. *The Thirteen Books of Euclid's Elements*, 2nd ed.; Dover Publications: New York, NY, USA, 1956.
123. Freni, D. Strongly Transitive Geometric Spaces: Applications to Hypergroups and Semigroups Theory. *Commun. Algebra* **2004**, *32*, 969–988. [CrossRef]
124. Mirvakili, S.; Davvaz, B. Strongly transitive geometric spaces: Applications to hyperrings. *Revista UMA* **2012**, *53*, 43–53.
125. Anvariyeh, S.M.; Davvaz, B. Strongly transitive geometric spaces associated with (m, n)-ary hypermodules. *An. Sti. Univ. Al. I. Cuza Iasi* **2013**, *59*, 85–101. [CrossRef]
126. Mittas, J. Hyperanneaux et certaines de leurs propriétés, C.R. *Acad. Sci. Paris* **1969**, *269*, 623–626.
127. Mittas, J. Hypergroupes et hyperanneaux polysymétriques, C.R. *Acad. Sci. Paris* **1970**, *271*, 920–923.
128. Mittas, J. Contributions a la théorie des hypergroupes, hyperanneaux, et les hypercorps hypervalues, C.R. *Acad. Sci. Paris* **1971**, *272*, 3–6.
129. Mittas, J. Certains hypercorps et hyperanneaux définis à partir de corps et anneaux ordonnés. *Bull. Math. Soc. Sci. Math. R. S. Roum.* **1971**, *15*, 371–378.
130. Mittas, J. Sur les hyperanneaux et les hypercorps. *Math. Balk.* **1973**, *3*, 368–382.
131. Mittas, J. Contribution à la théorie des structures ordonnées et des structures valuées. *Proc. Acad. Athens* **1973**, *48*, 318–331.
132. Mittas, J. Hypercorps totalement ordonnes. *Sci. Ann. Polytech. Sch. Univ. Thessalon.* **1974**, *6*, 49–64.
133. Mittas, J. Espaces vectoriels sur un hypercorps. Introduction des hyperspaces affines et Euclidiens. *Math. Balk.* **1975**, *5*, 199–211.
134. Viro, O. On basic concepts of tropical geometry. *Proc. Steklov Inst. Math.* **2011**, *273*, 252–282. [CrossRef]
135. Viro, O. Hyperfields for Tropical Geometry I. Hyperfields and dequantization. *arXiv* **2010**, arXiv:1006.3034.
136. Baker, M.; Bowler, N. Matroids over hyperfields. *arXiv* **2017**, arXiv:1601.01204.
137. Jun, J. Geometry of hyperfields. *arXiv* **2017**, arXiv:1707.09348.
138. Lorscheid, O. Tropical geometry over the tropical hyperfield. *arXiv* **2019**, arXiv:1907.01037.
139. Connes, A.; Consani, C. The hyperring of adèle classes. *J. Number Theory* **2011**, *131*, 159–194. [CrossRef]
140. Connes, A.; Consani, C. From monoids to hyperstructures: In search of an absolute arithmetic. *arXiv* **2010**, arXiv:1006.4810.
141. Massouros, C.G. Constructions of hyperfields. *Math. Balk.* **1991**, *5*, 250–257.

142. Massouros, C.G. Methods of constructing hyperfields. *Int. J. Math. Math. Sci.* **1985**, *8*, 725–728. [CrossRef]
143. Massouros, C.G. A class of hyperfields and a problem in the theory of fields. *Math. Montisnigri* **1993**, *1*, 73–84.
144. Massouros, C.G. On the theory of hyperrings and hyperfields. *Algebra Logika* **1985**, *24*, 728–742. [CrossRef]
145. Nakassis, A. Recent results in hyperring and hyperfield theory. *Int. J. Math. Math. Sci.* **1988**, *11*, 209–220. [CrossRef]
146. Massouros, C.G. A Field Theory Problem Relating to Questions in Hyperfield Theory. *AIP Conf. Proc.* **2011**, *1389*, 1852–1855. [CrossRef]
147. Massouros, C.G. Free and cyclic hypermodules. *Ann. Mat. Pura Appl.* **1988**, *150*, 153–166. [CrossRef]
148. Vahedi, V.; Jafarpour, M.; Aghabozorgi, H.; Cristea, I. Extension of elliptic curves on Krasner hyperfields. *Comm. Algebra* **2019**, *47*, 4806–4823. [CrossRef]
149. Vahedi, V.; Jafarpour, M.; Cristea, I. Hyperhomographies on Krasner Hyperfields. *Symmetry* **2019**, *11*, 1442. [CrossRef]
150. Vahedi, V.; Jafarpour, M.; Hoskova-Mayerova, S.; Aghabozorgi, H.; Leoreanu-Fotea, V.; Bekesiene, S. Derived Hyperstructures from Hyperconics. *Mathematics* **2020**, *8*, 429. [CrossRef]

© 2020 by the authors. Licensee MDPI, Basel, Switzerland. This article is an open access article distributed under the terms and conditions of the Creative Commons Attribution (CC BY) license (http://creativecommons.org/licenses/by/4.0/).

Article

Fuzzy Reduced Hypergroups

Milica Kankaras [1] **and Irina Cristea** [2,*]

[1] Department of Mathematics, University of Montenegro, 81000 Podgorica, Montenegro; milica.k@ucg.ac.me
[2] Centre for Information Technologies and Applied Mathematics, University of Nova Gorica, 5000 Nova Gorica, Slovenia
* Correspondence: irina.cristea@ung.si or irinacri@yahoo.co.uk; Tel.: +386-0533-15-395

Received: 10 January 2020; Accepted: 14 February 2020; Published: 17 February 2020

Abstract: The fuzzyfication of hypercompositional structures has developed in several directions. In this note we follow one direction and extend the classical concept of reducibility in hypergroups to the fuzzy case. In particular we define and study the fuzzy reduced hypergroups. New fundamental relations are defined on a crisp hypergroup endowed with a fuzzy set, that lead to the concept of fuzzy reduced hypergroup. This is a hypergroup in which the equivalence class of any element, with respect to a determined fuzzy set, is a singleton. The most well known fuzzy set considered on a hypergroup is the *grade fuzzy set*, used for the study of the fuzzy grade of a hypergroup. Based on this, in the second part of the paper, we study the fuzzy reducibility of some particular classes of crisp hypergroups with respect to the grade fuzzy set.

Keywords: fuzzy set; reducibility; grade fuzzy set

1. Introduction

In the algebraic hypercompositional structures theory, the most natural link with the classical algebraic structures theory is assured by certain equivalences, that work as a bridge between both theories. More explicitly, the quotient structure modulo the equivalence β defined on a hypergroup is always a group [1], the quotient structure modulo the equivalence Γ defined on a hyperring is a ring [2], while a commutative semigroup can be obtained factorizing a semihypergroup by the equivalence γ [3]. A more completed list of such equivalences is very clearly presented in [4]. All the equivalences having this property, i.e., they are the smallest equivalence relations defined on a hypercompositional structure such that the corresponding quotient (modulo that relation) is a classical structure with the same behaviour, are called *fundamental* relations, while the associated quotients are called *fundamental structures*. The study of the fundamental relations represents an important topic of research in Hypercompositional Algebra also nowadays [5–7]. But this is not the unique case when the name "fundamental" is given to an equivalence defined on a hyperstructure. Indeed, there exist three other equivalences, called *fundamental* by Jantosciak [8], who observed that, unlike what happens in classical algebraic structures, two elements can play interchangeable roles with respect to a hyperoperation. In other words, the hyperoperation does not distinguish between the action of the given elements. These particular roles have been translated by the help of the following three equivalences. We say that two elements x and y in a hypergroup $(H, \circ)$ are:

1. operationally equivalent, if their hyperproducts with all elements in H are the same;
2. inseparable, if x belongs to the same hyperproducts as y belongs to;
3. essentially indistinguishable, if they are both operationally equivalent and inseparable.

Then a hypergroup is called *reduced* [8] if the equivalence class of any element with respect to the essentially indistinguishable relation is a singleton. Besides the associated quotient structure (with respect to the same fundamental relation) is a reduced hypergroup, called the *reduced form* of the original hypergroup. Jantosciak explained the role of these fundamental relations by a very simple and well known result. Define on the set $H = \mathbb{Z} \times \mathbb{Z}^*$, where $\mathbb{Z}$ is the set of integers and $\mathbb{Z}^* = \mathbb{Z} \setminus \{0\}$, the equivalence $\sim$ that assigns equivalent fractions in the same class: $(x,y) \sim (u,v)$ if and only if $xv = yu$, for $(x,y), (u,v) \in H$. Endow H with a hypercompositional structure, considering the hyperproduct $(w,x) \circ (y,z) = \widehat{(wz+xy, xz)}_\sim$, where by $\widehat{x}_\sim$ we mean the equivalence class of the element x with respect to the equivalence $\sim$. It follows that the equivalence class of the element $(x,y) \in H$ with respect to all three fundamental relations defined above is equal to the equivalence class of (x,y) with respect to the equivalence $\sim$. Therefore H is not a reduced hypergroup, while its reduced form is isomorphic with $\mathbb{Q}$, the set of rationals.

Motivated by this example, in the same article [8], Jantosciak proposed a method to obtain all hypergroups having a given reduced hypergroup as their reduced form. This aspect of reducibility of hypergroups has been later on considered by Cristea et al. [9–11], obtaining necessary and sufficient condition for a hypergroup associated with a binary or n−ary relation to be reduced. The same author studied the regularity aspect of these fundamental relations and started the study of fuzzyfication of the concept of reducibility [12]. This study can be considered in two different directions, corresponding to the two different approaches of the theory of hypergroups associated with fuzzy sets. An overview of this theory is covered by the monograph [13] written by Davvaz and Cristea, the only one on this topic published till now. The fuzzy aspect of reducibility could be investigated in two ways: by studying the indistinguishability between the elements of a fuzzy hypergroup (i.e., a structure endowed with a fuzzy hyperoperation) or by studying the indistinguishability between the images of the elements of a crisp hypergroup through a fuzzy set. By a crisp hypergroup we mean a hypergroup, but the attribute "crisp" is used to emphasize that the structure is not fuzzy. The study conducted in this article follows the second direction, while the first direction will be developed in our future works.

The aim of this note is to study the concept of *fuzzy reduced hypergroup* as a crisp hypergroup which is fuzzy reduced with respect to the associated fuzzy set. One of the most known fuzzy sets associated with a hypergroup is the grade fuzzy set $\widetilde{\mu}$, introduced by Corsini [14]. It was studied by Corsini and Cristea [15] in order to define the *fuzzy grade of a hypergroup* as the length of the sequence of join spaces and fuzzy sets associated with the given hypergroup. For any element x in a hypergroup H, the value $\widetilde{\mu}(x)$ is defined as the average value of the reciprocals of the sizes of all hyperproducts containing x. The properties of this particular fuzzy set, in particular those related to the fuzzy grade, have been investigated for several classes of finite hypergroups, as: complete hypergroups, non-complete 1-hypergroups or i.p.s. hypergroups (i.e., hypergroups with partial scalar identities). Inspired by all these studies, first we introduce the definition of fuzzy reduced hypergroups and present some combinatorial aspects related to them. Then we focus on the fuzzy reducibility of i.p.s. hypergroups, complete hypergroups and non-complete 1-hypergroups with respect to the grade fuzzy set $\widetilde{\mu}$. Theorem 3 states that any proper complete hypergroup is not reduced, either fuzzy reduced with respect to $\widetilde{\mu}$. Regarding the i.p.s. hypergroups, we show that they are reduced, but not fuzzy reduced with respect to $\widetilde{\mu}$ (see Theorems 4 and 5). Finally, we present a general method to construct a non-complete 1-hypergroup, that is not reduced either fuzzy reduced with respect to $\widetilde{\mu}$. Some conclusions and new research ideas concerning this study are gathered in the last section.

2. Review of Reduced Hypergroups

In this section we briefly recall the basic definitions which will be used in the following, as well as the main properties of reduced hypergroups. We start with the three fundamental relations defined by

Jantosciak [8] on an arbitrary hypergroup. Throughout this note, by a hypergroup $(H, \circ)$, we mean a non-empty set H endowed with a hyperoperation, usually denoted as $\circ : H \times H \to \mathcal{P}^*(H)$ satisfying the associativity, i.e., for all $a, b, c \in H$ there is $(a \circ b) \circ c = a \circ (b \circ c)$ and the reproduction axiom, i.e., for all $a \in H$, there is $a \circ H = H \circ a = H$, where $\mathcal{P}^*(H)$ denotes the set of all non-empty subsets of H.

Definition 1 ([8]). *Two elements x, y in a hypergroup $(H, \circ)$ are called:*

1. *operationally equivalent or by short o-equivalent, and write $x \sim_o y$, if $x \circ a = y \circ a$, and $a \circ x = a \circ y$, for any $a \in H$;*
2. *inseparable or by short i-equivalent, and write $x \sim_i y$, if, for all $a, b \in H$, $x \in a \circ b \iff y \in a \circ b$;*
3. *essentially indistinguishable or by short e-equivalent, and write $x \sim_e y$, if they are operationally equivalent and inseparable.*

Definition 2 ([8]). *A reduced hypergroup has the equivalence class of any element with respect to the essentially indistinguishable relation $\sim_e$ a singleton, i.e., for any $x \in H$, there is $\hat{x}_e = \{x\}$.*

Example 1. *Let $(H, \circ)$ be a hypergroup, where the hyperoperation "$\circ$" is defined by the following table.*

$\circ$	a	b	c	d
a	a	a	a, b, c	a, b, d
b	a	a	a, b, c	a, b, d
c	a, b, c	a, b, c	a, b, c	c, d
d	a, b, d	a, b, d	c, d	a, b, d

One notices that $a \sim_o b$, because the lines (and columns) corresponding to a and b are exactly the same, thereby: $\hat{a}_o = \hat{b}_o = \{a, b\}$, while $\hat{c}_o = \{c\}$ and $\hat{d}_o = \{d\}$. But, on the other side, the equivalence class of any element in H with respect to the relation $\sim_i$ is a singleton, as well as with respect to the relation $\sim_e$, by consequence $(H, \circ)$ is reduced.

In [11] Cristea et al. discussed about the regularity of these fundamental relations, proving that in general none of them is strongly regular. This means that the corresponding quotients modulo these equivalences are not classical structures, but hypergroups. Moreover, Jantosciak [8] established the following result.

Theorem 1 ([8]). *For any hypergroup H, the associated quotient hypergroup $H/\sim_e$ is a reduced hypergroup, called the reduced form of H.*

Motivated by this property, Jantosciak concluded that the study of the hypergroups can be divided into two parts: the study of the reduced hypergroups and the study of all hypergroups having the same reduced form [8].

3. Fuzzy Reduced Hypergroups

As already mentioned in the introductory part of this article, the extension of the concept of reducibility to the fuzzy case can be performed on a crisp hypergroup endowed with a fuzzy set, by defining, similarly to the classical case, three equivalences as follows.

Definition 3. *Let $(H, \circ)$ be a crisp hypergroup endowed with a fuzzy set μ. For two elements $x, y \in H$, we say that*

1. *x and y are fuzzy operationally equivalent and write $x \sim_{fo} y$ if, for any $a \in H$, $\mu(x \circ a) = \mu(y \circ a)$ and $\mu(a \circ x) = \mu(a \circ y)$;*

2. x and y are fuzzy inseparable and write $x \sim_{fi} y$ if $\mu(x) \in \mu(a \circ b) \iff \mu(y) \in \mu(a \circ b)$, for $a, b \in H$;
3. x and y are fuzzy essentially indistinguishable and write $x \sim_{fe} y$, if they are fuzzy operationally equivalent and fuzzy inseparable.

Definition 4. *The crisp hypergroup $(H, \circ)$ is called fuzzy reduced if the equivalence class of any element in H with respect to the fuzzy essentially indistinguishable relation is a singleton, i.e.,*

$$\text{for all } x \in H, \hat{x}_{fe} = \{x\}.$$

Notice that the notion of fuzzy reducibility of a hypergroup is strictly connected with the definition of the involved fuzzy set.

Remark 1. *It is easy to see that, for each hypergroup H endowed with an arbitrary fuzzy set μ, the following implication holds: for any $a, b \in H$,*

$$a \sim_o b \Rightarrow a \sim_{fo} b.$$

Remark 2. *(i) First of all, let us better explain the meaning of $\mu(a \circ b)$, for any $a, b \in H$ and any arbitrary fuzzy set μ defined on H. Generally, $a \circ b$ is a subset of H, so $\mu(a \circ b)$ is the direct image of this subset through the fuzzy set μ, i.e., $\mu(a \circ b) = \{\mu(x) \mid x \in a \circ b\}$.*

From here, two things need to be stressed on. Firstly, if $a \circ b$ is a singleton, i.e., $a \circ b = \{c\}$, then $\mu(a \circ b)$ is a set containing the real number $\mu(c)$. Therefore we can write $\mu(c) \in \mu(a \circ b)$, but it is not correct writing $\mu(c) = \mu(a \circ b)$, because the first member is a real number, while the second one is a set containing the real number $\mu(c)$.

Secondly, if $a \in x \circ y$, then, clearly $\mu(a) \in \mu(x \circ y)$, but not also viceversa because it could happen that $\mu(a) \in \mu(x' \circ y')$ for $a \notin x' \circ y'$.

(ii) Generally, $a \sim_i b \not\Rightarrow a \sim_{fi} b$, as illustrated in the next example. Indeed, if $a \sim_i b$ then $a \in x \circ y$ if and only if $b \in x \circ y$. But it can happen that $\mu(a) \in \mu(x' \circ y')$ with $a \notin x' \circ y'$, so also $b \notin x' \circ y'$. And if $\mu(a) \neq \mu(b)$, then $\mu(b) \notin \mu(x' \circ y')$, thus $a \not\sim_i b$.

(iii) Finally, it is simple to see the implication $\mu(a) = \mu(b) \Rightarrow a \sim_{fi} b$.

The following example illustrates all the issues in the above mentioned remark.

Example 2 ([16])**.** *Let $(H, \circ)$ be a hypergroup represented by the following commutative Cayley table:*

$\circ$	e	a_1	a_2	a_3
e	e	a_1	a_2, a_3	a_2, a_3
a_1		a_2, a_3	e	e
a_2			a_1	a_1
a_3				a_1

One notices immediately that $a_2 \sim_i a_3$, while $a_1 \not\sim_i a_2$.

(a) Define now on H the fuzzy set μ as follows: $\mu(e) = 1, \mu(a_1) = \mu(a_2) = 0.3, \mu(a_3) = 0.5$. Since $\mu(a_1) = \mu(a_2)$, it follows that $a_1 \sim_{fi} a_2$. Moreover, since $e \circ a_1 = \{a_1\}$, we have

$$\mu(e \circ a_1) = \{\mu(a_1)\} = \{0.3\} \ni \mu(a_2),$$

while it is clear that $\mu(a_3) \notin \mu(e \circ a_1)$, so $a_2 \not\sim_{fi} a_3$.

(b) If we define on H the fuzzy set μ by taking $\mu(e) = \mu(a_1) = 1$, $\mu(a_2) = \mu(a_3) = \frac{1}{3}$, it follows that $e \sim_{fi} a_1$ and $a_2 \sim_{fi} a_3$.

Once again, it is evident that the three equivalences $\sim_{fo}$, $\sim_{fi}$, and $\sim_{fe}$ are strictly related with the definition of the fuzzy set considered on the hypergroup.

Now we will present an example of an infinite hypergroup and study its fuzzy reducibility.

Example 3. Consider the partially ordered group $(\mathbb{Z}, +, \leq)$ with the usual addition and orderings of integers. Define on $\mathbb{Z}$ the hyperoperation $a * b = \{x \in \mathbb{Z} \mid a + b \leq x\}$. Then $(\mathbb{Z}, *)$ is a hypergroup [17]. Define now on $\mathbb{Z}$ the fuzzy set μ as follows: $\mu(0) = 0$ and $\mu(x) = \frac{1}{|x|}$, for any $x \neq 0$. We obtain $\hat{x}_{f.o.} = \{x\}$, for any $x \in \mathbb{Z}$, therefore $(\mathbb{Z}, *)$ is fuzzy reduced with respect to μ. Indeed, for two arbitrary elements x and y in $\mathbb{Z}$, we have $x \sim_{fo} y$ if and only if $\mu(x * a) = \mu(y * a)$, for any $a \in \mathbb{Z}$, where $\mu(x * a) = \{\frac{1}{|x+a|}, \frac{1}{|x+a+1|}, \frac{1}{|x+a+2|}, \ldots\}$ and similarly, $\mu(y * a) = \{\frac{1}{|y+a|}, \frac{1}{|y+a+1|}, \ldots\}$. Since a is an arbitrary integer, for any x and y we always find a suitable integer a such that $x + a > 0$ and $y + a > 0$. This means that the sets $\mu(x * a)$ and $\mu(y * a)$ contain descending sequences of positive integers, so they are equal only when $x = y$. Therefore $x \sim_{fo} y$.

In the following, we will study the fuzzy reducibility of some particular types of finite hypergroups, with respect to the *grade fuzzy set* $\widetilde{\mu}$, defined by Corsini [14]. We recall here its definition. With any crisp hypergroupoid $(H, \circ)$ (not necessarily a hypergroup) we may associate the fuzzy set $\widetilde{\mu}$ considering, for any $u \in H$,

$$\widetilde{\mu}(u) = \frac{\sum_{(x,y) \in Q(u)} \frac{1}{|x \circ y|}}{q(u)}, \quad (1)$$

where $Q(u) = \{(a, b) \in H^2 \mid u \in a \circ b\}$ and $q(u) = |Q(u)|$. By convention, we take $\widetilde{\mu}(u) = 0$ anytime when $Q(u) = \emptyset$. In other words, the value $\widetilde{\mu}(u)$ is the average value of reciprocals of the sizes of all hyperproducts $x \circ y$ containing the element u in H. In addition, sometimes when we will refer to formula (1), we will denote its numerator by $A(u)$, while the denominator is already denoted by $q(u)$.

Remark 3. As already explained in Remark 2 (ii), generally, for an arbitrary fuzzy set, $x \sim_i y \not\Rightarrow x \sim_{fi} y$, while the implication holds if we consider the grade fuzzy set $\widetilde{\mu}$. Indeed, if $x \sim_i y$, then $x \in a \circ b$ if and only if $y \in a \circ b$ and therefore $Q(x) = Q(y)$, implying that $q(x) = q(y)$, and moreover $A(x) = A(y)$. This leads to the equality $\widetilde{\mu}(x) = \widetilde{\mu}(y)$. By consequence, based on Remark 2 (iii), it holds $x \sim_{fi} y$, with respect to $\widetilde{\mu}$.

Example 4. Let us consider now a total finite hypergroup H, i.e., $x \circ y = H$, for all $x, y \in H$. It is easy to see that $x \sim_e y$ for any $x, y \in H$, meaning that $\hat{x}_e = H$, for any $x \in H$. Thus, a total hypergroup is not reduced. What can we say about the fuzzy reducibility with respect to the grade fuzzy set $\widetilde{\mu}$?

For any $u \in H$, there is

$$\widetilde{\mu}(u) = \frac{|H|^2 \frac{1}{|H|}}{|H|^2} = \frac{1}{|H|}.$$

Since, $x \sim_\circ y$ for any $x, y \in H$, it follows that $x \sim_{fo} y$, for any $x, y \in H$. Then, it is clear that $\widetilde{\mu}(x) = \widetilde{\mu}(y)$, for all $x, y \in H$, implying that $x \sim_{fi} y$, for all $x, y \in H$. Concluding, it follows that any total finite hypergroup is neither reduced, nor fuzzy reduced.

Based now on Remarks 1 and 3, the following assertion is clear.

Corollary 1. *If $(H, \circ)$ is a not reduced hypergoup, then it is also not fuzzy reduced with respect to the grade fuzzy set $\widetilde{\mu}$.*

3.1. Fuzzy Reducibility in Complete Hypergroups

The complete hypergroups form a particular class of hypergroups, strictly related with the join spaces and the regular hypergroups. Their definition is based on the notion of *complete part*, introduced in Koskas [1], with the main role to characterize the equivalence class of an element under the relation β^*. More exactly, a nonempty set A of a semihypergroup $(H, \circ)$ is called a complete part of H, if for any natural number n and any elements $a_1, a_2, \ldots, a_n$ in H, the following implication holds:

$$A \cap \prod_{i=1}^{n} a_i \neq \emptyset \Rightarrow \prod_{i=1}^{n} a_i \subseteq A.$$

We may say, as it was mentioned in the review written by Antampoufis et al. [18], that a complete part A absorbs all hyperproducts of the elements of H having non-empty intersection with A. The intersection of all complete parts of H containing the subset A is called the *complete closure* of A in H and denoted by $C(A)$.

Definition 5. *A hypergroup $(H, \circ)$ is called a complete hypergroup if, for any $x, y \in H$, there is $C(x \circ y) = x \circ y$.*

As already explained in the fundamental book on hypergroups theory [19] and the other manuscripts related with complete hypergroups [16,20], in practice, it is more useful to use the following characterization of the complete hypergroups.

Theorem 2 ([19]). *Any complete hypergroup may be constructed as the union $H = \bigcup_{g \in G} A_g$ of its subsets, where*

(1) $(G, \cdot)$ *is a group.*
(2) *The family $\{A_g \,|\, g \in G\}$ is a partition of G, i.e., for any $(g_1, g_2) \in G^2$, $g_1 \neq g_2$, there is $A_{g_1} \cap A_{g_2} = \emptyset$.*
(3) *If $(a, b) \in A_{g_1} \times A_{g_2}$, then $a \circ b = A_{g_1 g_2}$.*

Example 5. *The hypergroup presented in Example 2 is complete, where the group $G = (\mathbb{Z}_3, +)$ and the partition set contains $A_0 = \{e\}$, $A_1 = \{a_1\}$, and $A_2 = \{a_2, a_3\}$. It is clear that all conditions in Theorem 2 are fulfilled.*

Let $(H, \circ)$ be a proper complete hypergroup (i.e., H is not a group). Define now on H the equivalence "$\sim$" by:

$$x \sim y \iff \exists g \in G \text{ such that } x, y \in A_g. \tag{2}$$

Proposition 1. *On a proper complete hypergroup $(H, \circ)$, the equivalence $\sim$ in (2) is a representation of the essentially indistinguishability equivalence $\sim_e$.*

Proof. By Theorem 2, one notices that, for any element $x \in H$, there exists a unique $g \in G$ such that $x \in A_{g_x}$. In the following, we will denote this element by g_x. First, suppose that $x \sim y$, i.e., there exists $g_x = g_y \in G$ such that $x, y \in A_{g_x}$. For any arbitrary element $a \in H$, we can say that $a \in A_{g_a}$, with $g_a \in G$, and by the definition of the hyperproduct in the complete hypergroup $(H, \circ)$, there is $x \circ a = A_{g_x g_a} = A_{g_y g_a} = y \circ a$, (and similarly, $a \circ x = a \circ y$,) implying that $x \sim_o y$ (i.e., x and y are operationally equivalent). Secondly, for any $x \in a \circ b = A_{g_a g_b} \cap A_{g_x}$, it follows that $g_a g_b = g_x$; but $g_x = g_y$, so $y \in A_{g_a g_b} = a \circ b$. Thereby,

$x \in a \circ b$ if and only if $y \in a \circ b$, meaning that $x \sim_i y$ (i.e., x and y are inseparable). We have proved that $\sim \subseteq \sim_e$.

Conversely, let us suppose that $x \sim_e y$. Since x and y are inseparable, i.e., $x \in a \circ b$ if and only if $y \in a \circ b$, we may write $x, y \in A_{g_a g_b}$. Therefore there exists $g_x = g_a \cdot g_b \in G$ such that $x, y \in A_{g_x}$, so $x \sim y$. □

Example 6. *If we continue with Example 2, we notice that the equivalence classes of the elements of H with respect to the equivalence~defined in (2) are: $\hat{e} = \{e\}$, $\hat{a}_1 = \{a_1\}$, $\hat{a}_2 = \{a_2, a_3\} = \hat{a}_3$. Regarding now the essentially indistinguishability equivalence $\sim_e$, we have the same equivalences classes: $\hat{e}_e = \{e\}$, $\hat{a}_{1e} = \{a_1\}$, $\hat{a}_{2e} = \{a_2, a_3\} = \hat{a}_{3e}$. This is clear from the Cayley table of the hypergroup: the lines corresponding to a_2 and a_3 are the same and every time a_2 and a_3 are in the same hyperproduct, so they are equivalent. Since these properties are not satisfied for a_1 and e, their equivalence classes are singletons.*

Proposition 2. *Let $(H, \circ)$ be a proper complete hypergroup and consider on H the grade fuzzy set $\widetilde{\mu}$. Then $\sim \subset \sim_{fe}$ (with respect to the fuzzy set $\widetilde{\mu}$).*

Proof. By the definition of the grade fuzzy set $\widetilde{\mu}$, one obtains that

$$\widetilde{\mu}(x) = \frac{1}{|A_{g_x}|}, \text{ for any } x \in H. \tag{3}$$

Take now $x, y \in H$ such that $x \sim y$. There exists $g_x \in G$ such that $x, y \in A_{g_x}$, thereby $\widetilde{\mu}(x) = \widetilde{\mu}(y)$ and by Remark 2 (ii) we have $x \sim_{fi} y$. Moreover, by Proposition 1 there is $x \sim_o y$ and by Remark 1 we get that $x \sim_{fo} y$. Concluding, we have proved that $x \sim y \Rightarrow x \sim_{fe} y$, with respect to $\widetilde{\mu}$. □

Theorem 3. *Any proper complete hypergroup is not reduced, either fuzzy reduced with respect to the grade fuzzy set.*

Proof. Since $(H, \circ)$ is a proper complete hypergroup, there exists at least one element g in G such that $|A_g| \geq 2$, i.e., there exist two distinct elements a and b in H with the property that $a \sim b$. Then $a \sim_e b$ and $a \sim_{fe} b$, meaning that $(H, \circ)$ is not reduced, either fuzzy reduced with respect to the grade fuzzy set $\widetilde{\mu}$. □

3.2. Fuzzy Reducibility in i.p.s. Hypergroups

An i.p.s. hypergroup is a canonical hypergroup with partial scalar identities. The name, given by Corsini [21], originally comes from Italian, the abbreviation "i.p.s." standing for "identità parziale scalare" (in English, partial scalar identity). First we dwell on the terminology connected with the notion of identity in a hypergroup $(H, \circ)$. An element $x \in H$ is called a *scalar*, if $|x \circ y| = |y \circ x| = 1$, for any $y \in H$. An element $e \in H$ is called *partial identity* of H if it is a *left identity* (i.e., there exists $x \in H$ such that $x \in e \circ x$) or a *right identity* (i.e., there exists $y \in H$ such that $y \in y \circ e$). We denote by I_p the set of all partial identities of H. Besides, for a given element $x \in H$, a *partial identity of x* is an element $u \in H$ such that $x \in x \circ u \cup u \circ x$. The element $u \in H$ is a *partial scalar identity of x* whenever from $x \in x \circ u$ it follows $x = x \circ u$ and from $x \in u \circ x$ it follows that $x = u \circ x$. We denote by $I_p(x)$ the set of all partial identities of x, by $I_{ps}(x)$ the set of all partial scalar identities of x, and by $Sc(H)$ the set of all scalars of H. It is obvious that $I_{ps}(x) = I_p(x) \cap Sc(H)$.

Remark 4. *The term "partial" here must not be confused with the "left or right" (identity), but it must be connected with the fact that the element u has a partial behaviour of identity with respect to the element x. So, u is not a left/right (i.e., partial) identity for the hypergroup H. Besides, an i.p.s. hypergroup is a commutative hypergroup, so the concept of partial intended as left/right element satisfying a property (i.e., left/right unit) has no sense. It is* probably better to understand an element u having the property of being partial identity for x as an element having a similar behaviour as an identity but only with respect to x, so a partial role of being identity.

Let us recall now the definition of an i.p.s. hypergroup. All finite i.p.s. hypergroups of order less than 9 have been determined by Corsini [21–23].

Definition 6. *A hypergroup $(H, \circ)$ is called i.p.s. hypergroup, if it satisfies the following conditions.*

1. *It is canonical, i.e.*

   - *it is commutative;*
   - *it has a scalar identity 0 such that $0 \circ x = x$, for any $x \in H$;*
   - *every element $x \in H$ has a unique inverse $x^{-1} \in H$, that is $0 \in x \circ x^{-1}$;*
   - *it is reversible, so $y \in a \circ x \Longrightarrow x \in a^{-1} \circ y$, for any $a, x, y \in H$.*

2. *It satisfies the relation: for any $a, x \in H$, if $x \in a \circ x$, then $a \circ x = x$.*

The most useful properties of i.p.s. hypergroups are gathered in the following result.

Proposition 3 ([21]). *Let $(H, \circ)$ be an i.p.s. hypergroup.*

1. *For any element x in H, the set $x \circ x^{-1}$ is a subhypergroup of H.*
2. *For any element x in H, different from $\{0\}$, we have: or x is a scalar of H, or there exists $u \in Sc(H) \setminus \{0\}$ such that $u \in x \circ x^{-1}$. Moreover $|Sc(H)| \geq 2$.*
3. *If x is a scalar of H, then the set of all partial scalar identities of x contains just 0.*
   If x is not a scalar of H, then $I_{ps}(x) \subset Sc(H) \cap x \circ x^{-1}$ and therefore $|I_{ps}(x)| \geq 2$.

Proposition 4. *Let $(H, \circ)$ be an i.p.s. hypergroup. For any scalar $u \in H$ and for any element $x \in H$, there exists a unique $y \in H$ such that $u \in x \circ y$.*

Proof. The existence part immediately follows from the reproducibility property of H. For proving the unicity, assume that there exist $y_1, y_2 \in H, y_1 \neq y_2$ such that $u \in x \circ y_1 \cap x \circ y_2$. Then, by reversibility, it follows that $y_1, y_2 \in x^{-1} \circ u$. Since u is a scalar element, we get $|x^{-1} \circ u| = 1$ and then $y_1 = y_2 = x^{-1} \circ u$. □

Example 7 ([21]). *Let us consider the following i.p.s. hypergroup $(H, \circ)$.*

H	0	1	2	3
0	0	1	2	3
1	1	2	0,3	1
2	2	0,3	1	2
3	3	1	2	0

Here we can notice that 0 is the only one identity of H. In addition, $Sc(H) = \{0, 3\}$, and $0 \in 0 \circ u$ only for $u = 0$, so $I_{ps}(0) = \{0\}$. Also, only for $u = 0$ there is $3 \in 3 \circ u$, thus $I_{ps}(3) = \{0\}$. (In general, if $x \in Sc(H)$ then $I_{ps}(x) = \{0\}$, according with Proposition 3). Similarly, one gets $I_p(1) = I_p(2) = \{0, 3\}$ and since $Sc(H) = \{0, 3\}$, it follows that $I_{ps}(1) = I_{ps}(2) = \{0, 3\}$.

Note that, in an i.p.s. hypergroup, the Jantosciak fundamental relations have a particular meaning, in the sense that, for any two elements there is

$$a \sim_o b \iff a \sim_i b \iff a \sim_e b \iff a = b.$$

By consequence, one obtains the following result.

Theorem 4. *Any i.p.s. hypergroup is reduced.*

In the following we will discuss the fuzzy reducibility of an i.p.s. hypergroup with respect to the grade fuzzy set $\tilde{\mu}$.

Theorem 5. *Any i.p.s. hypergroup is not fuzzy reduced with respect to the fuzzy set $\tilde{\mu}$.*

Proof. Since any i.p.s. hypergroup contains at least one non-zero scalar, take arbitrary such a $u \in Sc(H)$. We will prove that $u \sim_{fi} 0$ and $u \sim_{fo} 0$, therefore $|\hat{0}_{fe}| \geq 2$, meaning that H is not fuzzy reduced.

First we will prove that, for any $u \in Sc(H)$, there is $\tilde{\mu}(0) = \tilde{\mu}(u)$, equivalently with $u \sim_{fi} 0$. For doing this, based on the fact that $\tilde{\mu}(x) = \frac{A(x)}{q(x)}$, for all $x \in H$, we show that $A(0) = A(u)$ and $q(0) = q(u)$.

Let us start with the computation of $q(0)$ and $q(u)$. If $0 \in x \circ y$, it follows that $y \in x^{-1} \circ 0$, that is $y = x^{-1}$ and then $Q(0) = \{(x,y) \in H^2 \mid 0 \in x \circ y\} = \{(x, x^{-1}) \mid x \in H\}$. Thereby $q(0) = |Q(0)| = n = |H|$. On the other hand, by Proposition 4, we have $q(u) = n = |H|$ (since for any $x \in H$ there exists a unique $y \in H$ such that $u \in x \circ y$.)

Let us calculate now $A(0)$. By formula (1), we get that

$$A(0) = \sum_{(x,y) \in Q(0)} \frac{1}{|x \circ y|} = \sum_{x \in H} \frac{1}{|x \circ x^{-1}|} =$$

$$\sum_{a \in Sc(H)} \frac{1}{|a \circ a^{-1}|} + \sum_{x \notin Sc(H)} \frac{1}{|x \circ x^{-1}|} = |Sc(H)| + \sum_{x \notin Sc(H)} \frac{1}{|x \circ x^{-1}|}.$$

Since, for $u \in Sc(H), \exists x \notin Sc(H)$ such that $u \in x \circ x^{-1} \cap I_{ps}(x)$, we similarly get that

$$A(u) = |Sc(H)| + \sum_{x \notin Sc(H)} \frac{1}{|x \circ x^{-1}|}$$

and it is clear that $A(0) = A(u)$, so $\tilde{\mu}(0) = \tilde{\mu}(u)$. Therefore $0 \sim_{fi} u$.

It remains to prove the second part of the theorem, that is $0 \sim_{fo} u$, equivalently with $\tilde{\mu}(0 \circ x) = \tilde{\mu}(u \circ x), \forall x \in H$.

If $x \in Sc(H)$, then $u \circ x \in Sc(H)$ and by the first part of the theorem, there is $\tilde{\mu}(u \circ x) = \tilde{\mu}(0) = \tilde{\mu}(x) = \tilde{\mu}(0 \circ x)$.

If $x \notin Sc(H)$, then since $Sc(H) \subset I_{ps}(a)$, for any $a \notin Sc(H)$, it follows that $Sc(H) \subset I_{ps}(x)$, so $u \circ x = x$, and then $\tilde{\mu}(u \circ x) = \tilde{\mu}(x) = \tilde{\mu}(0 \circ x)$. Now the proof is complete. □

3.3. Fuzzy Reducibility in Non-Complete 1-Hypergroups

In this subsection we will study the reducibility and fuzzy reducibility of some particular finite non-complete 1-hypergroups defined and investigated by Corsini, Cristea [24] with respect to their fuzzy grade. A hypergroup H is called 1-*hypergroup* if the cardinality of its heart ω_H is 1.

The general construction of this particular hypergroup is the following one. Consider the set $H = H_n = \{e\} \cup A \cup B$, where $A = \{a_1, \ldots, a_\alpha\}$ and $B = \{b_1, \ldots, b_\beta\}$, with $\alpha, \beta \geq 2$ and $n = \alpha + \beta + 1$, such that $A \cap B = \emptyset$ and $e \notin A \cup B$. Define on H the hyperoperation "$\circ$" by the following rule:

- for all $a \in A$, $a \circ a = b_1$,
- for all $(a_1, a_2) \in A^2$ such that $a_1 \neq a_2$, set $a_1 \circ a_2 = B$,
- for all $(a, b) \in A \times B$, set $a \circ b = b \circ a = e$,
- for all $(b, b') \in B^2$, there is $b \circ b' = A$,
- for all $a \in A$, set $a \circ e = e \circ a = A$,
- for all $b \in B$, $b \circ e = e \circ b = B$ and
- $e \circ e = e$.

H_n is an 1-hypergroup which is not complete.

We will discuss the (fuzzy) reducibility of this hypergroup for different cardinalities of the sets A and B.

(1) Let us suppose $n = |H_6| = 6$, where $H = H_6 = e \cup A \cup B$, $\alpha = |A| = 2$, $\beta = |B| = 3$, $A \cap B = \emptyset$, $e \notin A \cup B$ with $A = \{a_1, a_2\}$, $B = \{b_1, b_2, b_3\}$. Thus the Cayley table of $(H_6, \circ)$ is the following one

H	e	a_1	a_2	b_1	b_2	b_3
e	e	A	A	B	B	B
a_1		b_1	B	e	e	e
a_2			b_1	e	e	e
b_1				A	A	A
b_2					A	A
b_3						A

From the table, we notice immediately that the elements b_2 and b_3 are essentially indistinguishable, while the equivalence class with respect to the $\sim_e$ relation of all other elements is a singleton. Thereby H is not reduced.

Calculating now the values of the grade fuzzy set $\widetilde{\mu}$, one obtains $\widetilde{\mu}(e) = 1, \widetilde{\mu}(a_1) = \widetilde{\mu}(a_2) = 0.5, \widetilde{\mu}(b_1) = 0.467, \widetilde{\mu}(b_2) = \widetilde{\mu}(b_3) = 0.333$. Since $\widetilde{\mu}(a_1) = \widetilde{\mu}(a_2)$, it follows that $a_1 \sim_{fi} a_2$. But $\widetilde{\mu}(a_1 \circ a_1) = \widetilde{\mu}(\{b_1\}) = \{\widetilde{\mu}(b_1)\}$, while $\widetilde{\mu}(a_1 \circ a_2) = \widetilde{\mu}(B) = \{\widetilde{\mu}(b_1), \widetilde{\mu}(b_2)\}$, so $\widetilde{\mu}(a_1 \circ a_1) \neq \widetilde{\mu}(a_1 \circ a_2)$, meaning that $a_1 \not\sim_{fo} a_2$, that is $a_1 \not\sim_{fe} a_2$.

On the other side, we have $b_2 \sim_{fo} b_3$ because they are also operationally equivalent, and $b_2 \sim_{fi} b_3$ because $\widetilde{\mu}(b_2) = \widetilde{\mu}(b_3)$. This is equivalent with $b_2 \sim_{fe} b_3$ and therefore H is not reduced with respect to $\widetilde{\mu}$.

(2) Consider now the most general case. The Cayley table of the hypergroup H is the following one:

H	e	a_1	a_2	$\cdots$	a_α	b_1	b_2	$\cdots$	b_β
e	e	A	A	$\cdots$	A	B	B	$\cdots$	B
a_1		b_1	B	$\cdots$	B	e	e	$\cdots$	e
a_2			b_1	$\cdots$	B	e	e	$\cdots$	e
$\vdots$				$\ddots$					$\vdots$
a_α					b_1	e	e	$\cdots$	e
b_1						A	A	$\cdots$	A
b_2							A	$\cdots$	A
$\vdots$								$\ddots$	$\vdots$
b_β									A

As already calculated in [24], there is $\widetilde{\mu}(e) = 1$, $\widetilde{\mu}(a_i) = \frac{1}{\alpha}$, for any $i = 1, \ldots, \alpha$, while $\widetilde{\mu}(b_1) = \frac{\alpha^2 + \alpha\beta + 2\beta - \alpha}{\beta(\alpha^2 + 2\beta)}$, and $\widetilde{\mu}(b_j) = \frac{1}{\beta}$, for any $j = 2, \ldots, \beta$. As in the previous case, we see that any two elements in $B \setminus \{b_1\}$ are operational equivalent, by consequence also fuzzy operational equivalent, and indistinguishable and fuzzy indistinguishable (because their values under the grade fuzzy set $\widetilde{\mu}$ are the same). Concluding, this non complete 1-hypergroup is always not reduced, either not fuzzy reduced with respect to $\widetilde{\mu}$.

4. Conclusions and Open Problems

The pioneering paper of Rosenfeld [25] on fuzzy subgroups has opened a new perspective of the study of algebraic structures (hyperstructures) using the combinatorial properties of fuzzy sets. Regarding this, there are two distinct views: one concerns the crisp (hyper)structures endowed with fuzzy sets, and the other one is related to fuzzy (hyper)structures, that are sets on which fuzzy (hyper)operations are defined. Several classical algebraic concepts have then been extended in the framework of these two lines of research, *reducibility in hypergroups*, defined by Jantosciak [8] already in 1990, being one of them. In this paper, we have focussed on the study of *fuzzy reducibility* in hypergroups, extending the concept of reducibility to hypergroups endowed with fuzzy sets. For a better understanding of this subject, we have chosen the grade fuzzy set $\widetilde{\mu}$ [14] and have investigated the fuzzy reducibility of some particular classes of hypergroups: those with scalar partial identities, complete hypergroups, and special non-complete 1-hypergroups. Several connections between reducibility and fuzzy reducibility (with respect to $\widetilde{\mu}$) have been underlined and motivated by several examples.

In our future work, following the other connection between fuzzy sets and (hyper)structures, we aim to describe the reducibility of fuzzy hypergroups [26], by studying the *reduced fuzzy hypergroups*: they are fuzzy hypergroups which are reduced. Moreover, these aspects will also be investigated in the case of the mimic fuzzy hypergroups [27].

Author Contributions: Conceptualization, M.K. and I.C.; methodology, M.K. and I.C.; investigation, M.K. and I.C.; writing—original draft preparation, M.K.; writing—review and editing, I.C.; supervision, I.C.; funding acquisition, I.C. All authors have read and agreed to the published version of the manuscript.

Funding: The second author acknowledges the financial support from the Slovenian Research Agency (research core funding No. P1-0285).

Conflicts of Interest: The authors declare no conflict of interest.

References

1. Koskas, M. Groupoides, demi-hypergroupes et hypergroupes. *J. Math. Pure Appl.* **1970**, *49*, 155–192.
2. Vougiouklis, T. The Fundamental Relation in Hyperrings. the General Hyperfield. In *Algebraic Hyperstructures and Applications (Xanthi, 1990)*; World Sci. Publishing: Teaneck, NJ, USA, 1991; pp. 203–211.
3. Freni, D. A new characterization of the derived hypergroup via strongly regular equivalences. *Commun. Algebra* **2002**, *30*, 3977–3989. [CrossRef]
4. Norouzi, M.; Cristea, I. Fundamental relation on m-idempotent hyperrings. *J. Open Math.* **2017**, *15*, 1558–1567. [CrossRef]
5. De Salvo, M.; Fasino, D.; Freni, D.; Lo Faro, G. On hypergroups with a β-class of finite height. *Symmetry* **2020**, *12*, 168. [CrossRef]
6. Zadeh, A.; Norouzi, M.; Cristea, I. The commutative quotient structure of m-idempotent hyperrings. *An. Stiintifice Univ. Ovidius Constanta Ser. Mat.* **2020**, *28*, 219–236.
7. Ameri, R.; Eyvazi, M.; Hoskova-Mayerova, S. Superring of Polynomials over a Hyperring. *Mathematics* **2019**, *7*, 902. [CrossRef]

8. Jantosciak, J. Reduced hypergroups. In Proceedings of the Algebraic Hyperstructures and Applications Proceedings of 4th International Congress, Xanthi, Greece, 27–30 June 1990; World Scientific: Singapore, 1991; pp. 119–122.
9. Cristea, I. Several aspects on the hypergroups associated with *n*-ary relations. *An. Stiintifice Univ. Ovidius Constanta Ser. Mat.* **2009**, *17*, 99–110.
10. Cristea, I.; Ştefănescu, M. Hypergroups and *n*-ary relations. *Eur. J. Comb.* **2010**, *31*, 780–789. [CrossRef]
11. Cristea, I.; Ştefănescu, M.; Angheluţa, C. About the fundamental relations defined on the hypergroupoids associated with binary relations. *Eur. J. Comb.* **2011**, *32*, 72–81. [CrossRef]
12. Cristea, I. Reducibility in hypergroups (crisp and fuzzy case). *Int. J. Algebr. Hyperstructures Its Appl.* **2015**, *2*, 67–76.
13. Davvaz, B.; Cristea, I. Fuzzy Algebraic Hyperstructures-An Introduction. In *Studies in Fuzziness and Soft Computing*; Springer: Cham, Switzerland, 2015; Volume 321.
14. Corsini, P. A new connection between hypergroups and fuzzy sets. *Southeast Asian Bull. Math.* **2003**, *27*, 221–229.
15. Corsini, P.; Cristea, I. Fuzzy grade of i.p.s. hypergroups of order 7. *Iran. J. Fuzzy Syst.* **2004**, *1*, 15–32.
16. Cristea, I. Complete hypergroups, 1-hypergroups and fuzzy sets. *An. Stiintifice Univ. Ovidius Constanta Ser. Mat.* **2002**, *10*, 25–37.
17. Novak, M.; Cristea, I. Cyclicity in EL-hypergroups. *Symmetry* **2018**, *10*, 611. [CrossRef]
18. Antampoufis, N.; Hošková–Mayerová, Š. A brief survey on the two different approaches of fundamental equivalence relations on hyperstructures. *Ratio Math.* **2017**, *33*, 47–60.
19. Corsini, P. *Prolegomena of Hypergroup Theory*; Aviani Editore: Tricesimo, Italy, 1993.
20. Davvaz, B.; Hassani Sadrabadi, E.; Cristea, I. Atanassov's intuitionistic fuzzy grade of the complete hypergroups of order less than or equal to 6. *Hacet. J. Math. Stat.* **2015**, *44*, 295–315.
21. Corsini, P. Sugli ipergruppi canonici finiti con identità parziali scalari. *Rend. Circ. Mat. Palermo* **1987**, *36*, 205–219. [CrossRef]
22. Corsini, P. (i.p.s.) Ipergruppi di ordine 6. *Ann. Sc. De L'Univ. Blaise Pascal, Clermont-II* **1987**, *24*, 81–104.
23. Corsini, P. (i.p.s.) Ipergruppi di ordine 7. *Atti Sem. Mat. Fis. Univ. Modena* **1986**, *34*, 199–216.
24. Corsini, P.; Cristea, I. Fuzzy sets and non complete 1-hypergroups. *An. Stiintifice Univ. Ovidius Constanta Ser. Mat.* **2005**, *13*, 27–54.
25. Rosenfeld, A. Fuzzy subgroups. *J. Math. Anal. Appl.* **1971**, *35*, 512–517. [CrossRef]
26. Sen, M.K.; Ameri, R.; Chowdhury, G. Fuzzy hypersemigroups. *Soft Comput.* **2008**, *12*, 891–900. [CrossRef]
27. Massouros, C.G.; Massouros, G.G. On 2-element fuzzy and mimic fuzzy hypergroups. *J. AIP Conf. Proc.* **2012**, *1479*, 2213–2216.

© 2020 by the authors. Licensee MDPI, Basel, Switzerland. This article is an open access article distributed under the terms and conditions of the Creative Commons Attribution (CC BY) license (http://creativecommons.org/licenses/by/4.0/).

Article
Regular Parameter Elements and Regular Local Hyperrings

Hashem Bordbar and Irina Cristea *

Centre for Information Technologies and Applied Mathematics, University of Nova Gorica,
5000 Nova Gorica, Slovenia; hashem.bordbar@ung.si
* Correspondence: irina.cristea@ung.si or irinacri@yahoo.co.uk; Tel.: +386-0533-15-395

Abstract: Inspired by the concept of regular local rings in classical algebra, in this article we initiate the study of the regular parameter elements in a commutative local Noetherian hyperring. These elements provide a deep connection between the dimension of the hyperring and its primary hyperideals. Then, our study focusses on the concept of regular local hyperring R, with maximal hyperideal M, having the property that the dimension of R is equal to the dimension of the vectorial hyperspace $\frac{M}{M^2}$ over the hyperfield $\frac{R}{M}$. Finally, using the regular local hyperrings, we determine the dimension of the hyperrings of fractions.

Keywords: Krasner hyperring; prime/maximal hyperideal; length of hypermodules; regular local hyperring; parameter elements

1. Introduction

A regular local ring is known as a Noetherian ring with just one maximal ideal generated by n elements, where n is the Krull dimension of the ring. This is equivalent with the condition that the ring R has a system of parameters that generate the maximal ideal, called regular system of parameters or regular parameter elements [1]. In commutative algebra, the theory of regular local rings plays a fundamental role and it has been developed since the late 30s and early 40s of the last century, thanks to the studies of two great mathematicians, Wolfgang Krull and Oscar Zariski. This happened just some years after that Frederic Marty introduced hypergroups as a generalization of groups in such a way that the single-valued group operation was extended to a hyperoperation, i.e., to a multi-valued operation. It is important to stress the fact that not all the properties of the group, such as the existence of the neutral element and the inverse, have been exactly transferred to the hypergroups, meaning that it is not obligatory that a hypergroup contains a neutral element or inverses. These requirements were requested later on, for so called canonical hypergroups, that were defined as the additive parts of the Krasner hyperrings and hyperfields [2,3].

The hyperring is a hypercompositional structure endowed with one hyperoperation, namely the addition, and one binary operation, namely the multiplication, satisfying certain properties. Krasner [2] introduced the concept of hyperring for the first time in 1957 and investigated its applicability to the theory of valued fields. There are some other types of hyperrings: if the multiplication is a hyperoperation and the addition is a binary operation, we talk about the multiplicative hyperrings, defined by Rota [4]. If both, the addition and the multiplication, are hyperoperations with the additive part being a canonical hypergroup, then we have superrings [5], which were introduced by Mittas in 1973 [6]. Until now, the most well known and studied type of hyperrings is the Krasner hyperring, that has a plentitude of applications in algebraic geometry [7,8], tropical geometry [9], theory of matroids [10], schemes theory [11], algebraic hypercurves [12,13], hypermomographies [14]. In addition, the theory of hypermodules was extensively investigated by Massouros [15]. In this article, the free and cyclic hypermodules are studied and several examples are provided such as the one obtained as a quotient of a P-module over a unitary ring P. Recently, Bordbar and his collaborators in [16–18] have introduced the length and the support of hypermodules and studied some properties of them, that are used also in this paper. Moreover, the connection between the hyperrings/hyperfields theory and geometry is very clearly explained by Massouros in [19]. This paper is a pylon in the current literature on algebraic hypercompositional algebra, because it describes the development of this theory (with a lot of examples and explanation of the terminology) from the first definition of hypergroup proposed by F. Marty to the hypergroups endowed with more axioms and used now a days.

In this paper, the theory of regular local rings is applied in the context of commutative Noetherian Krasner hyperrings, with only one maximal hyperideal, namely the local hyperrings. We first define the regular parameter elements in an arbitrary commutative local hyperring of finite dimension. These elements come up from the nice and deep relation existing between the dimension of a local hyperring with maximal hyperideal M and the set of the generators of its M-primary hyperideals (see Theorem 4). More precisely, since in a local hyperring R with maximal hyperideal M, the dimension of the ring R is equal with the height of the hyperideal M, i.e., $dim R = ht_R M$ as stated in Corolarry 1, we can say that the regular parameter elements are a consequence of the investigation of the height of M and the set of the generators for M-primary hyperideals in R. The other main objective of Section 3 is expressed by the result regarding the dimension of the quotient hyperrings (see Proposition 5). In Section 4, using the local hyeprring R with maximal hyperideal M and the structure of the quotient hyperring, we introduce the hypermodule $\frac{M}{M^2}$ over the hyperrings R and $\frac{R}{M}$, respectively. Since $\frac{R}{M}$ is a hyperfield, we conclude that $\frac{M}{M^2}$ is a vectorial hyperspace over the hyperfield $\frac{R}{M}$, as a direct consequence of Theorem 6. Moreover, the investigation on the relation between the dimension of the hyperring R (equivalently, the height $ht_R M$ of the maximal hyperideal M) and the dimension of the vectorial hyperspace $\frac{M}{M^2}$ conducts us to the definition of the regular local hyperrings. They are exactly local Noetherian hyperrings with the property that the maximal hyperideal M can be generated by d elements, where d is the dimension of the hyperring. This follows from the main result of this section, i.e., Theorem 7, saying that the dimension of a local Noetherian hyperring R is the smallest number of elements that generate an M-primary hyperideal of R. Finally we apply these results in the class of the hyperrings of fractions, i.e., hyperrings of the form $R_P = S^{-1}R$, where $S = R \setminus P$, with P a prime hyperideal of S, is a multiplicatively closed subset of R. We prove that the height of the prime hyperideal P is equal to the height of the hyperideal $S^{-1}P$ in the hyperring of fractions S_P (see Theorem 8). Final conclusions and some future works on this topic are gathered in the last section of the paper.

2. Preliminaries

In this section we collect some fundamental results regarding hyperrings, but for more details we refer the readers to [16–18]. Throughout the paper, R denotes a *Krasner hyperring*, unless stated otherwise and we call it by short a *hyperring*. It was introduced by Krasner [2] as follows.

Definition 1. *A (Krasner) hyperring is a hyperstructure* $(R, +, \cdot)$ *where*
1. $(R, +)$ *is a canonical hypergroup, i.e.,*
    (a) $(a, b \in R \Rightarrow a + b \subseteq R)$,
    (b) $(\forall a, b, c \in R)\ (a + (b + c) = (a + b) + c)$,
    (c) $(\forall a, b \in R)\ (a + b = b + a)$,
    (d) $(\exists 0 \in R)(\forall a \in R)\ (a + 0 = \{a\})$,
    (e) $(\forall a \in R)(\exists - a \in R)\ (0 \in a + x \Leftrightarrow x = -a)$,
    (f) $(\forall a, b, c \in R)(c \in a + b \Rightarrow a \in c + (-b))$.
2. $(R, \cdot)$ *is a semigroup with a bilaterally absorbing element 0, i.e.,*
    (a) $(a, b \in R \Rightarrow a \cdot b \in R)$,

(b) $(\forall a, b, c \in R) \, (a \cdot (b \cdot c) = (a \cdot b) \cdot c)$,

(c) $(\forall a \in R) \, (0 \cdot a = a \cdot 0 = 0)$.

3. The product distributes from both sides over the sum

(a) $(\forall a, b, c \in R) \, (a \cdot (b + c) = a \cdot b + a \cdot c \text{ and } (b + c) \cdot a = b \cdot a + c \cdot a)$.

Moreover, if $(R, \cdot)$ is commutative, i.e.,

4. $(\forall a, b \in R) \, (a \cdot b = b \cdot a)$,

the hyperring is called commutative.

Definition 2. *If every nonzero element in a hyperring R with multiplicative identity 1 is invertible, i.e.,*

(i) $(\forall a \in R)(\exists a^{-1} \in R) \, (a \cdot a^{-1} = 1_R)$,

then R is called a hyperfield.

Definition 3. *A nonempty set I of a hyperring R is called a hyperideal if, for all $a, b \in I$ and $r \in R$, we have $a - b \subseteq I$ and $a \cdot r \in I$. A proper hyperideal M of a hyperring R is called a maximal hyperideal of R if the only hyperideals of R that contain M are M itself and R. A hyperideal P of a hyperring R is called a prime hyperideal of R if, for every pair of elements a and b of R, the fact that $ab \in P$, implies either $a \in P$ or $b \in P$. In addition, a nonzero hyperring R having exactly one maximal hyperideal is called a local hyperring.*

Definition 4. *A hyperring homomorphism is a mapping f from a hyperring R_1 to a hyperring R_2 with units elements 1_{R_1} and 1_{R_2}, respectively, such that*

1. $(\forall a, b \in R) \, (f(a +_{R_1} b) = f(a) +_{R_2} f(b))$.
2. $(\forall a, b \in R) \, (f(a \cdot_{R_1} b) = f(a) \cdot_{R_2} f(b))$.
3. $f(1_{R_1}) = 1_{R_2}$.

Definition 5. *Let $f : R \to S$ be a hyperring homomorphism, I be a hyperideal of R and J be a hyperideal of S.*

(i) *The hyperideal $< f(I) >$ of S generated by the set $f(I)$ is called the extension of I and it is denoted by I^e.*

(ii) *The hyperideal $f^{-1}(J) = \{a \in R \mid f(a) \in J\}$ is called the contraction of J and it is denoted by J^c. It is known that, if J is a prime hyperideal in S, then J^c is a prime hyperideal in R.*

Definition 6. *A prime hyperideal P of R is called a minimal prime hyperideal over a hyperideal I of R if it is minimal (with respect to inclusion) among all prime hyperideals of R containing I. A prime hyperideal P is called a minimal prime hyperideal if it is a minimal prime hyperideal over the zero hyperideal of R.*

Definition 7. *A hyperring R is called Noetherian if it satisfies the ascending chain condition on hyperideals of R: for every ascending chain of hyperideals $I_1 \subseteq I_2 \subseteq I_3 \subseteq \ldots$ there exists $N \in \mathbb{N}$ such that $I_n = I_N$, for every natural number $n \geq N$ (this is equivalent to saying that every ascending chain of hyperideals has a maximal element). A hyperring R is called Artinian if it satisfies the descending chain condition on hyperideals of R: for every descending chain of hyperideals $I_1 \supseteq I_2 \supseteq I_3 \supseteq \ldots$ there exists $N \in \mathbb{N}$ such that $I_n = I_N$, for every natural number $n \geq N$ (this is equivalent to saying that every descending chain of hyperideals has a minimal element).*

Definition 8. *Let R be a hyperring with unit element 1. An R-hypermodule M is a commutative hypergroup $(M, +)$ together with a map $R \times M \longrightarrow M$ defined by*

$$(a, m) \mapsto a \cdot m = am \in M \tag{1}$$

such that for all $a, b \in R$ and $m_1, m_2 \in M$ we have:

1. $(a+b)m_1 = am_1 + bm_1$.
2. $a(m_1 + m_2) = am_1 + am_2$.
3. $(ab)m_1 = a(bm_1)$.
4. $a0_M = 0_R m_1 = 0_M$.
5. $1m_1 = m_1$, where 1 is the multiplicative identity in R.

Moreover, if R is a hyperfield, then M is called a *vectorial hyperspace* [20].

The next few results concern the concept of the *radical of a hyperideal* [16].

Definition 9. *The radical of a hyperideal I of a hyperring R, denoted by $r(I)$, is defined as*

$$r(I) = \{x \mid x^n \in I, \text{ for some } n \in \mathbb{N}\}.$$

It can be proved that the radical of I is the intersection of all prime hyperideals of R containing I. In addition, a hyperideal P in a hyperring R is called primary if $P \neq R$ and the fact that $xy \in P$ implies either $x \in P$ or $y \in r(P)$.

Lemma 1. *Let I be a hyperideal of the hyperring R, where $r(I)$ is a maximal hyperideal of R. Then I is a primary hyperideal of R.*

Proposition 1. *Let R be a Noetherian hyperring, M a maximal hyperideal of R and let P be a hyperideal of R. Then the following statements are equivalent:*

1. *P is primary.*
2. $r(P) = M$.
3. $M^n \subseteq P \subseteq M$, *for some $n \in \mathbb{N}$.*

Definition 10. *Let R be a commutative hyperring. An expression of the type*

$$P_0 \subset P_1 \subset \ldots \subset P_n \qquad (2)$$

(note the strict inclusions), where $P_0, \ldots, P_n$ are prime hyperideals of R, is called a chain of prime hyperideals of R; the length of such a chain is the number of the "links" between the terms of the chain, that is, 1 less than the number of prime hyperideals in the sequence.

Thus, the chain in (2) has length n. Note that, for a prime hyperideal P, we consider P to be a chain, with just one prime hyperideal of R, of length 0. Since R is non-trivial, it contains at least one prime hyperideal, so there certainly exists at least one chain of prime hyperideals of R of length 0.

Definition 11. *The supremum of the lengths of all chains of prime hyperideals of R is called the dimension of R, denoted by dimR.*

Definition 12. *Let P be a prime hyperideal of a commutative hyperring R. The height of P, denoted by $ht_R P$, is defined to be the supremum of the lengths of all chains*

$$P_0 \subset P_1 \subset \ldots \subset P_n$$

of prime hyperideals of R, for which $P_n = P$, if this supremum exists, and it is ∞, otherwise.

We conclude this preliminary section recalling the notion of *vectorial hyperspace*.

Definition 13 ([20]). *Let V be a vectorial hyperspace over a hyperfield F. A linear combination of vectors $v_1, v_2, \ldots, v_n$ is a set of the form $\sum_{i=1}^{n} r_i v_i$. Moreover, for a vectorial hyperspace V over a hyperfield F and $A \subseteq V$, A is called linearly independent if for every finite set of vectors*

$\{v_1, v_2, \ldots, v_n\} \subseteq A$, $0 \in \sum_{i=1}^{n} r_i v_i$ implies $r_i = 0$ for $1 \leq i \leq n$. If A is not linearly independent then A is called linearly dependent.

Definition 14 ([20]). *Let V be a vectorial hyperspace over F. The set $A \subseteq V$ is called a spanning set if each vector of V is contained in a linear combination of vectors from A. We say V is the span of A and write $V = Span[A]$. In addition, a basis for a vectorial hyperspace V is a subset B of V such that B is both spanning and linearly independent set. A vectorial hyperspace is finite dimensional if it has a finite basis. Moreover, the number of the elements in an arbitrary basis of a vectorial hyperspace is called the dimension of the vectorial hyperspace, denoted here by $vdimV$.*

Note that we denote the dimension of the vectorial hyperspace by $vdimV$ (where "v" stays for "vectorial") in order to not confuse it with the dimension of a hyperring, denoted here as $dimR$.

3. Regular Parameter Elements in Local Hyperrings

In this section, we introduce the notion of the *regular parameter elements* in a local hyperring and by using representative examples, we present some of their properties connected with the length and the dimension of the related vectorial hyperspace.

Definition 15. *A unitary hyperring R is called principal hyperideal hyperdomain when it has no zero divisors and all its hyperideals are generated by a single element.*

M. Krasner had the great idea to construct hyperrings and hyperfields as quotients of rings and fields, respectively, called by him *quotient hyperrings* and *quotient hyperfields*, while this construction is known in literature as *Krasner's construction* [3]. The importance of these hyperrings and hyperfields in Krasner's studies is very clear explained by G. Massouros and Ch. Massouros in [19], as well as their different names given by some authors [7–11], with the risk of creating confusions. Therefore, in order to keep the original terminology, we recommend to read the papers of Nakasis [21] and Massouros [15,22,23].

Example 1 ([3,21]). *Let $(R, +, \cdot)$ be a ring and G a normal subset of R (which means that $rG = Gr$ for every $r \in R$) such that $(G, \cdot)$ is a group and the unit element of G is a unit element of R. Define an equivalence relation $\cong$ on R as follows: $r \cong s$ if and only if $rG = sG$. Then, the equivalence class represented by r is $P(r) = \{s \in R \mid sG = rG\} = rG$. Define now a hyperoperation $\oplus$ on the set of all equivalence classes R/G as follows: $P(r) \oplus P(s) = \{P(t) \mid P(t) \cap (P(r) + P(s)) \neq \emptyset\}$ $= \{tG \mid \exists g_1, g_2 \in G \text{ such that } t = rg_1 + sg_2\} = \{tG \mid tG \subseteq rG + sG\}$, and define a binary operation on R/G as $rG \cdot sG = rsG(P(r) \cdot P(s) = P(rs))$. Then, $(R/G, \oplus, \cdot)$ forms a hyperring. Moreover, if we choose R to be a field, then we get that $(R/G, \oplus, \cdot)$ is a hyperfield.*

Notice that the condition on the normality of G can be substituted with a more general one, i.e., $rGsG = rsG$, for every $r, s \in R^*$, which is practically equivalent to the normality of G only when the multiplicative semigroup is a group, so only when R is a field, as Massouros proved in [22].

Proposition 2. *Let R be a principal hyperideal hyperdomain and let $p \in R - \{0\}$. Then the following statements are equivalent:*

(i) pR is a maximal hyperideal of R.
(ii) pR is a non-zero prime hyperideal of R.

Proof. $(i) \longrightarrow (ii)$ This implication is clear because $p \neq 0$ and every maximal hyperideal of a hyperring is also a prime hyperideal.

$(ii) \longrightarrow (i)$ Since pR is a prime hyperideal of R, i.e., p is not a unit element of R, we have $pR \subset R$. Let I be a hyperideal of R such that $pR \subseteq I \subset R$. Since R is a principal hyperideal hyperdomain, there exists an element $a \in R$ such that $I = aR$. In addition, a can not be a unit element because I is a proper hyperideal. Since R is a unitary hyperring, it follows that $p \in I$ and so $p = ab$ for some $b \in R$. Since pR is a prime hyperideal, it follows that p is irreducible and because a is not a unit, we have that b is a unit element of R. Thus, $pR = aR = I$ and therefore pR is a maximal hyperideal. □

Example 2 ([16]). *Consider the set of integers $\mathbb{Z}$ and its multiplicative subgroup $G = \{-1, 1\}$. Based on Example 1, the Krasner's construction $\frac{\mathbb{Z}}{G}$ is a principal hyperideal hyperdomain. In addition, the prime (also maximal) hyperideals of $\frac{\mathbb{Z}}{G}$ have the form $< pG >$, where p is a prime number. In $\frac{\mathbb{Z}}{G}$ we have $< 0\mathbb{Z} > \subset < 2\mathbb{Z} >$ a chain of prime hyperideals of length 1. Since every nonzero prime hyperideal of $\frac{\mathbb{Z}}{G}$ is maximal, there does not exist a chain of prime hyperideals of $\frac{\mathbb{Z}}{G}$ of length 2, therefore $\dim \frac{\mathbb{Z}}{G} = 1$.*

Remark 1. *Let R be a non-trivial commutative hyperring. By [16], every prime hyperideal of R is contained in a maximal hyperideal of R (and every maximal hyperideal is prime). Moreover, every prime hyperideal of R contains a minimal prime hyperideal. It follows that $\dim R$ is equal to the supremum of lengths of chains $P_0 \subset P_1 \subset \ldots \subset P_n$ of prime hyperideals of R, with P_n a maximal and P_0 a minimal prime hyperideal. Indeed, if we have an arbitrary chain of prime hyperideals of R with length h, like $P'_0 \subset P'_1 \subset \ldots \subset P'_h$, then it is bounded above by the length of a special chain as it follows. If P'_0 is not a minimal prime hyperideal, then another prime hyperideal can be inserted before it. On the other hand, if P'_h is not a maximal hyperideal of R, then another prime hyperideal can be inserted above it. Thus, if $\dim R$ is finite, then*

$$\dim R = \sup\{ht_R M \mid M \text{ is a maximal hyperideal of } R\}$$

$$= \sup\{ht_R P \mid P \text{ is a prime hyperideal of } R\}.$$

As a consequence, we have the following result.

Corollary 1. *If R is a local commutative hyperring with the maximal hyperideal M, then $\dim R = ht_R M$.*

The first property on the height of a hyperideal is highlighted in the next result.

Theorem 1 ([16]). *Let R be a commutative Noetherian hyperring and let $a \in R$ be a non-unit element. Let P be a minimal prime hyperideal over the principal hyperideal $\langle a \rangle$ of R. Then, $ht_R P \leq 1$.*

Now we can extend Theorem 1 to the case when P is a minimal prime hyperideal over a hyperideal I generated not by one element, but by n elements.

Theorem 2 ([16]). *Let R be a commutative Noetherian hyperring. Suppose that I is a proper hyperideal of R generated by n elements and P is a minimal prime hyperideal over I. Then, $ht_R P \leq n$.*

Lemma 2. *In a commutative Noetherian hyperring R, let I be a hyperideal and P a prime hyperideal of R, such that $I \subseteq P$ and $ht_R I = ht_R P$. Then, P is a minimal prime hyperideal over I.*

Proof. Suppose that P is not a minimal prime hyperideal over I. Then, by Remark 1, there exists a prime hyperideal Q of R which is minimal and $I \subseteq Q \subset P$. Thus, $ht_R I \leq ht_R Q < ht_R P$, obtaining a contradiction. So P is a minimal prime hyperideal over I. □

The next theorem states the conditions under which there exists a proper hyperideal of height n and generated by n elements.

Theorem 3 ([17]). *Let R be a commutative Noetherian hyperring and P a prime hyperideal of R, with $ht_R P = n$. Then there exists a proper hyperideal I of R having the following properties:*

(i) $I \subseteq P$.

(ii) I is generated by n elements.
(iii) $ht_R I = n$.

We have now all the elements to determine the dimension of a commutative local Noetherian hyperring.

Theorem 4. *Suppose that R is a commutative local Noetherian hyperring, having M as its unique maximal hyperideal. Then, dimR is equal to the smallest number of elements of R that generate an M-primary hyperideal. In other words,*

$$dimR = min\{i \in \mathbb{N}, \exists a_1, a_2, \ldots, a_i \in R, \sum_{j=1}^{i} Ra_j \text{ is an } M - primary\ hyperideal\}.$$

Proof. Let

$$d = min\{i \in \mathbb{N}, \exists\ a_1, a_2, \ldots, a_i \in R, \sum_{j=1}^{i} Ra_j \text{ is an } M - primary\ hyperideal\}.$$

By using Corollary 1 and Theorem 2, we have $dimR = ht_R M \leq d$, because an M-primary hyperideal must have M as its minimal prime hyperideal. On the other hand, by using Lemma 2 and Theorem 3, there exists a hyperideal P of R which has M as a minimal prime hyperideal and which can be generated by $dimR = ht_R M$ elements. In addition, every prime hyperideal of R is contained in M. Hence, M must be the only one associated prime hyperideal of P. Thus, P is an M-primary hyperideal.

As a result that there exists an M-primary hyperideal of R which can be generated by $dimR$ elements, we have $d \leq dimR$. Therefore, $d = dimR$ and this completes the proof. □

Definition 16. *Let R be a local hyperring of dimension d, with M as its unique maximal hyperideal. By regular parameter elements of R we mean a set of d elements of R that generate an M-primary hyperideal of R.*

Remark 2. *Based on Theorem 4, we conclude that each local hyperring of dimension at least 1 possesses a set of regular parameter elements.*

Example 3. *On the set $R = \{0, 1, 2\}$ define the hyperaddition + and the multiplication · by the following tables*

+	0	1	2
0	0	1	2
1	1	R	1
2	2	1	M

·	0	1	2
0	0	0	0
1	0	1	2
2	0	2	0

Then, R is a commutative local Noetherian hypering and $M = \{0, 2\}$ is the only maximal hyperideal of R. Thus, $dimR = ht_R M$ and M is an M-primary hyperideal. One can check that $ht_R M = 0$, so $dimR = 0$. Therefore this hyperring has no regular parameter elements.

Proposition 3. *Let R be a local hyperring of dimension d, with the unique maximal hyperideal M, and let $a_1, a_2, \ldots, a_d$ be a set of regular parameter elements for R. For any $n_1, n_2, \ldots, n_d \in \mathbb{N}$, we have that $a_1^{n_1}, a_2^{n_2}, \ldots, a_d^{n_d}$ form a set of regular parameter elements for R, too.*

Proof. The proof is straightforward. □

Proposition 4. *Let $f : R \to S$ be a surjective homomorphism of commutative hyperrings. Suppose that $Q_1, Q_2, \ldots, Q_n, P_1, P_2, \ldots, P_n$ are hyperideals of R, all of which containing $Ker f$, with $r(Q_i) = P_i$, for $i = 1, 2, \ldots, n$. Then*

$$I = Q_1 \cap Q_2 \cap \ldots \cap Q_n$$

is a (minimal) primary decomposition of I if and only if

$$I^e = Q_1^e \cap Q_2^e \cap \ldots \cap Q_n^e$$

with $r(Q_i^e) = P_i^e$ for $i = 1, 2, \ldots, n$ is a (minimal) primary decomposition of I^e.

Proof. The proof is straightforward. □

We can conclude that I is a decomposable hyperideal of R (meaning that it has a primary decomposition) if and only if I^e is a decomposable hyperideal of S. Moreover, if

$$I = Q_1 \cap Q_2 \cap \ldots \cap Q_n$$

where $r(Q_i) = P_i$, for $i = 1, 2, \ldots, n$, then we call the set $\{P_1, P_2, \ldots, P_n\}$, the set of the *associated prime hyperideals related to I* and denoted by $ass_R I$. Using Proposition 4, for the hyperideal I^e of S the related associated prime hyperideals form the set

$$ass_S I^e = \{P^e \mid P \in ass_R I\} = \{P_1^e, P_2^e, \ldots, P_n^e\}.$$

Corollary 2. *Let I be a proper hyperideal of the commutative hyperring R. Using the canonical hyperring homomorphism from R to the quotient $\frac{R}{I}$, we conclude that if J is a hyperideal of R such that $I \subseteq J$, then J is a decomposable hyperideal of R if and only if $\frac{J}{I}$ is a decomposable hyperideal of $\frac{R}{I}$, and we have*

$$ass_{\frac{R}{I}}(\frac{J}{I}) = \{\frac{P}{I} : P \in ass_R J\}.$$

Theorem 5 ([17]). *Let R be a commutative Noetherian hyperring, P a prime hyperideal of R and I a proper hyperideal of R generated by n elements, such that $I \subseteq P$. Then:*

$$ht_{\frac{R}{I}} \frac{P}{I} \leq ht_R P \leq ht_{\frac{R}{I}} \frac{P}{I} + n.$$

Proposition 5. *Let R be a local commutative hyperring of dimention d with its maximal hyperideal M, and let $a_1, a_2, \ldots, a_t \in M$. Then,*

$$dim R - t \leq dim\, R/(a_1, a_2, \ldots, a_t) \leq dim R.$$

Moreover, $dim R/(a_1, a_2, \ldots, a_t) = dim R - t$ if and only if $a_1, a_2, \ldots, a_t$ are all distinct elements and form a set of regular parameter elements of R.

Proof. Using Corollary 1, we have $dim R = ht_R M$ and therefore

$$dim R/(a_1, a_2, \ldots, a_t) = ht_{R/(a_1, a_2, \ldots, a_t)} M/(a_1, a_2, \ldots, a_t).$$

By using Theorem 5, we conclude that

$$dim R - t \leq dim\, R/(a_1, a_2, \ldots, a_t) \leq dim R.$$

Now let $can : R \to R/(a_1, a_2, \ldots, a_t)$ be the canonical hyperring homomorphism such that $can(r) = r + (a_1, a_2, \ldots, a_t)$. Set $\overline{M} = \frac{M}{(a_1, a_2, \ldots, a_t)}$.

Suppose that $dim R/(a_1, a_2, \ldots, a_t) = d - t$. Then, $t \leq d$, and by using Theorem 4, there exist $\overline{a_{t+1}}, \overline{a_{t+2}}, \ldots, \overline{a_d} \in R/(a_1, a_2, \ldots, a_t)$, where $\overline{a_j} = a_j + (a_1, a_2, \ldots, a_t)$ for $j =$

$t+1,\ldots,d$, such that $(\overline{a_{t+1}}, \overline{a_{t+2}}, \ldots, \overline{a_d})$ is an $\overline{M}$-primary hyperideal of $R/(a_1, a_2, \ldots, a_t)$. Therefore, $a_{t+1}, a_{t+2}, \ldots, a_d \in M$. Thus,

$$\frac{(a_1, a_2, \ldots, a_t, a_{t+1}, \ldots, a_d)}{(a_1, a_2, \ldots, a_t)}$$

is an $\overline{M}$-primary hyperideal of $R/(a_1, a_2, \ldots, a_t)$. By using Proposition 4 and Corollary 2, we conclude that $(a_1, a_2, \ldots, a_d)$ is an M-primary hyperideal of R. It now follows from Theorem 4, that $a_1, a_2, \ldots, a_d$ are all distinct and $\{a_1, a_2, \ldots, a_d\}$ is a set of regular parameter elements of R. Therefore, $a_1, a_2, \ldots, a_t$ are all distinct and form a set of regular parameter elements of R.

Now suppose that $t \leq d$ and there exist $a_{t+1}, \ldots, a_d \in M$ such that

$$a_1, a_2, \ldots, a_t, a_{t+1}, \ldots, a_d$$

form a set of regular parameter elements of R. Thus, $(a_1, a_2, \ldots, a_d)$ is an M-primary hyperideal of R. Therefore, by using Proposition 4 and Corollary 2, $(\overline{a_{t+1}}, \overline{a_{t+2}}, \ldots, \overline{a_d})$ is an $\overline{M}$-primary hyperideal of $R/(a_1, a_2, \ldots, a_t)$. Thus, by using Theorem 4, we have $d - t \geq \dim \overline{R}$. But, it follows from the first part that $d - t \leq \dim \overline{R}$, and so the proof is complete. □

4. Regular Local Hyperrings

The aim of this section is to define the *regular local hyperrings*. For doing this, we will first prove that, in a local Noetherian hyperring R with the unique maximal hyperideal M, the R-hypermodule $\frac{M}{M^2}$ is a vectorial hyperspace over the hyperfield $\frac{R}{M}$. Then we will establish a relation between the dimension of the hyperring R and the dimension of the vectorial hyperspace $\frac{M}{M^2}$. Finally, we will present a new characterization of the dimension of the hyperring of fractions.

Definition 17 ([18]). *Let M be a hypermodule over the commutative hyperring R, N be a subhypermodule of M and $I \subseteq M$ with $I \neq 0$. We define the hyperideal*

$$(N :_R I) = \{r \in R \mid r \cdot x \in N \text{ for all } x \in I\}.$$

For any element $m \in M$, we denote $(N :_R m)$ instead of $(N :_R \{m\})$. In addition, in the special case when $N = 0$, the hyperideal

$$(0 :_R I) = \{r \in R \mid r \cdot x = 0 \text{ for all } x \in I\}$$

is called the annihilator of I and is denoted by $Ann_R(I)$. Moreover, for any element $m \in M$, we call the hyperideal $(0 :_R m)$ the annihilator of the element m.

Example 4. *Let us continue with Example 3, where R is an R-hypermodule and its hyperideal M is a subhypermodule. We have*

$$Ann_R(M) = (0 :_R M) = \{r \in R \mid r \cdot m = 0 \text{ for all } m \in M\} = M.$$

Moreover, for $m = 2$ we have

$$(0 :_R 2) = \{r \in R \mid r \cdot 2 = 0\} = \{0, 2\}.$$

Proposition 6. *Let M be a hypermodule over the commutative hyperring R and I be a hyperideal of R such that $I \subseteq Ann_R(M)$. Then, M is a hypermodule over $\frac{R}{I}$.*

Proof. Suppose that $r, r' \in R$ such that $r + I = r' + I$. We have $r - r' \subseteq I \subseteq Ann_R(M)$, and so $(r - r')m = 0$ for any arbitrary element $m \in M$. Thus, $rm = r'm$. Hence, we can define a mapping $\frac{R}{I} \times M \longrightarrow M$ such that $(r+I, m) \longrightarrow rm$. It is a routine to check that M has an $\frac{R}{I}$-hypermodule structure. □

Proposition 7. *Let R be a local hyperring with its maximal hyperideal M and consider the hyperfield $F = \frac{R}{M}$. Let N be a finitely generated R-hypermodule. Then the R-hypermodule $\frac{N}{MN}$ has a natural structure as a hypermodule over $\frac{R}{M}$ as an F-vectorial hyperspace.*

Moreover, let $n_1, n_2, \ldots, n_t \in N$. Then the following statements are equivalent.

(i) *N is generated by $n_1, n_2, \ldots, n_t$.*
(ii) *The R-hypermodule $\frac{N}{MN}$ is generated by $n_1 + MN, n_2 + MN, \ldots, n_t + MN$.*
(iii) *The F-vectorial hyperspace $\frac{N}{MN}$ is generated by $n_1 + MN, n_2 + MN, \ldots, n_t + MN$.*

Proof. Since the R-hypermodule $\frac{N}{MN}$ is annihilated by M, i.e., $M \subseteq Ann_R(\frac{N}{MN})$, by using Proposition 6, it has a natural structure as hypermodule over $\frac{R}{M}$. In addition, since $F = \frac{R}{M}$ is a hyperfield, $\frac{N}{MN}$ is also an F-vectorial hyperspace.

It is clear that (i) implies (ii).

The R-hypermodule and F-vectorial hyperspace structures of $\frac{N}{MN}$ are related by the formula
$$r(n + MN) = (r + M)(n + MN)$$
for all $r \in R$ and $n \in N$. Thus, the equivalence of statements (ii) and (iii) is clear.

It remains to prove that (ii) implies (i). Assume that (ii) holds, so the R-hypermodule $\frac{N}{MN}$ is generated by the elements
$$n_1 + MN, n_2 + MN, \ldots, n_t + MN.$$

Let $G = Rn_1 + Rn_2 + \ldots + Rn_t$. First, we will show that $N = G + MN$. Let $n \in N$. Then there exist $r_1, r_2, \ldots, r_t \in R$ such that
$$n + MN = r_1(n_1 + MN) + r_2(n_2 + MN) + \ldots + r_t(n_t + MN).$$

Hence, $n - \sum_{i=1}^{n} r_i n_i \in MN$. It follows that $N \subseteq G + MN$. On the other side, it is clear that $G + MN \subseteq N$. Thus, we have $N = G + MN$. Now by using Corollary 2.9 in [16], we conclude that $N = G$, so N is generated by the elements $n_1, n_2, \ldots, n_t$. □

Theorem 6. *The F-vectorial hyperspace $\frac{N}{MN}$ in Proposition 7 has finite dimension and the number of the elements in each minimal generating set for the R-hypermodule N is equal to $vdim_F \frac{N}{MN}$.*

Proof. Since N is a finitely generated R-hypermodule, it follows from Proposition 7 that $\frac{N}{MN}$ is a finitely generated F-vectorial hyperspace. Therefore, its dimension is finite.

Let $\{n'_1, n'_2, \ldots, n'_p\}$ be a minimal generating set for N. By Proposition 7, we know that the set
$$\{n'_1 + MN, n'_2 + MN, \ldots, n'_p + MN\}$$
is a generating set for the F-vectorial hyperspace $\frac{N}{MN}$ and since $\{n'_1, n'_2, \ldots, n'_p\}$ is a minimal generating set for N, it follows that no proper subset of $\{n'_1 + MN, n'_2 + MN, \ldots, n'_p + MN\}$ generates $\frac{N}{MN}$. Thus, the set $\{n'_1 + MN, n'_2 + MN, \ldots, n'_p + MN\}$ is a basis for the F-vectorial hyperspace $\frac{N}{MN}$ and so $vdim_F \frac{N}{MN} = p$. □

Note that the R-hypermodule $\frac{M}{M^2}$ is annihilated by M. If R is a Noetherian hyperring, then by Proposition 6, $\frac{M}{M^2}$ has a natural structure as a vectorial hyperspace over the hyperfield $\frac{R}{M}$. In addition, using Proposition 7, the dimension of the vectorial hyperspace $\frac{M}{M^2}$ is equal to the number of the elements in an arbitrary minimal generating set for M. Therefore, we have the following result.

Theorem 7. *Let R be a local Noetherian hyperring with maximal hyperideal M. Then,*

$$\dim R \leq vdim_{\frac{R}{M}} \frac{M}{M^2}.$$

Proof. The inequality is clear because M is an M-primary hyperideal of the hyperring R and as mentioned before $vdim_{\frac{R}{M}} \frac{M}{M^2}$ is the number of the elements of an arbitrary minimal generating set for M. Based on Theorem 4, $\dim R$ is the smallest number of elements that generate an M-primary hyperideal of R. □

Definition 18. *Let R be a local Noetherian hyperring with maximal hyperideal M. Then, R is called a regular hyperring when $\dim R = vdim_{\frac{R}{M}} \frac{M}{M^2}$.*

Remark 3. *For a local Noetherian hyperring R with one maximal hyperideal M and $\dim R = d$, we have the following statements:*

(i) *The dimension of the $\frac{R}{M}$-vectorial hyperspace $\frac{M}{M^2}$ is the number of the elements in each minimal generating set for the hyperideal M. By using Theorem 4, at least d elements are needed to generate M, and R is a regular hyperring when the hyperideal M can be generated by exactly d elements.*

(ii) *Suppose that R is a regular hyperring and $a_1, a_2, \ldots, a_d \in M$. By using Proposition 7, the elements $a_1, a_2, \ldots, a_d$ generate M if and only if $a_1 + M^2, a_2 + M^2, \ldots, a_d + M^2$ in $\frac{M}{M^2}$ form a basis for this $\frac{R}{M}$-vectorial hyperspace, equivalently if and only if $a_1 + M^2, a_2 + M^2, \ldots, a_d + M^2$ form a linearly independent set.*

We conclude this section with a new characterization of the dimension of the hyperring of fractions $R_P = S^{-1}R$, where $S = R \setminus P$, with P a prime hyperideal of S, is a multiplicatively closed subset of R. As proved in [16,17], R_P is a local Noetherian hyperring. First, we recall the main properties of the hyperring of fractions.

Proposition 8 ([16]). *Let S be a multiplicatively closed subset of a hyperring R.*

(i) *Every hyperideal in $S^{-1}R$ is an extended hyperideal.*
(ii) *If I is a hyperideal in R, then $I^e = S^{-1}R$ if and only if $I \cap S \neq \emptyset$.*
(iii) *A hyperideal I is a contracted hyperideal of R if and only if no element of S is a zero divisor in R/I.*
(iv) *The prime hyperideals of $S^{-1}R$ are in one to one correspondence with the prime hyperideals of R that don't meet S, with the correspondence given by $P \leftrightarrow S^{-1}P$.*

Theorem 8. *Let S be a multiplicatively closed subset of R and P be a prime hyperideal of R such that $P \cap S = \emptyset$. Then, $ht_R P = ht_{S^{-1}R} S^{-1}P$.*

Proof. By Proposition 8, it follows that $S^{-1}P$ is a prime hyperideal of $S^{-1}R$. Let

$$P_0 \subset P_1 \subset \ldots \subset P_n = P$$

be a chain of prime hyperideals of R. Again by Proposition 8, it follows that

$$P_0^e \subset P_1^e \subset \ldots \subset P_n^e$$

is a chain of prime hyperideals of $S^{-1}R$ with $P_n^e = P^e = S^{-1}P$, and therefore $ht_R P \leq ht_{S^{-1}R} S^{-1}P$. On the other side, if

$$Q_0 \subset Q_1 \subset \ldots \subset Q_n$$

is a chain of prime hyperideals of $S^{-1}R$ with $Q_n = P^e$, then using Propositions 8, we get that
$$Q_0^c \subset Q_1^c \subset \ldots \subset Q_n^c$$
is a chain of prime hyperideals of R with $Q_n^c = P^{ec} = P$. So we have $ht_{S^{-1}R}S^{-1}P \leq ht_R P$. Therefore, it follows that $ht_R P = ht_{S^{-1}R}S^{-1}P$. □

Combining the previous results, we get now the following important consequence.

Corollary 3. *For a prime hyperideal P of a commutative hyperring R, it follows that*
$$ht_R P = ht_{R_P} S^{-1} P = dim(R_P).$$

In the following we will illustrate the previous results by several examples.

Example 5. *Let R be a commutative Noetherian hyperring. Suppose that there exists a prime hyperideal P of R with $htP = n$ and that can be generated by n elements $a_1, a_2, \ldots, a_n$. Consider the localisation hyperring R_P, that is a local hyperring. According with Theorem 8 and Corollary 3, it has dimension n. Since the dimension of localization hyperring R_P is n, the maximal hyperideal of this hyperring, i.e.,*
$$PR_P = (\sum_{i=1}^{n} Ra_i) R_P = \sum_{i=1}^{n} R_P \frac{a_i}{1},$$
can be generated by n elements, that are also regular parameter elements, it follows that R_P is a regular hyperring.

Example 6. *Let p be a prime number. Based on Example 2, we have $ht_{\mathbb{Z}} p\mathbb{Z} = 1$. Since $p\mathbb{Z}$ is a prime hyperideal of $\mathbb{Z}$ which can be generated by one element, it follows from Example 5 that $\mathbb{Z}_{p\mathbb{Z}}$ is a regular hyperring of dimension 1 and p is a regular parameter element.*

Example 7. *Let R be a principal hyperideal hyperdomain which is not a hyperfield, and let M be a maximal hyperideal of R. By using Proposition 2, M is a prime hyperideal of R and also principal hyperideal with height 1. Using Example 5, it follows that R_M is a regular hyperring of dimension 1.*

5. Conclusions

In this paper, we have to define the regular parameter elements in a commutative local Noetherian hyperring R with maximal hyperideal M and present some properties related to them. After investigating some results concerning the quotient hypermodule $\frac{M}{M^2}$ over the hyperfield $\frac{R}{M}$, our study has focused on regular local hyperrings. We have studied the relation between the dimension of a commutative local Noetherian hyperring and the dimension of the vectorial hyperspace $\frac{M}{M^2}$ over the hyperfield $\frac{R}{M}$.

Our future work will include new results regarding the regular local hyperrings. In particular we will investigate whether they are hyperdomains or not. In addition, we will study the properties of the hyperideals generated by a subset of the regular parameter elements and the relation between the length of these hyperideals and the dimension of the hyperring.

Author Contributions: Conceptualization, H.B. and I.C.; methodology, H.B. and I.C.; investigation, H.B. and I.C.; writing—original draft preparation, H.B.; writing—review and editing, I.C.; funding acquisition, I.C. All authors have read and agreed to the published version of the manuscript.

Funding: The second author acknowledges the financial support from the Slovenian Research Agency (research core funding No. P1-0285).

Conflicts of Interest: The authors declare no conflict of interest.

References

1. Kemper, G. Regular local rings. In *A Course in Commutative Algebra*; Graduate Texts in Mathematics; Springer: Berlin/ Heildeberg, Germany, 2011; Volume 256.
2. Krasner, M. *Approximation des Corps Values Complets de Caracteristique p, p > 0, Par ceux de Caracteristique Zero*; Colloque d' Algebre Superieure (Bruxelles, Decembre 1956); CBRM: Bruxelles, Belgium, 1957.
3. Krasner, M. A class of hyperrings and hyperfields. *Int. J. Math. Math. Sci.* **1983**, *6*, 307–312.
4. Rota, R. Strongly distributive multiplicative hyperrings. *J. Geom.* **1990**, *1–2*, 130–138.
5. Ameri, R.; Eyvazi, M.; Hoskova-Mayerova, S. Superring of Polynomials over a Hyperring. *Mathematics* **2019**, *7*, 902.
6. Mittas, J.D. Sur certaines classes de structures. hypercompositionnelles. *Proc. Acad. Athens* **1973**, *48*, 298–318.
7. Connes, A.; Consani, C. On the notion of geometry over $\mathbf{F}_1$. *J. Algebraic Geom.* **2011**, *20*, 525–557.
8. Jun, J. Algebraic geometry over hyperrings. *Adv. Math.* **2018**, *323*, 142–192.
9. Viro, O. Hyperfields for tropical geometry I. Hyperfields and dequantization. *arXiv* **2010**, arXiv: 1006.3034.
10. Baker, M.; Bowler, N. Matroids over partial hyperstructures. *Adv. Math.* **2019**, *343*, 821–863.
11. Jun, J. Hyperstructures of affine algebraic group schemes. *J. Number Theory* **2016**, *167*, 336–352.
12. Vahedi, V.; Jafarpour, M.; Aghabozorgi, H.; Cristea, I. Extension of elliptic curves on Krasner hyperfields. *Comm. Algebra* **2019**, *47*, 4806–4823.
13. Vahedi, V.; Jafarpour, M.; Hoskova-Mayerova, S.; Aghabozorgi, H.; Leoreanu-Fotea, V.; Bekesiene, S. Derived Hyperstructures from Hyperconics. *Mathematics* **2020**, *8*, 429.
14. Vahedi, V.; Jafarpour, M.; Cristea, I. Hyperhomographies on Krasner hyperfields. *Symmetry* **2019**, *11*, 1442.
15. Massouros, C.G. Free and cyclic hypermodules. *Ann. Mat. Pura Appl.* **1988**, *150*, 153–166.
16. Bordbar, H.; Cristea, I. Height of prime hyperideals in Krasner hyperrings. *Filomat* **2017**, *31*, 6153–6163.
17. Bordbar, H.; Cristea, I.; Novak, M. Height of hyperideals in Noetherian Krasner hyperrings. *Politehn. Univ. Bucharest Sci. Bull. Ser. A Appl. Math. Phys* **2017**, *79*, 31–42.
18. Bordbar, H.; Novak, M.; Cristea, I. A note on the support of a hypermodule. *J. Algebra Appl.* **2020**, *19*, 2050019.
19. Massouros, G.; Massouros, C. Hypercompositional Algebra, Computer Science and Geometry. *Mathematics* **2020**, *8*, 1338.
20. Mittas, J. Espaces vectoriels sur un hypercorps, Introduction des hyperspaces affines et Euclidiens. *Math. Balk.* **1975**, *5*, 199–211.
21. Nakassis, A. Recent results in hyperring and hyperfield theory. *Int. J. Math. Math. Sci.* **1988**, *11*, 209–220.
22. Massouros, C.G. On the theory of hyperrings and hyperfields. *Algebra Logika* **1985**, *24*, 728–742.
23. Massouros, C.G. Constructions of hyperfields. *Math. Balk.* **1991**, *5*, 250–257.

Article

Fuzzy Multi-Hypergroups

Sarka Hoskova-Mayerova [1,*], Madeline Al Tahan [2] and Bijan Davvaz [3]

[1] Department of Mathematics and Physics, University of Defence in Brno, Kounicova 65, 662 10 Brno-střed, Czech Republic
[2] Department of Mathematics, Lebanese International University, Beqaa Valley 1803, Lebanon; madeline.tahan@liu.edu.lb
[3] Department of Mathematics, Yazd University, Yazd 89139, Iran; davvaz@yazd.ac.ir
* Correspondence: sarka.mayerova@seznam.cz

Received: 27 January 2020; Accepted: 12 February 2020; Published: 14 February 2020

Abstract: A fuzzy multiset is a generalization of a fuzzy set. This paper aims to combine the innovative notion of fuzzy multisets and hypergroups. In particular, we use fuzzy multisets to introduce the concept of fuzzy multi-hypergroups as a generalization of fuzzy hypergroups. Different operations on fuzzy multi-hypergroups are defined and discussed and some results known for fuzzy hypergroups are generalized to fuzzy multi-hypergroups.

Keywords: hypergroup; multiset; fuzzy multiset; fuzzy multi-hypergroup

MSC: 20N25; 20N20

1. Introduction

In 1934, Frederic Marty [1] defined the concept of a hypergroup as a natural generalization of a group. It is well known that the composition of two elements in a group is an element, whereas the composition of two elements in a hypergroup is a non-empty set. The law characterizing such a structure is called the multi-valued operation, or hyperoperation and the theory of the algebraic structures endowed with at least one multi-valued operation is known as the Hyperstructure Theory. Marty's motivation to introduce hypergroups is that the quotient of a group modulo of any subgroup (not necessarily normal) is a hypergroup. Significant progress in the theory of hyperstructures has been made since the 1970s, when its research area was enlarged and different hyperstructures were introduced (e.g., hyperrings, hypermodules, hyperlattices, hyperfields, etc.). Many types of hyperstructures have been used in different contexts, such as automata theory, topology, cryptography and code theory, geometry, graphs and hypergraphs, analysis of the convex systems, finite groups' character theory, theory of fuzzy and rough sets, probability theory, ethnology, and economy [2]. An overview of the most important works and results in the field of hyperstructures up to 1993 is given in the book by Corsini [3], this book was followed in 1994 by the book by Vougiouklis [4]. An overview regarding the applications of hyperstructure theory is given in the book by Corsini and Leoreanu [5]. The book by Davvaz and Leoreanu-Fotea [6] deals with the hyperring theory and applications. A more recent book [7] gives an introduction into fuzzy algebraic hyperstructures.

A set is a well-defined collection of distinct objects, i.e., every element in a set occurs only once. A generalization of the notion of set was introduced by Yager [8]. He introduced the bag (multiset) structure as a set-like object in which repeated elements are significant. He discussed operations on multisets, such as intersection and union, and he showed the usefulness of the new defined structure in relational databases. These new structures have many applications in mathematics and computer science [9]. For example, the prime factorization of a positive integer is a multiset in which its elements are primes (e.g., 90 has the multiset $\{2, 3, 3, 5\}$). Moreover, the eigenvalues of a matrix (e.g., the

5×5 lower triangular matrix (a_{ij}) with $a_{11} = a_{22} = -5$, $a_{33} = 1$, $a_{44} = a_{55} = 0$ has the multiset $\{-5, -5, 1, 0, 0\}$) and roots of a polynomial (e.g., the polynomial $(x+2)^2(x-1)x^3$ over the field of complex numbers has the multiset $\{-2, -2, 1, 0, 0, 0\}$) can be considered as multisets.

As an extension of the classical notion of sets, Zadeh [10] introduced a concept similar to that of a set but whose elements have degrees of membership and he called it fuzzy set. In classical sets, the membership function can take only two values: 0 and 1. It takes 0 if the element belongs to the set, and 1 if the element does not belong to the set. In fuzzy sets, there is a gradual assessment of the membership of elements in a set which is assigned a number between 0 and 1 (both included). Several applications for fuzzy sets appear in real life. We refer to the papers [11–13]. Yager, in [8], generalized the fuzzy set by introducing the concept of fuzzy multiset (fuzzy bag) and he discussed a calculus for them in [14]. An element of a fuzzy multiset can occur more than once with possibly the same or different membership values. If every element of a fuzzy multiset can occur at most once, we go back to fuzzy sets [15].

In [16], Onasanya and Hoskova-Mayerova defined multi-fuzzy groups and in [17,18], the authors defined fuzzy multi-polygroups and fuzzy multi-H_v-ideals and studied their properties. Moreover, Davvaz in [19] discussed various properties of fuzzy hypergroups. Our paper generalizes the work in [16,17,19] to combine hypergroups and fuzzy multisets. More precisely, it is concerned with fuzzy multi-hypergroups and constructed as follows: After the Introduction, Section 2 presents some preliminary definitions and results related to fuzzy multisets and hypergroups that are used throughout the paper. Section 3 introduces, for the first time, fuzzy multi-hypergroups as a generalization of fuzzy hypergroups and studies its properties. Finally, Section 4 defines some operations (e.g., intersection, selection, product, etc.) on fuzzy multi-hypergroups and discusses them.

2. Preliminaries

In this section, we present some basic definitions and results related to both fuzzy multisets and hypergroups [20,21] that are used throughout the paper.

2.1. Fuzzy Multisets

A multiset is a collection of objects that can be repeated. Yager in his paper [8] introduced, under the name *bag*, a structure similar to a set in which repeated elements are allowed. He studied some operations on bags, such as intersection, union, and addition. Moreover, he introduced the operation of selecting elements from a bag based upon their membership in a set. Furthermore, he suggested a definition for fuzzy multisets (fuzzy bags). A multiset (bag) M is characterized by a count function $C_M : X \to \mathbb{N}$, where $\mathbb{N}$ is the set of natural numbers.

In a multiset, and unlike a set, the multiple occurrences for each of its elements is allowed and it is called multiplicity. For example, the multiset $\{\alpha, \beta\}$ contains two elements α and β, each having multiplicity 1, whereas the multiset $\{\alpha, \beta, \beta\}$ contains two elements α having multiplicity 1 and β having multiplicity 2. The two multisets $\{\alpha, \beta, \beta\}$ and $\{\alpha, \beta\}$ are not equal although they are considered equal as sets.

Definition 1 ([17,22])**.** *Let X be a set. A multiset M drawn from X is represented by a function $C_M : X \to \{0, 1, 2, \ldots\}$. For each $x \in X$, $C_M(x)$ denotes the number of occurrences of x in M.*

Assume $X = \{a_1, \ldots, a_k\}$. For a multiset M on X with count function C_M, the following two equivalent expressions are used:
$$M = \{a_1/n_1, \ldots, a_k/n_k\}$$
and
$$M = \{\underbrace{a_1, \ldots, a_1}_{n_1}, \ldots, \underbrace{a_k, \ldots, a_k}_{n_k}\}.$$

Here, x_i has multiplicity n_i for all $i = 1, \ldots, k$, or equivalently, $C_M(x_i) = n_i$ for all $i = 1, \ldots, k$.

One potential useful application for the theory of multisets lies in the field of relational databases [8].

Example 1. *Consider a class of eight students and a relation of students over the scheme (student name, grade) given as follows:*
R ={(Sam, 90), (Nader, 85), (Mady, 90), (Tala, 60), (Joe, 70), (Ziad, 95), (Lune, 85), (Bella, 60)}. Assume we are interested in finding the grades of students in this class. We can take the projection of R on grades and get:

$$Proj_{grade} = \{60, 70, 85, 90, 95\}.$$

The set $Proj_{grade}$ does not give the set of grades of all students in the class but it gives the set of different grades of the students. If we need the set of grades of all students in the class then we need to consider the following multiset M:

$$M = \{60, 60, 70, 85, 85, 90, 90, 95\} = \{60/2, 70/1, 85/2, 90/2, 95/1\}.$$

Or equivalently, we can say that M is a multiset with count function C_M defined as follows: $C_M(60) = C_M(85) = C_M(90) = 2$ and $C_M(70) = C_M(95) = 1$. The notation used here for the multiset M is the same as that used by Yager in his pioneering paper [8] about multisets.

Definition 2 ([23]). *Let X be a non-empty set. A fuzzy multiset A drawn from X is represented by a function $CM_A : X \to Q$, where Q is the set of all multisets drawn from the unit interval $[0, 1]$.*

In the above definition, the value $CM_A(x)$ is a multiset drawn from $[0, 1]$. For each $x \in X$, the membership sequence is defined as the decreasingly ordered sequence of elements in $CM_A(x)$ and it is denoted by:

$$\{\mu_A^1(x), \mu_A^2(x), \ldots, \mu_A^p(x)\} : \mu_A^1(x) \geq \mu_A^2(x) \geq \ldots \geq \mu_A^p(x).$$

Fuzzy sets introduced by Zadeh [10] can be considered as a special case of fuzzy multisets by setting $p = 1$ so that $CM_A(x) = \mu_A^1(x)$.

Example 2. *Let $X = \{0, 1, 2, 3\}$. Then $A = \{(0.5, 0.5, 0.3, 0.1, 0.1)/1, (0.65, 0.2, 0.2, 0.2)/3\}$ is a fuzzy multiset of X with fuzzy count function CM_A. Or equivalently, we can write it as:*

$$CM_A(0) = CM_A(2) = 0, CM_A(1) = (0.5, 0.5, 0.3, 0.1, 0.1), CM_A(3) = (0.65, 0.2, 0.2, 0.2).$$

Let A, B be fuzzy multisets of X with fuzzy count functions CM_A, CM_B respectively. Then $L(x; A) = \max\{j : \mu_A^j(x) \neq 0\}$ and $L(x) = L(x; A, B) = \max\{L(x; A), L(x; B)\}$. When we define an operation between two fuzzy multisets, the length of their membership sequences should be set as equal. In case two membership sequences have different lengths then the shorter sequence is extended with zeros. As an illustration, we consider the following example.

Example 3. *Let $X = \{0, 1, 2, 3\}$ and define the fuzzy multisets A, B of X as follows:*

$$A = \{(0.4, 0.2)/0, (0.5, 0.3, 0.1)/1, (0.65, 0.2, 0.2, 0.2)/3\},$$

$$B = \{(0.5, 0.3, 0.1, 0.1)/1, (0.5, 0.1)/2, (0.65, 0.2, 0.2)/3\}.$$

In order to define any operations on A, B, we can rewrite A, B as follows:

$$A = \{(0.4, 0.2)/0, (0.5, 0.3, 0.1, 0)/1, (0, 0)/2, (0.65, 0.2, 0.2, 0.2)/3\},$$

$$B = \{(0, 0)/0, (0.5, 0.3, 0.1, 0.1)/1, (0.5, 0.1)/2, (0.65, 0.2, 0.2, 0)/3\}.$$

Definition 3 ([24]). *Let X be a set and A, B be fuzzy multisets of X. Then*

1. $A \subseteq B$ if and only if $CM_A(x) \leq CM_B(x)$ for all $x \in X$. i.e., $\mu_A^j(x) \leq \mu_B^j(x)$, $j = 1, \ldots, L(x)$ for all $x \in X$,
2. $A = B$ if and only if $CM_A(x) = CM_B(x)$ for all $x \in X$. i.e., $\mu_A^j(x) = \mu_B^j(x)$, $j = 1, \ldots, L(x)$ for all $x \in X$,
3. $A \cap B$ is defined by $\mu_{A \cap B}^j(x) = \mu_A^j(x) \wedge \mu_B^j(x)$, $j = 1, \ldots, L(x)$ for all $x \in X$,
4. $A \cup B$ is defined by $\mu_{A \cap B}^j(x) = \mu_A^j(x) \vee \mu_B^j(x)$, $j = 1, \ldots, L(x)$ for all $x \in X$.

Example 4. *Let $X = \{0, 1, 2, 3\}$ and define the fuzzy multisets A, B, C of X as follows:*

$$A = \{(0.5, 0.3, 0.1, 0.1)/1, (0.65, 0.2, 0.2, 0.2)/3\},$$

$$B = \{(0.5, 0.3, 0.1, 0.1)/1, (0.5, 0.1)/2, (0.65, 0.2, 0.2, 0.2)/3\},$$

$$C = \{(0.5, 0.3)/0, (0.5, 0.1)/2, (0.35, 0.3, 0.1)/3\}.$$

Then it is clear that:

1. $A \subseteq B$ and $A \neq B$,
2. $A \cap C = \{(0.35, 0.2, 0.1)/3\}$, and
3. $A \cup C = \{(0.5, 0.3)/0, (0.5, 0.3, 0.1, 0.1)/1, (0.5, 0.1)/2, (0.65, 0.3, 0.2, 0.2)/3\}$.

Definition 4 ([25]). *Let X, Y be non-empty sets, $f : X \to Y$ be a mapping, and A, B be fuzzy multisets of X, Y respectively. Then*

1. *The image of A under f is denoted by $f(A)$ with fuzzy count function $CM_{f(A)}$ defined as follows: For all $y \in Y$,*

$$CM_{f(A)}(y) = \begin{cases} \vee_{f(x)=y} CM_A(x) & \text{if } f^{-1}(y) \neq \emptyset \\ 0 & \text{otherwise} \end{cases}$$

2. *The inverse image of B under f is denoted by $f^{-1}(B)$ with fuzzy count function $CM_{f^{-1}(B)}$ defined as: $CM_{f^{-1}(B)}(x) = CM_B(f(x))$ for all $x \in X$.*

2.2. Hypergroups

Let H be a non-empty set and $\mathcal{P}^*(H)$ be the family of all non-empty subsets of H. Then, a mapping $\circ : H \times H \to \mathcal{P}^*(H)$ is called a *binary hyperoperation* on H and $(H, \circ)$ is called a *hypergroupoid*.

In the above definition, if A and B are two non-empty subsets of H and $x \in H$, then we define:

$$A \circ B = \bigcup_{\substack{a \in A \\ b \in B}} a \circ b, \quad x \circ A = \{x\} \circ A \text{ and } A \circ x = A \circ \{x\}.$$

A hypergroupoid $(H, \circ)$ is called a *semihypergroup* if "$\circ$" is associative, i.e., if $x \circ (y \circ z) = (x \circ y) \circ z$ for all $x, y, z \in H$ and is called a *quasihypergroup* if the reproduction axiom is satisfied, i.e., if $x \circ H = H = H \circ x$ for all $x \in H$. The couple $(H, \circ)$ is called a *hypergroup* if it is a semihypergroup and a quasihypergroup. A hypergroup $(H, \circ)$ is called *commutative* if $x \circ y = y \circ x$ for all $x, y \in H$. A subset S of a hypergroup $(H, \circ)$ is called a *subhypergroup* of H if $(S, \circ)$ is a hypergroup. To prove that S is a subhypergroup of H, it suffices to show that the reproduction axiom is satisfied for S.

Example 5. *In the finite hypergroup, we can represent the hyperoperation by the Cayley square table in the same way as for the finite group. The only difference is that various places in the table may contain several elements instead of a single element, i.e.,*

*	e	a	b
e	{e}	{a,b}	{a,b}
a	{a}	{e,b}	{e,b}
b	{b}	{e,a}	{e,a}

is a hypergroup.

Definition 5 ([17,26]). *Let $(H_1, \circ_1)$ and $(H_2, \circ_2)$ be hypergroups and $f : H_1 \to H_2$ be a mapping. Then f is:*

1. *a homomorphism if $f(x \circ_1 y) \subseteq f(x) \circ_2 f(y)$ for all $x, y \in H_1$;*
2. *a strong homomorphism if $f(x \circ_1 y) = f(x) \circ_2 f(y)$ for all $x, y \in H_1$;*
3. *an isomorphism if f is a bijective strong homomorphism.*

Example 6. *Let H be any non-empty set and define $\circ$ on H as follows:*

$$x \circ y = H \text{ for all } x, y \in H.$$

Then $(H, \circ)$ is a hypergroup known as total hypergroup.

Example 7. *Let H be any non-empty set and define $\circ$ on H as follows:*

$$x \circ y = \{x, y\} \text{ for all } x, y \in H.$$

Then $(H, \circ)$ is a hypergroup known as biset hypergroup.

Example 8. *Let $\mathbb{Z}$ be the set of integers and define $\star$ on $\mathbb{Z}$ as follows: For all $x, y \in \mathbb{Z}$,*

$$x \star y = \begin{cases} 2\mathbb{Z} & \text{if } x, y \text{ have same parity;} \\ 2\mathbb{Z} + 1 & \text{otherwise.} \end{cases}$$

Then $(\mathbb{Z}, \star)$ is a commutative hypergroup.

For more examples and details about hypergroups, we refer to the books [3–6,26,27] and to the papers [20,21,28–32].

3. Construction of Fuzzy Multi-Hypergroups

Inspired by the definition of the multi-fuzzy group [25] and the fuzzy multi-polygroup [17], we introduce the concept of the fuzzy multi-hypergroup. Further, we investigate their properties. It is well known that groups and polygroups [26] are considered as special cases of hypergroups. Hence, the results in this section can be considered as more general than that in [17,25].

Definition 6 ([7]). *Let $(H, \circ)$ be a hypergroup and μ be a fuzzy subset of H. Then μ is called a fuzzy subhypergroup of H if for all $x, y \in H$, the following conditions hold.*

1. $\mu(x) \wedge \mu(y) \leq \inf_{z \in x \circ y} \mu(z)$;
2. *for every $x, a \in H$ there exists $y \in H$ such that $x \in a \circ y$ and $\mu(x) \wedge \mu(a) \leq \mu(y)$;*
3. *for every $x, a \in H$ there exists $z \in H$ such that $x \in z \circ a$ and $\mu(x) \wedge \mu(a) \leq \mu(z)$.*

Definition 7. *Let $(H, \circ)$ be a hypergroup. A fuzzy multiset A over H is a fuzzy multi-hypergroup of H if for all $x, y \in H$, the following conditions hold.*

1. $CM_A(x) \wedge CM_A(y) \leq \inf_{z \in x \circ y} CM_A(z)$ *(or equivalently, $CM_A(x) \wedge CM_A(y) \leq CM_A(z)$ for all $z \in x \circ y$);*
2. *for every $x, a \in H$ there exists $y \in H$ such that $x \in a \circ y$ and $CM_A(x) \wedge CM_A(a) \leq CM_A(y)$;*
3. *for every $x, a \in H$ there exists $z \in H$ such that $x \in z \circ a$ and $CM_A(x) \wedge CM_A(a) \leq CM_A(z)$.*

Remark 1. *It is clear, using Definitions 6 and 7, that if $(H, \circ)$ is a hypergroup and μ is a fuzzy subhypergroup of H then μ is a fuzzy multi-hypergroup of H. Hence, the results of fuzzy subhypergroups are considered a special case of our work.*

Remark 2. *Let A be a fuzzy multiset over a commutative hypergroup $(H, \circ)$. Then conditions 2. and 3. of Definition 7 are equivalent. Hence, to show that A is a fuzzy multi-hypergroup of H, it suffices to show that either conditions 1. and 2. of Definition 7 are valid or conditions 1. and 3. of Definition 7 are valid.*

We present different examples on fuzzy multi-hypergroups.

Example 9. *Let $(H, \circ)$ be the hypergroup defined by the following table:*

$\circ$	a	b
a	H	a
b	a	b

It is clear that $A = \{(0.2, 0.1)/a, (0.5, 0.4, 0.4)/b\}$ is a fuzzy multi-hypergroup of H.

Example 10. *Let $(H_1, \circ_1)$ be the hypergroup defined by the following table:*

$\circ_1$	0	1	2
0	$\{0, 1\}$	$\{0, 1\}$	H_1
1	$\{0, 1\}$	$\{0, 1\}$	H_1
2	H_1	H_1	2

It is clear that $A = \{(0.2, 0.1)/0, (0.2, 0.1)/1, (0.5, 0.4, 0.4)/2\}$ is a fuzzy multi-hypergroup of H_1.

Example 11. *Let $(\mathbb{Z}, \star)$ be the hypergroup defined in Example 8. It is clear that A, with the fuzzy count function CM_A, is a fuzzy multi-hypergroup of $\mathbb{Z}$. Where*

$$CM_A(x) = \begin{cases} (0.7, 0.5, 0.5) & \text{if } x \text{ is an even integer;} \\ (0.7, 0.3, 0.2) & \text{otherwise.} \end{cases}$$

Proposition 1. *Let $(H, \circ)$ be a hypergroup and A be a fuzzy multi-hypergroup of H. Then the following assertions are true.*

1. *$CM_A(z) \geq CM_A(x_1) \wedge \ldots \wedge CM_A(x_n)$ for all $z \in x_1 \circ \ldots \circ x_n$ and $n \geq 2$;*
2. *$CM_A(z) \geq CM_A(x)$ for all $z \in x^n$.*

Proof.

- Proof of 1. By mathematical induction on the value of n, $CM_A(z) \geq CM_A(x_1) \wedge \ldots \wedge CM_A(x_n)$ for all $z \in x_1 \circ \ldots \circ x_n$ is true for $n = 2$. Assume that $CM_A(z) \geq CM_A(x_1) \wedge \ldots \wedge CM_A(x_n)$ for all $z \in x_1 \circ \ldots \circ x_n$ and let $z' \in x_1 \circ \ldots \circ x_n \circ x_{n+1}$. Then there exists $x \in x_1 \circ \ldots \circ x_n$ such that $z' \in x \circ x_{n+1}$. Having A a fuzzy multi-hypergroup implies that $CM_A(z') \geq CM_A(x) \wedge CM_A(x_{n+1})$. And using our assumption that $CM_A(x) \geq CM_A(x_1) \wedge \ldots \wedge CM_A(x_n)$ implies that our statement is true for $n + 1$.
- Proof of 2. The proof follows from 1. by setting $x_i = x$ for all $i = 1, \ldots, n$. □

Example 12. *Let $(H, \circ)$ be any hypergroup and $a \in H$ be a fixed element. We define a fuzzy multiset A of H with fuzzy count function CM_A as $CM_A(x) = CM_A(a)$ for all $x \in H$. Then A is a fuzzy multi-hypergroup of H (the constant fuzzy multi-hypergroup).*

Remark 3. *Let $(H, \circ)$ be a hypergroup. Then we can define at least one fuzzy multi-hypergroup of H, which is mainly the one that is described in Example 12.*

Proposition 2. Let $(H, \circ)$ be the biset hypergroup and A be a fuzzy multiset of H. Then A is a fuzzy multi-hypergroup of H.

Proof. Let A be a fuzzy multiset of H. Since $(H, \circ)$ is a commutative hypergroup, it suffices to show that conditions 1. and 2. of Definition 7 are satisfied. (See Remark 2). For condition 1., let $x, y \in H$ and $z \in x \circ y = \{x, y\}$. It is clear that $CM_A(z) \geq CM_A(x) \wedge CM_A(y)$. For condition 2., let $a, x \in H$. Then there exists $y = x \in H$ such that $x \in a \circ y$ and $CM_A(y) = CM_A(x) \geq CM_A(a) \wedge CM_A(x)$. Therefore, A is a fuzzy multi-hypergroup of H. □

Proposition 3. Let $(H, \circ)$ be the total hypergroup and A be a fuzzy multiset of H. Then A is a fuzzy multi-hypergroup of H if and only if A is the fuzzy multiset described in Example 12.

Proof. If A is the fuzzy multiset described in Example 12 then it is clear that A is a fuzzy multi-hypergroup of H. Let A be a fuzzy multi-hypergroup of H and $a \in H$. For all $x \in H$, we have $x \in a \circ a = H$ and $a \in x \circ x$. The latter and having A a fuzzy multi-hypergroup of H implies that $CM_A(x) \geq CM_A(a)$ and $CM_A(a) \geq CM_A(x)$. Thus, $CM_A(x) = CM_A(a)$ for all $x \in H$. □

Notation 1. Let $(H, \circ)$ be a hypergroup, A be a fuzzy multiset of H and $CM_A(x) = (\mu_A^1(x), \mu_A^2(x), \ldots, \mu_A^p(x))$. We say that $CM_A(x) > 0$ if $\mu_A^1(x) > 0$.

Definition 8. Let $(H, \circ)$ be a hypergroup and A be a fuzzy multiset of H. Then $A_\star = \{x \in X : CM_A(x) > 0\}$.

Proposition 4. Let $(H, \circ)$ be a hypergroup and A be a fuzzy multi-hypergroup of H. Then $A_\star$ is either the empty set or a subhypergroup of H.

Proof. Let $a \in A_\star \neq \emptyset$. We need to show that the reproduction axiom is satisfied for $A_\star$. We show that $a \circ A_\star = A_\star$ and $A_\star \circ a = A_\star$ is done similarly. For all $x \in A_\star$ and $z \in a \circ x$, we have $CM_A(z) \geq CM_A(a) \wedge CM_A(x) > 0$. The latter implies that $z \in A_\star$ and hence, $A_\star \circ a \subseteq A_\star$. Moreover, for all $x \in A_\star$, Condition 2. of Definition 7 implies that there exist $y \in H$ such that $x \in a \circ y$ and $CM_A(y) \geq CM_A(x) \wedge CM_A(a) > 0$. The latter implies that $y \in A_\star$ and $x \in a \circ A_\star$. Thus, $A_\star \subseteq a \circ A_\star$. □

Definition 9. Let $(H, \circ)$ be a hypergroup and A, B be fuzzy multisets of H. Then $A \circ B$ is defined by the following fuzzy count function.

$$CM_{A \circ B}(x) = \vee \{CM_A(y) \wedge CM_B(z) : x \in y \circ z\}.$$

Theorem 1. Let $(H, \circ)$ be a hypergroup and A be a fuzzy multiset of H. If A is a fuzzy multi-hypergroup of H then $A \circ A = A$.

Proof. Let $z \in H$. Then $CM_A(z) \geq CM_A(x) \wedge CM_A(y)$ for all $z \in x \circ y$. The latter implies that $CM_A(z) \geq \vee \{CM_A(x) \wedge CM_B(y) : z \in x \circ y\} \geq CM_{A \circ A}(z)$. Thus, $A \circ A \subseteq A$. Having $(H, \circ)$ a hypergroup and A a fuzzy multi-hypergroup of H implies that for every $x \in H$ there exist $y \in H$ such that $x \in x \circ y$ and $CM_A(y) \geq CM_A(x)$. Moreover, we have $CM_{A \circ A}(x) = \vee \{CM_A(y) \wedge CM_B(z) : x \in y \circ z\} \geq CM_A(x) \wedge CM_A(y) = CM_A(x)$. Thus, $A \subseteq A \circ A$. □

Notation 2. Let $(H, \circ)$ be a hypergroup, A be a fuzzy multiset of H and $CM_A(x) = (\mu_A^1(x), \mu_A^2(x), \ldots, \mu_A^p(x))$. We say that $CM_A(x) \geq (t_1, \ldots, t_k)$ if $p \geq k$ and $\mu_A^i(x) \geq t_i$ for all $i = 1, \ldots, k$. If $CM_A(x) \not\geq (t_1, \ldots, t_k)$ and $(t_1, \ldots, t_k) \not\geq CM_A(x)$ then we say that $CM_A(x)$ and $(t_1, \ldots, t_k)$ are not comparable.

In [19], Davvaz studied fuzzy hypergroups and proved some important results about them related to level subhypergroups of fuzzy hypergroups. In what follows, we do suitable changes to extend the results of [19] to fuzzy multi-hypergroups.

Theorem 2. *Let $(H, \circ)$ be a hypergroup, A a fuzzy multiset of H with fuzzy count function CM and $t = (t_1, \ldots, t_k)$ where $t_i \in [0,1]$ for $i = 1, \ldots, k$ and $t_1 \geq t_2 \geq \ldots \geq t_k$. Then A is a fuzzy multi-hypergroup of H if and only if CM_t is either the empty set or a subhypergroup of H.*

Proof. Let CM_t be a subhypergroup of H and $x, y \in H$. By setting $t_0 = CM(x) \wedge CM(y)$, we get that $x, y \in CM_{t_0}$. Having CM_{t_0} a subhypergroup of H implies that for all $z \in x \circ y$, $CM(z) \geq t_0 = CM(x) \wedge CM(y)$. We prove condition 2. of Definition 7 and condition 3. is done similarly. Let $a, x \in H$ and $t_0 = CM(x) \wedge CM(a)$. Then $a, x \in CM_{t_0}$. Having CM_{t_0} a subhypergroup of H implies that $a \circ CM_{t_0} = CM_{t_0}$. The latter implies that there exist $y \in CM_{t_0}$ such that $x \in a \circ y$. Thus, $CM(y) \geq t_0 = CM(x) \wedge CM(y)$.

Conversely, let A be a fuzzy multi-hypergroup of H and $CM_t \neq \emptyset$. We need to show that $CM_t = a \circ CM_t = CM_t \circ a$ for all $a \in CM_t$. We prove that $CM_t = a \circ CM_t$ and $CM_t = CM_t \circ a$ is done similarly. Let $x \in CM_t$. Then $CM(z) \geq CM(x) \wedge CM(a) \geq t$ for all $z \in a \circ x$. The latter implies that $z \in CM_t$. Thus, $a \circ CM_t \subseteq CM_t$. Let $x \in CM_t$. Since A is a fuzzy multi-hypergroup of H, it follows that there exist $y \in H$ such that $x \in a \circ y$ and $CM(y) \geq CM(x) \wedge CM(a) \geq t$. The latter implies that $y \in CM_t$ and hence, $CM_t \subseteq a \circ CM_t$. □

Proposition 5. *Let $(H, \circ)$ be a hypergroup, A a fuzzy multiset of H with fuzzy count function CM and $t = (t_1, \ldots, t_k), s = (s_1, \ldots, s_n)$ where $t_i, s_i \in [0,1]$ for $i = 1, \ldots, \max(k, n)$ and $t_1 \geq t_2 \geq \ldots \geq t_k$, $s_1 \geq s_2 \geq \ldots \geq s_n$. If $t < s$ and $CM_t = CM_s$ then there exist no $x \in H$ such that $t \leq CM(x) < s$.*

Proof. Let $CM_t = CM_s$ and suppose that there exist $x \in H$ such that $t \leq CM(x) < s$. Then $x \in CM_t$ and $x \notin CM_s$ which contradicts the given. □

Proposition 6. *Let $(H, \circ)$ be a hypergroup and S be a subhypergroup of H. Then $S = CM_t$ for some $t = (t_1, \ldots, t_k)$ where $t_i \in [0,1]$ for $i = 1, \ldots, k$, $t_1 \neq 0$, and $t_1 \geq t_2 \geq \ldots \geq t_k$.*

Proof. Let $t = (t_1, \ldots, t_k)$ where $t_i \in [0,1]$ for $i = 1, \ldots, k$ and define the fuzzy multiset A of H as follows:
$$CM(x) = \begin{cases} t & \text{if } x \in S \\ 0 & \text{otherwise.} \end{cases}$$

It is clear that $S = CM_t$. We still need to prove that CM is a fuzzy multi-hypergroup of H. Using Theorem 2, it suffices to show that $CM_\alpha \neq \emptyset$ is a subhypergroup of H for all $\alpha = (a_1, \ldots, a_s)$ with $a_i \in [0,1]$ for $i = 1, \ldots, s$. One can easily see that

$$CM_\alpha = \begin{cases} H & \text{if } \alpha = 0 \\ S & \text{if } 0 < \alpha \leq t \\ \emptyset & \text{if } (\alpha > t) \text{ or } (\alpha \text{ and } t \text{ are not comparable)}. \end{cases}$$

Thus, CM_α is either the empty set or a subhypergroup of H. □

4. Operations on Fuzzy Multi-Hypergroups

In this section, we define some operations on fuzzy multi-hypergroups, study them, and present some examples.

Proposition 7. Let $(H, \circ)$ be a hypergroup and A, B be fuzzy multisets of H. If A and B are fuzzy multi-hypergroups of H and one of them is the constant fuzzy multi-hypergroup then $A \cap B$ is a fuzzy multi-hypergroup of H.

Proof. We prove conditions of Definition 7 are satisfied for $A \cap B$. (1) Let $x, y \in H$ and $z \in x \circ y$. Then $CM_{A \cap B}(z) = CM_A(z) \wedge CM_B(z)$. Having A, B fuzzy multi-hypergroups of H implies that $CM_A(z) \geq CM_A(x) \wedge CM_A(y)$ and $CM_B(z) \geq CM_B(x) \wedge CM_B(y)$. The latter implies that $CM_{A \cap B}(z) \geq CM_A(x) \wedge CM_A(y) \wedge CM_B(x) \wedge CM_B(y) = CM_{A \cap B}(x) \wedge CM_{A \cap B}(y)$. (2) Without loss of generality, let B be the constant fuzzy multiset of H with $CM_B(x) = \alpha$ for all $x \in H$. Let $a, x \in H$. Then there exist $y \in H$ such that $x \in a \circ y$ and $CM_A(y) \geq CM_A(x) \wedge CM_A(a)$. The latter implies that $CM_{A \cap B}(y) = CM_A(y) \wedge \alpha \geq CM_A(x) \wedge CM_A(a) \wedge \alpha = (CM_A(x) \wedge \alpha) \wedge (CM_A(a) \wedge \alpha) = CM_{A \cap B}(x) \wedge CM_{A \cap B}(a)$. (3) is done in a similar way to (2). □

Example 13. Let $(H, \circ)$ be the hypergroup defined in Example 9 and A, B be fuzzy multisets of H defined as:

$$A = \{(0.2, 0.1)/a, (0.5, 0.4, 0.4)/b\}, B = \{(0.7, 0.05, 0.05)/a, (0.7, 0.05, 0.05)/b\}.$$

Since A is a fuzzy multi-hypergroup of H and B is a constant fuzzy multi-hypergroup of H, it follows that $A \cap B = \{(0.2, 0.05)/a, (0.5, 0.05, 0.05)/b\}$ is a fuzzy multi-hypergroup of H.

Definition 10. Let H be any set and A, B be fuzzy multisets of H with fuzzy count functions CM_A, CM_B respectively. Then the fuzzy multiset $A \uplus B$ is given as follows: For all $x \in H$,

$$CM(x) = \frac{CM_A(x) + CM_B(x)}{2}.$$

Example 14. Let $H = \{a, b\}$, A, B be fuzzy multisets of H given as:

$$A = \{(0.8, 0.4, 0.4)/a, (0.1, 0.1)/b\}, B = \{(0.6, 0.4, 0.4, 0.4)/a, (0.3, 0.1, 0.1)/b\}.$$

Then $A \uplus B = \{(0.7, 0.4, 0.4, 0.2)/a, (0.2, 0.1, 0.05)/b\}$.

Proposition 8. Let $(H, \circ)$ be a hypergroup and A, B be fuzzy multisets of H. If A and B are fuzzy multi-hypergroups of H and A or B is constant then $A \uplus B$ is a fuzzy multi-hypergroup of H.

Proof. Without loss of generality, let B be the constant fuzzy multiset of H with $CM_B(x) = \alpha$ for all $x \in H$. We prove that the conditions of Definition 7 are satisfied for $A \uplus B$. (1) Let $x, y \in H$ and $z \in x \circ y$. Having $CM_A(z) \geq CM_A(x) \wedge CM_A(y)$ implies that $\frac{CM_A(z) + \alpha}{2} \geq \frac{(CM_A(x) \wedge CM_A(y)) + \alpha}{2}$. We get that $CM(z) = \frac{CM_A(z) + \alpha}{2} \geq \frac{CM_A(x) + \alpha}{2} \wedge \frac{CM_A(y) + \alpha}{2} = CM(x) \wedge CM(y)$. (2) Let $a, x \in H$. Then there exist $y \in H$ such that $x \in a \circ y$ and $CM_A(y) \geq CM_A(x) \wedge CM_A(a)$. The latter implies that $CM(y) = \frac{CM_A(y) + \alpha}{2} \geq \frac{(CM_A(x) \wedge CM_A(a)) + \alpha}{2} = \frac{CM_A(x) + \alpha}{2} \wedge \frac{CM_A(a) + \alpha}{2} = CM(x) \wedge CM(a)$. (3) is done in a similar way to (2). □

Example 15. In Example 13, $A \uplus B = \{(0.45, 0.075, 0.025)/a, (0.6, 0.225, 0.225)/b\}$ is a fuzzy multi-hypergroup of H.

Definition 11. Let H be a non-empty set and A be a fuzzy multiset of H. We define A', the complement of A, to be the fuzzy multiset defined as: For all $x \in H$,

$$CM_{A'}(x) = (1 - \mu_A^p(x), \ldots, 1 - \mu_A^1(x)) \text{ when } CM_A(x) = (\mu_A^1(x), \mu_A^2(x), \ldots, \mu_A^p(x)).$$

Remark 4. Let $(H, \circ)$ be a hypergroup and A be the constant fuzzy multi-hypergroup of H defiend in Example 12. Then, A' is also a fuzzy multi-hypergroup of H.

We present an example of a (non-constant) fuzzy multiset A of H where A and A' are both fuzzy multi-hypergroups of H.

Example 16. Let $H = \{a, b\}$, $(H, \circ)$ be the biset hypergroup on H and A be a fuzzy multiset of H defined as: $A = \{(0.4, 0.3, 0.3)/a, (0.2, 0.1)/b\}$. Proposition 2 asserts that A and $A' = \{(0.7, 0.7, 0.6)/a, (0.9, 0.8)/b\}$ are fuzzy multi-hypergroups of H.

Remark 5. Let $(H, \circ)$ be a hypergroup and A be a fuzzy multi-hypergroup of H. Then, A' is not necessary a fuzzy multi-hypergroup of H.

The following example is an illustration for Remark 5.

Example 17. Let $(H, \circ)$ be the hypergroup defined in Example 9 and A be the fuzzy multi-hypergroup of H defined as: $A = \{(0.2, 0.1)/a, (0.5, 0.4, 0.4)/b\}$. Then $A' = \{(0.9, 0.8)/a, (0.6, 0.6, 0.5)/b\}$ is not a fuzzy multi-hypergroup of H as $b \in a \circ a$ and $CM_{A'}(b) \not\geq CM_{A'}(a)$.

Definition 12. Let X be any set, $S \subseteq X$ and A a fuzzy multiset of X with fuzzy count function CM_A. Then the selection operation $\otimes$ is defined by the fuzzy multiset $A \otimes S$ with the fuzzy count function CM as follows:

$$CM(x) = \begin{cases} CM_A(x) & \text{if } x \in S \\ 0 & \text{otherwise.} \end{cases}$$

Proposition 9. Let $(H, \circ)$ be a hypergroup and S be a subhypergroup of H. If A is a fuzzy multi-hypergroup of S then $A \otimes S$ is a fuzzy multi-hypergroup of H.

Proof. Let $z \in x \circ y$. If $x \notin S$ or $y \notin S$ then $CM(z) \geq 0 = CM(x) \wedge CM(y)$. If $x, y \in S$ then $z \in x \circ y$. Having A a fuzzy multi-hypergroup of S implies that $CM_A(z) \geq CM_A(x) \wedge CM_A(y)$. The latter implies that $CM(z) = CM_A(z) \geq CM_A(x) \wedge CM_A(y) = CM(x) \wedge CM(y)$. We prove condition 2. of Definition 7 and condition 3 is done similarly. Let $a, x \in H$. If $a \notin S$ or $x \notin S$ then for all $y \in H$ with $x \in a \circ y$, $CM(y) \geq 0 = CM(x) \wedge CM(a)$. If $a, x \in S$ then there exist $y \in S$ such that $x \in a \circ y$ and $CM_A(y) \geq CM_A(x) \wedge CM_A(a)$. The latter implies that $CM(y) \geq CM(x) \wedge CM(a)$. □

Definition 13. Let X be any set, $S \subseteq X$ and A a fuzzy multiset of X with fuzzy count function CM_A. Then, the selection operation $\odot$ is defined by the fuzzy multiset $A \otimes S$ with the fuzzy count function CM.

$$CM(x) = \begin{cases} CM_A(x) & \text{if } x \notin S \\ 0 & \text{otherwise.} \end{cases}$$

Proposition 10. Let $(H, \circ)$ be a hypergroup and $S \subset H$ with $H - S$ a subhypergroup of H. If A is a fuzzy multi-hypergroup of $H - S$ then $A \odot S$ is a fuzzy multi-hypergroup of H.

Proof. $A \odot S$ is given by the fuzzy count function CM where

$$CM(x) = \begin{cases} 0 & \text{if } x \in S \\ CM_A(x) & \text{otherwise.} \end{cases} = \begin{cases} 0 & \text{if } x \notin H - S \\ CM_A(x) & \text{otherwise.} \end{cases}$$

One can easily see that $A \odot S = A \otimes (H - S)$. Proposition 9 completes the proof. □

Proposition 11. $(H_1, \circ_1), (H_2, \circ_2)$ be hypergroups with fuzzy multisets A, B respectively. If A and B are fuzzy multi-hypergroups of H_1 and H_2 respectively then $A \times B$ is a fuzzy multi-hypergroup of the productional hypergrpup $(H_1 \times H_2, \circ)$, where for all $(x, y) \in H_1 \times H_2$, $CM_{A \times B}(x, y) = CM_A(x) \wedge CM_B(y)$.

Proof. Let $(x_3, y_3) \in (x_1, y_1) \circ (x_2, y_2)$. Then $x_3 \in x_1 \circ x_2$ and $y_3 \in y_1 \circ y_2$. Having A, B fuzzy multi-hypergroups of H_1, H_2 respectively implies that $CM_A(x_3) \geq CM_A(x_1) \wedge CM_A(x_2)$ and $CM_B(y_3) \geq CM_B(y_1) \wedge CM_B(y_2)$. We get now

$$CM_{A \times B}(x_3, y_3) = CM_A(x_3) \wedge CM_B(y_3) \geq CM_A(x_1) \wedge CM_A(x_2) \wedge CM_B(y_1) \wedge CM_B(y_2).$$

The latter implies that $CM_{A \times B}(x_3, y_3) \geq CM_{A \times B}(x_1, y_1) \wedge CM_{A \times B}(x_2, y_2)$. We now prove condition 2. of Definition 7 and condition 3. is done similarly. Let $(x, y), (a, b) \in H_1 \times H_2$. Having $x, a \in H_1, y, b \in H_2$ implies that there exist $z \in H_1, w \in H_2$ such that $x \in a \circ z, y \in b \circ w$ and $CM_A(z) \geq CM_A(x) \wedge CM_A(a), CM_B(w) \geq CM_B(y) \wedge CM_B(b)$. We get that $CM_{A \times B}(z, w) = CM_A(z) \wedge CM_B(w) \geq CM_A(x) \wedge CM_A(a) \wedge CM_B(w) \geq CM_B(y) \wedge CM_B(b) = CM_{A \times B}(x, y) \wedge CM_{A \times B}(a, b)$. The latter implies that there exist $(z, w) \in H_1, H_2$ such that $(x, y) \in (a, b) \circ (z, w)$ and $CM_{A \times B}(z, w) \geq CM_{A \times B}(x, y) \wedge CM_{A \times B}(a, b)$. □

Corollary 1. $(H_i, \circ_i)$ be a hypergroup with fuzzy multiset A_i for $i = 1, \ldots, n$. If A_i is a fuzzy multi-hypergroup of H_i then $A_1 \times \ldots \times A_n$ is a fuzzy multi-hypergroup of the productional hypergroup $(H_1 \times \ldots \times H_n, \circ)$, where for all $(x_1, \ldots, x_n) \in H_1 \times \ldots \times H_n$, $CM_{A_1 \times \ldots \times A_n}(x_1, \ldots, x_n) = CM_{A_1}(x_1) \wedge \ldots \wedge CM_{A_n}(x_n)$.

Proof. The proof follows by using mathematical induction and Proposition 11. □

Example 18. Let $(H, \circ)$ be the hypergroup defined in Example 9 and $(J, \star)$ be the biset hypergroup on the set $\{c, d\}$. Then, the productional hypergroup $(H \times J, \bullet)$ is given by the following table:

$\bullet$	(a, c)	(a, d)	(b, c)	(b, d)
(a, c)	(a, c)	$\{(a, c), (a, d)\}$	$\{(a, c), (b, c)\}$	$H \times J$
(a, d)	$\{(a, c), (a, d)\}$	(a, d)	$H \times J$	$\{(a, d), (b, d)\}$
(b, c)	$\{(a, c), (b, c)\}$	$H \times J$	(b, c)	$\{(b, c), (b, d)\}$
(b, d)	$H \times J$	$\{(a, d), (b, d)\}$	$\{(b, c), (b, d)\}$	(b, d)

Since $A = \{(0.2, 0.1)/a, (0.5, 0.4, 0.4)/b\}$ is a fuzzy multi-hypergroup of H and $B = \{(0.3, 0.05)/c, (0.2, 0.2, 0.1)/d\}$ is a fuzzy multi-hypergroup of J. Then

$$A \times B = \{(0.2, 0.05)/(a, c), (0.2, 0.1)/(a, d), (0.3, 0.05)/(b, c), (0.2, 0.2, 0.1)/(b, d)\}$$

is a fuzzy multi-hypergroup of $H \times J$.

The following propositions (Propositions 12 and 13) deal with the strong homomorphic image and pre-image of a fuzzy multi-hypergroup.

Proposition 12. Let $(H_1, \circ_1), (H_2, \circ_2)$ be hypergroups, A, B be fuzzy multisets of H_1, H_2 respectively and $f : H_1 \to H_2$ be a strong homomorphism. If A is a fuzzy multi-hypergroup of H_1 then $f(A)$ is a fuzzy multi-hypergroup of H_2.

Proof. Let $y_1, y_2 \in H_2$ and $y_3 \in y_1 \circ_2 y_2$. If $f^{-1}(y_1) = \emptyset$ or $f^{-1}(y_2) = \emptyset$ then $CM_{f(A)}(y_1) = 0$ or $CM_{f(A)}(y_2) = 0$. We get that $CM_{f(A)}(y_3) \geq 0 = CM_{f(A)}(y_1) \wedge CM_{f(A)}(y_2)$. If $f^{-1}(y_1) \neq \emptyset$ and $f^{-1}(y_2) \neq \emptyset$ then there exist $x_1, x_2 \in H_1$ such that $CM_A(x_1) = \vee_{f(x)=y_1} CM_A(x)$ and $CM_A(x_2) =$

$\bigvee_{f(x)=y_2} CM_A(x)$. Having f a homomorphism implies that $y_3 \in f(x_1) \circ_2 f(x_2) = f(x_1 \circ_1 x_2)$. The latter implies that there exists $x_3 \in x_1 \circ x_2$ such that $y_3 = f(x_3)$. Since A is a fuzzy multi-hypergroup of H_1, it follows that $CM_{f(A)}(y_3) \geq CM_A(x_3) \geq CM_A(x_1) \wedge CM_A(x_2) = CM_{f(A)}(y_1) \wedge CM_{f(A)}(y_2)$. We prove now condition 2. of Definition 7 and condition 3. is done similarly. Let $y, b \in H_2$. If $f^{-1}(y) = \emptyset$ or $f^{-1}(b) = \emptyset$ then for all $z \in H_2$ such that $y \in b \circ_2 z$, $CM_{f(A)}(z) \geq 0 = CM_{f(A)}(y) \wedge CM_{f(A)}(b)$. If $f^{-1}(y) \neq \emptyset$ and $f^{-1}(b) \neq \emptyset$ then there exist $x_1, a \in H_1$ such that $CM_A(x_1) = \bigvee_{f(x)=y} CM_A(x)$ and $CM_A(a) = \bigvee_{f(x)=b} CM_A(x)$. Having A a fuzzy multi-hypergroup of H_1 implies that there exist $x_2 \in H$ with $x_1 \in x_2 \circ_1 a$ and $CM_A(x_2) \geq CM_A(x_1) \wedge CM_A(a)$. Since f is a strong homomorphism, it follows that $y = f(x_1) \in f(x_2) \circ_2 b$ and $CM_{f(A)}(f(x_2)) \geq CM_A(x_2) \geq CM_A(x_1) \wedge CM_A(a) = CM_{f(A)}(y) \wedge CM_{f(A)}(b)$. □

We can use Proposition 12 to prove Proposition 9.

Corollary 2. *Let $(H, \circ)$ be a hypergroup and S be a subhypergroup of H. If A is a fuzzy multi-hypergroup of S then $A \otimes S$ is a fuzzy multi-hypergroup of H.*

Proof. Let $f : S \to H$ be the inclusion map defined by $f(x) = x$ for all $x \in S$. One can easily see that $CM_{f(A)}$ is the fuzzy count function of $A \otimes S$. □

Proposition 13. *Let $(H_1, \circ_1), (H_2, \circ_2)$ be hypergroups, A, B be fuzzy multisets of H_1, H_2 respectively and $f : H_1 \to H_2$ be a surjective strong homomorphism. If B is a fuzzy multi-hypergroup of H_2 then $f^{-1}(B)$ is a fuzzy multi-hypergroup of H_1.*

Proof. Let $x_1, x_2 \in H_1$ and $x_3 \in x_1 \circ_1 x_2$. Then $CM_{f^{-1}(B)}(x_3) = CM_B(f(x_3))$. Having $f(x_3) \in f(x_1 \circ_1 x_2) = f(x_1) \circ f(x_2)$ implies that $CM_{f^{-1}(B)}(x_3) = CM_B(f(x_3)) \geq CM_B(f(x_1)) \wedge CM_B(f(x_2)) = CM_{f^{-1}(B)}(x_1) \wedge CM_{f^{-1}(B)}(x_2)$. We prove now condition 2. of Definition 7 and condition 3. is done similarly. Let $x, a \in H_1$. Having $y = f(x), b = f(a) \in H_2$ and B a fuzzy multi-hypergroup of H_2 implies that there exist $z \in H_2$ such that $y \in b \circ_2 z$ and $CM_B(z) \geq CM_B(y) \wedge CM_B(b)$. Since f is a surjective strong homomorphism, it follows that there exist $w \in H_1$ such that $f(w) = z$ and $x \in z \circ_1 w$. We get now that $CM_{f^{-1}(B)}(z) = CM_B(z) \geq CM_B(y) \wedge CM_B(b) = CM_{f^{-1}(B)}(x) \wedge CM_{f^{-1}(B)}(w)$. □

5. Conclusions

In this paper, a new link between algebraic hyperstructures and fuzzy multisets was initiated and as a result fuzzy multi-hypergroups were defined and studied. In particular, different operations on fuzzy multi-hypergroups were defined and studied and several results and examples were obtained. The foundations that we made through this paper can be used to get an insight into other types of hyperstructures. As a result, different real life problems involving the concept of the fuzzy multiset can be dealt with from a different perspective.

Author Contributions: Conceptualization, S.H.-M., M.A.T. and B.D.; methodology, S.H.-M., M.A.T. and B.D.; investigation, S.H.-M., M.A.T. and B.D.; resources, S.H.-M., M.A.T. and B.D.; writing—original draft, S.H.-M., M.A.T. and B.D.; writing—review and editing, S.H.-M., M.A.T. and B.D.; project administration, S.H.-M.; funding acquisition, S.H-M. All authors have read and agreed to the published version of the manuscript.

Funding: The APC was funded by by the Ministry of Defence in the Czech Republic.

Acknowledgments: The work presented in this paper was supported within the project for development of basic and applied research developed in the long term by the departments of theoretical and applied bases FMT supported by the Ministry of Defence in the Czech Republic.

Conflicts of Interest: The authors declare no conflict of interest.

References

1. Marty, F. Sur une generalization de la notion de group. In Proceedings of the 8th Congress on Mathmatics Scandenaves, Stockholm, Sweden, 14–18 August 1934; pp. 45–49.
2. Antampoufis, N.; Hoskova-Mayerova, S. A Brief Survey on the two Different Approaches of Fundamental Equivalence Relations on Hyperstructures. *Ratio Mathematica* **2017**, *33*, 47–60. [CrossRef]
3. Corsini, P. *Prolegomena of Hypergroup Theory*, 2nd ed.; Udine: Tricesimo, Italy, 1993.
4. Vougiouklis, T. *Hyperstructures and Their Representations*; Hadronic Press Monographs: Palm Harbor, FL, USA, 1994.
5. Corsini, P.; Leoreanu, V. *Applications of Hyperstructures Theory*; Advances in Mathematics, Kluwer Academic Publisher: Dordrecht, The Netherlands, 2003.
6. Davvaz, B.; Leoreanu-Fotea, V. *Hyperring Theory and Applications*; International Academic Press: Cambridge, MA, USA, 2007.
7. Davvaz, B.; Cristea, I. *Fuzzy Algebraic Hyperstructures*; Studies in Fuzziness and Soft Computing 321; Springer International Publishing: Cham, Switzerland, 2015. [CrossRef]
8. Yager, R. On the theory of bags. *Int. J. Gen. Syst.* **1987**, *13*, 23–37. [CrossRef]
9. Onasanya, B.O.; Hoskova-Mayerova, S. Results on Functions on Dedekind Multisets. *Symmetry* **2019**, *11*, 1125. [CrossRef]
10. Zadeh, L.A. Fuzzy sets. *Inform. Control* **1965**, *8*, 338–353. [CrossRef]
11. Mei, Y.; Peng, J.; Yang, J. Convex aggregation operators and their applications to multi-hesitant fuzzy multi-criteria decision-making. *Information* **2018**, *9*, 207. [CrossRef]
12. Joshi, D.K.; Beg, I.; Kumar, S. Hesitant probabilistic fuzzy linguistic sets with applications in multi-criteria group decision making problems. *Mathematics* **2018**, *6*, 47. [CrossRef]
13. Yaqoob, N.; Gulistan, M.; Kadry, S.; Wahab, H.A. Complex intuitionistic fuzzy graphs with application in cellular network provider companies. *Mathematics* **2019**, *7*, 35. [CrossRef]
14. Miyamoto, S. *Fuzzy Multisets and Their Generalizations, Multiset Processing*; Springer: Berlin, Germany, 2001; pp. 225–235.
15. Onasanya, B.O.; Hoskova-Mayerova, S. Some Topological and Algebraic Properties of alpha-level Subsets' Topology of a Fuzzy Subset. *Analele St. Univ. Ovidius Constanta* **2018**, *26*, 213–227. [CrossRef]
16. Onasanya, B.O.; Hoskova-Mayerova, S. Multi-fuzzy group induced by multisets. *Ital. J. Pure Appl. Math.* **2019**, *41*, 597–604.
17. Al Tahan, M.; Hoskova-Mayerova, S.; Davvaz, B. Fuzzy multi-polygroups. *J. Intell. Fuzzy Syst.* **2019**. [CrossRef]
18. Al Tahan, M.; Hoskova-Mayerova, S.; Davvaz, B. Some results on (generalized) fuzzy multi-H_v-ideals of H_v-rings. *Symmetry* **2019**, *11*, 1376. [CrossRef]
19. Davvaz, B. Fuzzy H_v-groups. *Fuzzy Sets Syst.* **1999**, *101*, 191–195. [CrossRef]
20. Cristea, I. Several aspects on the hypergroups associated with n-ary relations. *Analele Stiint. Univ. Ovidius Constanta Ser. Mat.* **2009**, *17*, 99–110.
21. Yaqoob, N.; Cristea, I.; Gulistan, M. Left almost polygroups. *Ital. J. Pure Appl. Math.* **2018**, *39*, 465–474.
22. Jena, S.P.; Ghosh, S.K.; Tripathi, B.K. On theory of bags and lists. *Inform. Sci.* **2011**, *132*, 241–254. [CrossRef]
23. Shinoj, T.K.; John, S.J. Intutionistic fuzzy multisets. *Int. J. Eng. Sci. Innov. Technol. (IJESIT)* **2013**, *2*, 1–24.
24. Syropoulos, A. *Mathematics of Multisets, Multiset Processing*; Springer: Berlin, Germany, 2001; pp. 347–358.
25. Shinoj, T.K.; Baby, A.; John, S.J. On some algebraic structures of fuzzy multisets. *Ann. Fuzzy Math. Inform.* **2015**, *9*, 77–90.
26. Davvaz, B. *Polygroup Theory and Related Systems*; World Scientific Publishing Co. Pte. Ltd.: Hackensack, NJ, USA, 2013.
27. Davvaz, B. *Semihypergroup Theory*; Elsevier/Academic Press: London, UK, 2016.
28. Leoreanu-Fotea, V.; Rosenberg, I.; Davvaz, B.; Vougiouklis, T. A new class of n-ary hyperoperations. *Eur. J. Comb.* **2015**, *44*, 265–273. [CrossRef]
29. Cristea, I. Fuzzy Subhypergroups Degree. *J. Mult. Valued Log. Soft Comput.* **2016**, *27*, 75–88.
30. Cristea, I.; Hoskova, S. Fuzzy pseudotopological hypergroupoids. *Iran. J. Fuzzy Syst.* **2009**, *6*, 11–19.

31. Vougiouklis, T. On the hyperstructure theory. *Southeast Asian Bull. Math.* **2016**, *40*, 603–620.
32. De Salvo, M.; Fasino, D.; Freni, D.; Faro, G.L. On hypergroups with a β-class of finite height. *Symmetry* **2020**, *12*, 168. [CrossRef]

© 2020 by the authors. Licensee MDPI, Basel, Switzerland. This article is an open access article distributed under the terms and conditions of the Creative Commons Attribution (CC BY) license (http://creativecommons.org/licenses/by/4.0/).

Article

Sequences of Groups, Hypergroups and Automata of Linear Ordinary Differential Operators

Jan Chvalina [1], Michal Novák [1,*], Bedřich Smetana [2] and David Staněk [1]

1. Department of Mathematics, Faculty of Electrical Engineeering and Communication, Brno University of Technology, Technická 8, 616 00 Brno, Czech Republic; chvalina@feec.vutbr.cz (J.C.); xstanek41@stud.feec.vutbr.cz (D.S.)
2. Department of Quantitative Methods, University of Defence in Brno, Kounicova 65, 662 10 Brno, Czech Republic; bedrich.smetana@unob.cz
* Correspondence: novakm@feec.vutbr.cz; Tel.: +420-541146077

Abstract: The main objective of our paper is to focus on the study of sequences (finite or countable) of groups and hypergroups of linear differential operators of decreasing orders. By using a suitable ordering or preordering of groups linear differential operators we construct hypercompositional structures of linear differential operators. Moreover, we construct actions of groups of differential operators on rings of polynomials of one real variable including diagrams of actions–considered as special automata. Finally, we obtain sequences of hypergroups and automata. The examples, we choose to explain our theoretical results with, fall within the theory of artificial neurons and infinite cyclic groups.

Keywords: hyperstructure theory; linear differential operators; ODE; automata theory

Citation: Chvalina, J.; Novák, M.; Smetana, B.; Staněk, D. Sequences of Groups, Hypergroups and Automata of Linear Ordinary Differential Operators. *Mathematics* **2021**, *9*, 319. https://doi.org/10.3390/math9040319

Academic Editor: Christos G. Massouros
Received: 29 December 2020
Accepted: 2 February 2021
Published: 5 February 2021

Publisher's Note: MDPI stays neutral with regard to jurisdictional claims in published maps and institutional affiliations.

Copyright: © 2021 by the authors. Licensee MDPI, Basel, Switzerland. This article is an open access article distributed under the terms and conditions of the Creative Commons Attribution (CC BY) license (https://creativecommons.org/licenses/by/4.0/).

1. Introduction

This paper discusses sequences of groups, hypergroups and automata of linear differential operators. It is based on the algebraic approach to the study of linear ordinary differential equations. Its roots lie in the work of Otakar Borůvka, a Czech mathematician, who tied the algebraic, geometrical and topological approaches, and his successor, František Neuman, who advocated the algebraic approach in his book [1]. Both of them (and their students) used the classical group theory in their considerations. In several papers, published mainly as conference proceedings such as [2–4], the existing theory was extended by the use of hypercompositional structures in place of the usual algebraic structures. The use of hypercompositional generalizations has been tested in the automata theory, where it has brought several interesting results; see, e.g., [5–8]. Naturally, this approach is not the only possible one. For another possible approach, investigations of differential operators by means of orthognal polynomials, see, e.g., [9,10].

Therefore, in this present paper we continue in the direction of [2,4] presenting results parallel to [11]. Our constructions, no matter how theoretical they may seem, are motivated by various practical issues of signal processing [12–16]. We construct sequences of groups and hypergroups of linear differential operators. This is because, in signal processing (but also in other real-life contexts), two or more connecting systems create a standing higher system, characteristics of which can be determined using characteristics of the original systems. Cascade (serial) and parallel connecting of systems of signal transfers are used in this. Moreover, series of groups motivated by the Galois theory of solvability of algebraic equations and the modern theory of extensions of fields, are often discussed in literature. Notice also paper [11] where the theory of artificial neurons, used further on in some examples, has been studied.

Another motivation for the study of sequences of hypergroups and their homomorphisms can be traced to ideas of classical homological algebra which comes from the algebraic description of topological spaces. A homological algebra assigns to any topological space a family of abelian groups and to any continuous mapping of topological spaces a family of group homomorphisms. This allows us to express properties of spaces and their mappings (morphisms) by means of properties of the groups or modules or their homomorphisms. Notice that a substantial part of homology theory is devoted to the study of exact short and long sequences of the above mentiones structures.

2. Sequences of Groups and Hypergroups: Definitions and Theorems

2.1. Notation and Preliminaries

It is crucial that one understands the notation used in this paper. Recall that we study, by means of algebra, linear ordinary differential equations. Therefore, our notation, which follows the original model of Borůvka and Neuman [1], uses a mix of algebraic and functional notation.

First, we denote intervals by J and regard open intervals (bounded or unbounded). Systems of functions with continuous derivatives of order k on J are denoted by $\mathbb{C}^k(J)$; for $k = 0$ we write $\mathbb{C}(J)$ instead of $\mathbb{C}^0(J)$. We treat $\mathbb{C}^k(J)$ as a ring with respect to the usual addition and multiplication of functions. We denote by δ_{ij} the Kronecker delta, $i, j \in \mathbb{N}$, i.e., $\delta_{ii} = \delta_{jj} = 1$ and $\delta_{ij} = 0$, whenever $i \neq j$; by $\overline{\delta_{ij}}$ we mean $1 - \delta_{ij}$. Since we will be using some notions from the theory of hypercompositional structures, recall that by $\mathcal{P}(X)$ one means the power set of X while $(P)^*(X)$ means $\mathcal{P}(X) \setminus \emptyset$.

We regard linear homogeneous differential equations of order $n \geq 2$ with coefficients, which are real and continuous on J, and–for convenience reasons–such that $p_0(x) > 0$ for all $x \in J$, i.e., equations

$$y^{(n)}(x) + p_{n-1}(x)y^{(n-1)}(x) + \cdots + p_0(x)y(x) = 0. \tag{1}$$

By $\mathbb{A}_n$ we, adopting the notation of Neuman [1], mean the set of all such equations.

Example 1. *The above notation can be explained on an example taken from [17], in which Neuman considers the third-order linear homogeneous differential equation*

$$y'''(x) - \frac{q_1'(x)}{q_1(-x)}y''(x) + (q_1(x) - 1)^2 y'(x) - \frac{q_1'(x)}{q_1(x)}y(x) = 0$$

on the open interval $J \in \mathbb{R}$. One obtains this equation from the system

$$\begin{aligned} y_1' &= y_2 \\ y_2' &= -y_1 + q_1(x)y_3 \\ y_3' &= -q_1(x)y_2 \end{aligned}$$

Here $q_1 \in C^+(J)$ satisfies the condition $q_1(x) \neq 0$ on J. In the above differential equation we have $n = 3$, $p_0(x) = -\frac{q_1'(x)}{q_1(x)}$, $p_1(x) = (q_1(x) - 1)^2$ and $p_2(x) = -\frac{q_1'(x)}{q_1(-x)}$. It is to be noted that the above three equations form what is known as set of global canonical forms for the third-order equation on the interval J.

Denote $L_n(p_{n-1}, \ldots, p_0) : \mathbb{C}^n(J) \to \mathbb{C}^n(J)$ the above linear differential operator defined by

$$L_n(p_{n-1}, \ldots, p_0)y(x) = y^{(n)}(x) + \sum_{k=0}^{n-1} p_k(x)y^{(k)}(x), \tag{2}$$

where $y(x) \in \mathbb{C}^n(J)$ and $p_0(x) > 0$ for all $x \in J$. Further, denote by $\mathbb{LA}_n(J)$ the set of all such operators, i.e.,

$$\mathbb{LA}_n(J) = \{L(p_{n-1}, \ldots, p_0) \mid p_k(x) \in \mathbb{C}(J), p_0(x) > 0\}. \tag{3}$$

By $\mathbb{LA}_n(J)_m$ we mean subsets of $\mathbb{LA}_n(J)$ such that $p_m \in \mathbb{C}_+(J)$, i.e., there is $p_m(x) > 0$ for all $x \in J$. If we want to explicitly emphasize the variable, we write $y(x)$, $p_k(x)$, etc. However, if there is no specific need to do this, we write y, p_k, etc. Using vector notation $\vec{p}(x) = (p_{n-1}(x), \ldots, p_0(x))$, we can write

$$L_n(\vec{p})y = y^{(n)} + \sum_{k=0}^{n-1} p_k y^{(k)}. \tag{4}$$

Writing $L(\vec{p}) \in \mathbb{LA}_n(J)$ (or $L(\vec{p}) \in \mathbb{LA}_n(J)_m$) is a shortcut for writing $L_n(\vec{p})y \in \mathbb{LA}_n(J)$ (or, $L_n(\vec{p})y \in \mathbb{LA}_n(J)_m$).

On the sets of linear differential operators, i.e., on sets $\mathbb{LA}_n(J)$, or their subsets $\mathbb{LA}_n(J)_m$, we define some binary operations, hyperoperations or binary relations. This is possible because our considerations happen within a ring (of functions).

For an arbitrary pair of operators $L(\vec{p}), L(\vec{q}) \in \mathbb{LA}_n(J)_m$, where $\vec{p} = (p_{n-1}, \ldots, p_0)$, $\vec{q} = (q_{n-1}, \ldots, q_0)$, we define an operation "$\circ_m$" with respect to the m-th component by $L(\vec{p}) \circ_m L(\vec{q}) = L(\vec{u})$, where $\vec{u} = (u_{n-1}, \ldots, u_0)$ and

$$u_k(x) = p_m(x) q_k(x) + (1 - \delta_{km}) p_k(x) \tag{5}$$

for all $k = n-1, \ldots, 0$, $k \neq m$ and all $x \in J$. Obviously, such an operation is not commutative.

Moreover, apart from the above binary operation we can define also a relation "$\leq_m$" comparing the operators by their m-th component, putting $L(\vec{p}) \leq_m L(\vec{q})$ whenever, for all $x \in J$, there is

$$p_m(x) = q_m(x) \text{ and at the same time } p_k(x) \leq q_k(x) \tag{6}$$

for all $k = n-1, \ldots, 0$. Obviously, $(\mathbb{LA}_n(J)_m, \leq_m)$ is a partially ordered set.

At this stage, in order to simplify the notation, we write $\mathbb{LA}_n(J)$ instead of $\mathbb{LA}_n(J)_m$ because the lower index m is kept in the operation and relation. The following lemma is proved in [2].

Lemma 1. *Triads $(\mathbb{LA}_n(J), \circ_m, \leq_m)$ are partially ordered (noncommutative) groups.*

Now we can use Lemma 1 to construct a (noncommutative) hypergroup. In order to do this, we will need the following lemma, known as Ends lemma; for details see, e.g., [18–20]. Notice that a join space is a special case of a hypergroup–in this paper we speak of hypergroups because we want to stress the parallel with groups.

Lemma 2. *Let $(H, \cdot, \leq)$ be a partially ordered semigroup. Then $(H, *)$, where $* : H \times H \to \mathcal{H}$ is defined, for all $a, b \in H$ by*

$$a * b = [a \cdot b)_\leq = \{x \in H \mid a \cdot b \leq x\},$$

is a semihypergroup, which is commutative if and only if "$\cdot$" is commutative. Moreover, if $(H, \cdot)$ is a group, then $(H, *)$ is a hypergroup.

Thus, to be more precise, defining

$$\star_m : \mathbb{LA}_n(J) \times \mathbb{LA}_n(J) \to \mathcal{P}(\mathbb{LA}_n(J)), \tag{7}$$

by

$$L(\vec{p}) \star_m L(\vec{q}) = \{L(\vec{u}) \mid L(\vec{p}) \circ_m L(\vec{q}) \leq_m L(\vec{u})\} \tag{8}$$

for all pairs $L(\vec{p}), L(\vec{q}) \in \mathbb{LA}_n(J)_m$, lets us state the following lemma.

Lemma 3. *Triads $(\mathbb{LA}_n(J), \star_m)$ are (noncommutative) hypergroups.*

Notation 1. *Hypergroups* $(\mathbb{L}\mathbb{A}_n(J), \star_m)$ *will be denoted by* $\mathbb{H}\mathbb{L}\mathbb{A}_n(J)_m$ *for an easier distinction.*

Remark 1. *As a parallel to* (2) *and* (3) *we define*

$$\overline{L}(q_n, \ldots, q_0) y(x) = \sum_{k=0}^{n} q_k(x) y^{(k)}(x), q_0 \neq 0, q_k \in \mathbb{C}(J) \tag{9}$$

and

$$\overline{\mathbb{L}\mathbb{A}}_n(J) = \{q_n, \ldots, q_0) \mid q_0 \neq 0, q_k(x) \in \mathbb{C}(J)\} \tag{10}$$

and, by defining the binary operation "$\circ_m$" and "$\leq_m$" in the same way as for $\mathbb{L}\mathbb{A}_n(J)_m$, it is easy to verify that also $(\overline{\mathbb{L}\mathbb{A}}_n(J), \circ_m, \leq_m)$ are noncommutative partially ordered groups. Moreover, given a hyperoperation defined in a way parallel to (8), *we obtain hypergroups* $(\overline{\mathbb{L}\mathbb{A}}_n(J)_m, \star_m)$, *which will be, in line with Notation* 1, *denoted* $\overline{\mathbb{H}\mathbb{L}\mathbb{A}}_n(J)_m$.

2.2. Results

In this subsection we will construct certain mappings between groups or hypergroups of linear differential operators of various orders. The result will have a form of sequences of groups or hypergroups.

Define mappings $F_n : \mathbb{L}\mathbb{A}_n(J) \to \mathbb{L}\mathbb{A}_{n-1}(J)$ by

$$F_n(L(p_{n-1}, \ldots, p_0)) = L(p_{n-2}, \ldots, p_0)$$

and $\phi_n : \mathbb{L}\mathbb{A}_n(J) \to \overline{\mathbb{L}\mathbb{A}}_{n-1}(J)$ by

$$\phi_n : (L(p_{n-1}, \ldots, p_0)) = \overline{L}(p_{n-2}, \ldots, p_0).$$

It can be easily verify that both F_n and ϕ_n are, for an arbitrary $n \geq 2$, group homomorphisms.

Evidently, $\mathbb{L}\mathbb{A}_n(J) \subset \overline{\mathbb{L}\mathbb{A}}_n(J), \overline{\mathbb{L}\mathbb{A}}_{n-1}(J) \subset \overline{\mathbb{L}\mathbb{A}}_n(J)$ for all admissible $n \in \mathbb{N}$. Thus we obtain two complete sequences of ordinary linear differential operators with linking homomorphisms F_n and ϕ_n :

$$\begin{array}{c}
\overline{\mathbb{L}\mathbb{A}}_0(J) \xrightarrow{\overline{id}_{0,1}} \overline{\mathbb{L}\mathbb{A}}_1(J) \xrightarrow{\overline{id}_{1,2}} \overline{\mathbb{L}\mathbb{A}}_2(J) \xrightarrow{\overline{id}_{2,3}} \cdots \\
\uparrow id_0 \quad \phi_1 \quad \uparrow id_1 \quad \phi_2 \quad \uparrow id_2 \quad \phi_3 \\
\mathbb{L}\mathbb{A}_0(J) \xleftarrow{F_1} \mathbb{L}\mathbb{A}_1(J) \xleftarrow{F_2} \mathbb{L}\mathbb{A}_2(J) \xleftarrow{F_3} \cdots
\end{array} \tag{11}$$

$$\begin{array}{c}
\cdots \overline{\mathbb{L}\mathbb{A}}_{n-2}(J) \xrightarrow{\overline{id}_{n-2,n-1}} \overline{\mathbb{L}\mathbb{A}}_{n-1}(J) \xrightarrow{\overline{id}_{n-1,n}} \overline{\mathbb{L}\mathbb{A}}_n(J) \xrightarrow{\overline{id}_{n,n+1}} \cdots \\
\uparrow id_{n-2} \quad \phi_{n-2} \quad \uparrow id_{n-1} \quad \phi_n \quad \uparrow id_n \quad \phi_{n+1} \\
\cdots \mathbb{L}\mathbb{A}_{n-2}(J) \xleftarrow{F_{n-1}} \mathbb{L}\mathbb{A}_{n-1}(J) \xleftarrow{F_n} \mathbb{L}\mathbb{A}_n(J) \xleftarrow{F_{n+1}} \cdots
\end{array}$$

where $\overline{id}_{k,k+1}, id_k$ are corresponding inclusion embeddings.

Notice that this diagram, presented at the level of groups, can be lifted to the level of hypergroups. In order to do this, one can use Lemma 3 and Remark 1. However, this is not enough. Yet, as Lemma 4 suggests, it is possible to show that the below presented assignment is functorial, i.e., not only objects are mapped onto objects but also morphisms (isotone group homomorphisms) are mapped onto morphisms (hypergroup homomorphisms). Notice that Lemma 4 was originally proved in [4]. However, given the minimal impact of the proceedings and its very limited availability and accessibility, we include it here with a complete proof.

Lemma 4. *Let $(G_k, \cdot_k, \leq_k), k = 1, 2$ be preordered groups and $f : (G_1, \cdot_1, \leq_1) \to (G_2, \cdot_2, \leq_2)$ a group homomorphism, which is isotone, i.e., the mapping $f : (G_1, \leq_1) \to (G_2, \leq_2)$ is order-preserving. Let $(H_k, *_k), k = 1, 2$ be hypergroups constructed from $(G_k, \cdot_k, \leq_k), k = 1, 2$ by Lemma 2, respectively. Then $f : (H_1, *_1) \to (H_2, *_2)$ is a homomorphism, i.e., $f(a *_1 b) \subseteq f(a) *_2 f(b)$ for any pair of elements $a, b \in H_1$.*

Proof. Let $a, b \in H_1$ be a pair of elements and $c \in f(a *_1 b)$ be an arbitrary element. Then there is $d \in a *_1 b = [a \cdot_1 b)_{\leq_1}$, i.e., $a \cdot_1 b \leq_1 d$ such that $c = f(d)$. Since the mapping f is an isotone homomorphism, we have $f(a) \cdot_2 f(b) = f(a \cdot_1 b) \leq f(d) = c$, thus $c \in [f(a) \cdot_2 f(b))_{\leq_2}$. Hence

$$f(a *_1 b) = f([a \cdot_1 b)_{\leq_1}) \subseteq [f(a) \cdot_2 f(b))_{\leq} = f(a) *_2 f(b).$$

□

Consider a sequence of partially ordered groups of linear differential operators

$$\mathbb{LA}_0(J) \xleftarrow{F_1} \mathbb{LA}_1(J) \xleftarrow{F_2} \mathbb{LA}_2(J) \xleftarrow{F_3} \ldots$$
$$\ldots \xleftarrow{F_{n-2}} \mathbb{LA}_{n-2}(J) \xleftarrow{F_{n-1}} \mathbb{LA}_{n-1}(J) \xleftarrow{F_n} \mathbb{LA}_n(J) \xleftarrow{F_{n+1}} \mathbb{LA}_{n+1}(J) \leftarrow \ldots$$

given above with their linking group homomorphisms $F_k : \mathbb{LA}_k(J) \to \mathbb{LA}_{k-1}(J)$ for $k = 1, 2, \ldots$. Since mappings $F_n : \mathbb{LA}_n(J) \to \mathbb{LA}_{n-1}(J)$, or rather

$$F_n : (\mathbb{LA}_n(J), \circ_m, \leq_m) \to (\mathbb{LA}_{n-1}(J), \circ_m, \leq_m),$$

for all $n \geq 2$, are group homomorphisms and obviously mappings isotone with respect to the corresponding orderings, we immediately get the following theorem.

Theorem 1. *Suppose $J \subseteq \mathbb{R}$ is an open interval, $n \in \mathbb{N}$ is an integer $n \geq 2, m \in \mathbb{N}$ such that $m \leq n$. Let $(\mathbb{HLA}_n(J)_m, *_m)$ be the hypergroup obtained from the group $(\mathbb{LA}_n(J)_m, \circ_m)$ by Lemma 2. Suppose that $F_n : (\mathbb{LA}_n(J)_m, \circ_m) \to (\mathbb{LA}_{n-1}(J)_m, \circ_m)$ are the above defined surjective group-homomorphisms, $n \in \mathbb{N}, n \geq 2$. Then $F_n : (\mathbb{HLA}_n(J)_m, *_m) \to \mathbb{HLA}_{n-1}(J)_m, *_m)$ are surjective homomorphisms of hypergroups.*

Proof. See the reasoning preceding the theorem. □

Remark 2. *It is easy to see that the second sequence from (11) can be mapped onto the sequence of hypergroups*

$$\mathbb{HLA}_0(J)_m \xleftarrow{F_1} \mathbb{HLA}_1(J)_m \xleftarrow{F_2} \mathbb{HLA}_2(J)_m \xleftarrow{F_3} \ldots$$
$$\ldots \xleftarrow{F_{n-2}} \mathbb{HLA}_{n-1}(J)_m \xleftarrow{F_{n-1}} \mathbb{HLA}_n(J)_m \leftarrow \ldots$$

This mapping is bijective and the linking mappings are surjective homomorphisms F_n. Thus this mapping is functorial.

3. Automata and Related Concepts

3.1. Notation and Preliminaries

The concept of an automaton is mathematical interpretation of diverse real-life systems that work on a discrete time-scale. Various types of automata, called also machines, are applied and used in numerous forms such as money changing devices, various calculating machines, computers, telephone switch boards, selectors or lift switchings and other technical objects. All the above mentioned devices have one aspect in common–states are switched from one to another based on outside influences (such as electrical or mechanical impulses), called inputs. Using the binary operation of concatenation of chains of input symbols one obtains automata with input alphabets in the form of semigroups or a groups. In the case of our paper we work with input sets in the form of hypercompositional structures. When focusing on the structure given by transition function and simultaneously neglecting the output functions and output sets, one reaches a generalization of automata–quasi-automata (or semiautomata); see classical works such as, e.g., [3,18,21–24].

To be more precise, a quasi-automaton is a system (A, S, δ) which consists of a non-void set A, an arbitrary semigroup S and a mapping $\delta : A \times S \to A$ such that

$$\delta(\delta(a, r, s)) = \delta(a, r, s) \qquad (12)$$

for arbitrary $a \in A$ and $r, s \in S$. Notice that the concept of quasi-automaton has been introduced by S. Ginsberg as quasi-machine and was meant to be a generalization of the Mealy-type automaton. Condition (12) is sometimes called Mixed Associativity Condition (MAC). With most authors it is nameless, though.

For further reading on automata theory and its links to the theory of hypercompositional structures (also known as algebraic hyperstructures), see, e.g., [24–26]. Furthermore, for clarification and evolution of terminology, see [8]. For results obtained by means of quasi-multiautomata, see, e.g., [5–8,27].

Definition 1. *Let A be a nonempty set, $(H, \cdot)$ be a semihypergroup and $\delta : A \times H \to A$ a mapping satisfying the condition*

$$\delta(\delta(s, a), b) \in \delta(s, a \cdot b) \qquad (13)$$

for any triad $(s, a, b) \in A \times H \times H$, where $\delta(s, a \cdot b) = \{\delta(s, x); x \in a \cdot b\}$. Then the triad (A, H, δ) is called quasi-multiautomaton with the state set A and the input semihypergroups $(H, \cdot)$. The mapping $\delta : A \times H \to A$ is called the transition function (or the next-state function) of the quasi-multiautomaton (A, H, δ). Condition (13) is called Generalized Mixed Associativity Condition (or GMAC).

In this section, $\mathbb{R}_n[x]$ means, as usually, the ring of polynomials of degree at most n.

3.2. Results

Now, consider linear differential operators $L(m, p_{n-1}, \ldots, p_0) : C^\infty(\mathbb{R}) \to C^\infty(\mathbb{R})$ defined by

$$L(m, p_{n-1}, \ldots, p_0)f = m\frac{d^n f(x)}{dx^n} + \sum_{k=0}^{n-1} p_k(x)\frac{d^k f(x)}{dx^k}. \qquad (14)$$

Denote by $\mathbb{L}_{A1}\mathbb{A}_n(\mathbb{R})$ the additive abelian group of differential operators $L(m, p_{n-1}, \ldots, p_0)$, where for $L(m, p_{n-1}, \ldots, p_0), L(k, q_{n-1}, \ldots, q_0) \in \mathbb{L}_{A1}\mathbb{A}_n(\mathbb{R})$ we define

$$L(m, p_{n-1}, \ldots, p_0) + L(k, q_{n-1}, \ldots, q_0) = L(m+k, p_{n-1}+q_{n-1}, \ldots, p_0+q_0), \qquad (15)$$

where

$$L(m+k, p_{n-1}+q_{n-1}, \ldots, p_0+q_0)f = (m+k)\frac{d^n f(x)}{dx^n} + \sum_{k=0}^{n-1}(p_k(x)+q_k(x))\frac{d^k f(x)}{dx^k}. \qquad (16)$$

Suppose that $p_k \in \mathbb{R}_{n-1}[x]$ and define

$$\delta_n : R_n[x] \times \mathbb{L}_{A1}\mathbb{A}_n(\mathbb{R}) \to \mathbb{R}_n[x] \qquad (17)$$

by

$$\delta_n(f, L(m, p_{n-1}, \ldots, p_0)) = m\frac{d^n f(x)}{dx^n} + f(x) + m + \sum_{k=0}^{n-1} p_k(x), f \in \mathbb{R}_n[x]. \qquad (18)$$

Theorem 2. Let $\mathbb{L}_{A1}\mathbb{A}_n(\mathbb{R})$, $\mathbb{R}_n[x]$ be structures and $\delta_n : \mathbb{R}_n[x] \times \mathbb{L}_{A1}\mathbb{A}_n(\mathbb{R}) \to \mathbb{R}_n[x]$ the mapping defined above. Then the triad $(\mathbb{R}_n[x], \mathbb{L}_{A1}\mathbb{A}_n(\mathbb{R}), \delta_n)$ is a quasi-automaton, i.e., an action of the group $\mathbb{L}_{A1}\mathbb{A}_n(\mathbb{R})$ on the group $\mathbb{R}_n[x]$.

Proof. We are going to verify the mixed associativity condition (MAC) which should satisfy the above defined action:

Suppose $f \in \mathbb{R}_n[x]$, $f(x) = \sum_{k=0}^{n} a_k x^k$, $L(m, p_{n-1}, \ldots, p_0)$, $L(k, q_{n-1}, \ldots, q_0) \in \mathbb{L}_{A1}\mathbb{A}_n(\mathbb{R})$. Then

$$\delta_n(\delta_n(f, L(m, p_{n-1}, \ldots, p_0)), L(k, q_{n-1}, \ldots, q_0)) =$$
$$= \delta_n\left(m\frac{d^n f(x)}{dx^n} + f(x) + m + \sum_{k=0}^{n-1} p_k(x), L(k, q_{n-1}, \ldots, q_0) \right) =$$
$$= \delta_n\left(m \cdot n! \cdot a_n + m + f(x) + \sum_{k=0}^{n-1} p_k(x), L(k, q_{n-1}, \ldots, q_0) \right) =$$
$$= k\frac{d^n f(x)}{dx^n} + m \cdot n! \cdot a_n + m + f(x) + \sum_{k=0}^{n-1} p_k(x) + \sum_{k=0}^{n-1} q_k(x) + k =$$
$$= (m+k)n! \cdot a_n + (m+k) + f(x) + \sum_{k=0}^{n-1}(p_k(x) + q_k(x)) =$$
$$= (m+k)(n! \cdot a_n + 1) + f(x) + \sum_{k=0}^{n-1}(p_k(x) + q_k(x)) =$$
$$= (m+k)\frac{d^n f(x)}{dx^n} + f(x) + (m+k) + \sum_{k=0}^{n-1}(p_k(x) + q_k(x)) =$$
$$= \delta_n(f, L(m+k, p_{n-1} + q_{n-1}, \ldots, p_0 + q_0)) =$$
$$= \delta_n(f, L(m, p_{n-1}, \ldots, p_0) + L(k, q_{n-1}, \ldots, q_0)), \tag{19}$$

thus the mixed associativity condition is satisfied. □

Since $\mathbb{R}_n[x], \mathbb{L}_{A1}\mathbb{A}_n(\mathbb{R})$ are endowed with naturally defined orderings, Lemma 2 can be straightforwardly applied to construct semihypergroups from them.

Indeed, for a pair of polynomials $f, g \in \mathbb{R}_n[x]$ we put $f \leq g$, whenever $f(x) \leq g(x), z \in \mathbb{R}_n[x]$. In such a case $(\mathbb{R}_n[x], \leq)$ is a partially ordered abelian group. Now we define a binary hyperoperation

$$\# : \mathbb{R}_n[x] \times \mathbb{R}_n[x] \to \mathcal{P}^\star(\mathbb{R}_n[x]) \tag{20}$$

by

$$f \# g = \{h; h \in \mathbb{R}_n[x], f(x) + g(x) \leq h(x), x \in \mathbb{R}\} = [f + g]_\leq. \tag{21}$$

By Lemma 2 we have that $(\mathbb{R}_n[x], \#)$ is a hypergroup.
Moreover, defining

$$\# : \mathbb{L}_{A1}\mathbb{A}_n(\mathbb{R}) \times \mathbb{L}_{A1}\mathbb{A}_n(\mathbb{R}) \to \mathcal{P}^\star(\mathbb{L}_{A1}\mathbb{A}_n(\mathbb{R})) \tag{22}$$

by $L(m, \overrightarrow{p(x)}) \# L(k, \overrightarrow{q(x)}) = \left[L(m, \overrightarrow{p(x)}) + L(k, \overrightarrow{q(x)}) \right]_\leq = \left[L(m+k, \overrightarrow{p(x)} + \overrightarrow{q(x)}) \right]_\leq = \{ L(r, \overrightarrow{u(x)}); m+k \leq r, \overrightarrow{p(x)} + \overrightarrow{q(x)} \leq \overrightarrow{u(x)} \}$, which means

$$p_j(x) + q_j(x) \leq u_j(x),$$

where $j = 0, 1, \ldots, n-1$, we obtain, again by Lemma 2 that the hypergroupoid $(\mathbb{L}_{A1}\mathbb{A}_n(\mathbb{R}), \#)$ is a commutative semihypergroup.

Finally, define a mapping

$$\sigma_n : \mathbb{L}_{A1}\mathbb{A}_n(\mathbb{R}) \times \mathbb{R}_n[x] \to \mathbb{R}_n[x] \quad (23)$$

by

$$\sigma_n(L(m, p_{n-1}, \ldots, p_0, f)) = L(m, p_0 \circ f + p_{n-1}, \ldots, p_0 \circ f + p_1, p_0). \quad (24)$$

Below, in the proof of Theorem 3, we show that the mapping satisfies the GMAC condition.

This allows us to construct a quasi-multiautomaton.

Theorem 3. *Suppose* $(\mathbb{L}_{A_1}\mathbb{A}_n(\mathbb{R}), \#), (\mathbb{R}_n[x], \#)$ *are hypergroups constructed above and* $\sigma_n : \mathbb{L}_{A1}\mathbb{A}_n(\mathbb{R}) \times \mathbb{R}_n[x] \to \mathbb{R}_n[x]$ *is the above defined mapping. Then the structure*

$$((\mathbb{L}_{A1}\mathbb{A}_n(\mathbb{R}), \#), (\mathbb{R}_n[x], \#), \sigma_n)$$

is a quasi-multiautomaton.

Proof. Suppose $L(m, \vec{p}) \in \mathbb{L}_{A1}\mathbb{A}_n(\mathbb{R}), f, g \in \mathbb{R}_n[x]$. Then

$$\sigma_n(\sigma_n(L(m, \vec{p}), f), g) = \sigma_n(L(m, p \circ f + p_{n-1}, \ldots, p \circ f + p_1, p_0), g) =$$
$$= L(m, p \circ g + p \circ f + p_{n-1}, \ldots, p \circ g + p \circ f + p_1, p_0) =$$
$$= L(m, p \circ (g + f) + p_{n-1}, \ldots p \circ (g + f) + p_1, p_0) \in$$
$$\in \{\sigma_n(L(m, p \circ h + p_{n-1}, \ldots p \circ h + p_1, p_0); f, g, h \in \mathbb{R}_n[x], f + g \leq h\} =$$
$$= \sigma_n(L, m, p_{n-1}, \ldots, p_1, p_0), [f + g]_\leq) = \sigma_n(L(m, \vec{p}), f\#g), \quad (25)$$

hence the GMAC condition is satisfied. □

Now let us discuss actions on objects of different dimensions. Recall that a homomorphism of automaton (S, G, δ_S) into the automaton (T, H, δ_T) is a mapping $F = \phi \times \psi : S \times G \to T \times H$ such that $\phi : S \to T$ is a mapping and $\psi : G \to H$ is a homomorphism (of semigroups or groups) such that for any pair $[s, g] \in S \times G$ we have

$$\phi(\delta_S(s, g)) = \delta_T(\phi(s), \psi(g)), \text{ i.e., } \phi \circ \delta_S = \delta_T \circ (\phi \times \psi). \quad (26)$$

In order to define homomorphisms of our considered actions and especially in order to construct a sequence of quasi-automata with decreasing dimensions of the corresponding objects, we need a different construction of a quasiautomaton.

If $f \in \mathbb{R}_n[x], f(x) = \sum_{k=0}^n a_k x^k$ and $L(m, \vec{p}) \in \mathbb{L}_{A1}\mathbb{A}_n(\mathbb{R})$, we define

$$\tau_n(L(m, p_{n-1}, \ldots, p_0), f) = L(m, a_n + p_{n-1}, \ldots, a_1 + p_0 + a_0). \quad (27)$$

Now, if $g \in \mathbb{R}_n[x], g(x) = \sum_{k=0}^n b_k x^k$, we have

$$\tau_n(\tau_n(L(m, p_{n-1}, \ldots, p_0), f), g) =$$
$$= \tau_n(L(m, a_n + p_{n-1}, \ldots, a_1 + p_1 + a_0), g) =$$
$$= L(m, a_n + b_n + p_{n-1}, \ldots, a_1 + b_1 + p_0 + a_0 + b_0) =$$
$$= \tau_n\left(L(m, p_{n-1}, \ldots, p_0), \sum_{k=0}^n (a_k + b_k)x\right) =$$
$$= \tau_n(L(m, p_{n-1}, \ldots, p_0), f + g). \quad (28)$$

Hence $\tau_n : \mathbb{L}_{A1}\mathbb{A}_n(\mathbb{R}) \times \mathbb{R}_n[x] \to \mathbb{L}_{A1}\mathbb{A}_n(\mathbb{R})$ is the transition function (satisfying MAC) of the automaton $\mathcal{A} = (\mathbb{L}_{A1}\mathbb{A}_n(\mathbb{R}), \mathbb{R}_{n-1}[x], \tau_n)$.

Consider now two automata–$\mathcal{A}_{n-1} = (\mathbb{L}_{A1}\mathbb{A}_{n-1}(\mathbb{R}), \mathbb{R}_{n-1}[x], \tau_{n-1})$ and the above one. Define mappings

$$\phi_n : \mathbb{L}_{A1}\mathbb{A}_n(\mathbb{R}) \to \mathbb{L}_{A1}\mathbb{A}_{n-1}(\mathbb{R}), \quad \psi_n : \mathbb{R}_n[x] \to \mathbb{R}_{n-1}[x] \tag{29}$$

in the following way: For $L(m, p_{n-1}, \ldots, p_0) \in \mathbb{L}_{A1}\mathbb{A}_n(\mathbb{R})$ put

$$\phi_n(L(m, p_{n-1}, \ldots, p_0)) = L(m, p_{n-2}, \ldots, p_0) \in \mathbb{L}_{A1}\mathbb{A}_{n-1}(\mathbb{R}) \tag{30}$$

and for $f \in \mathbb{R}_n[x]$, $f(x) = \sum_{k=0}^{n} a_k x^k$ define

$$\psi_n(f) = \psi_n\left(\sum_{k=0}^{n} a_k x^k\right) = \sum_{k=0}^{n-1} a_k x^k \in \mathbb{R}_{n-1}[x]. \tag{31}$$

Evidently, there is $\psi_n(f+g) = \psi_n(f) + \psi(g)$ for any pair of polynomials $f, g \in \mathbb{R}_n[x]$.

Theorem 4. *Let $\phi_n : \mathbb{L}_{A1}\mathbb{A}_n(\mathbb{R}) \to \mathbb{L}_{A1}\mathbb{A}_{n-1}(\mathbb{R})$, $\psi_n : \mathbb{R}_n[x] \to \mathbb{R}_{n-1}[x]$, $\tau_n : \mathbb{L}_{A1}\mathbb{A}_n(\mathbb{R}) \times \mathbb{R}_n[x] \to \mathbb{L}_{A1}\mathbb{A}_n(\mathbb{R})$, $n \in \mathbb{N}, n \geq 2$, be mappings defined above. Define $F_n : \mathcal{A}_n \to \mathcal{A}_{n-1}$ as mapping*

$$F_n = \phi_n \times \psi_n : \mathbb{L}_{A1}\mathbb{A}_n(\mathbb{R}) \times \mathbb{R}_n[x] \to \mathbb{L}_{A1}\mathbb{A}_{n-1}(\mathbb{R}) \times \mathbb{R}_{n-1}[x].$$

Then the following diagram

$$\begin{array}{ccc}
\mathbb{L}_{A1}\mathbb{A}_n(\mathbb{R}) \times \mathbb{R}_n[x] & \xrightarrow{\tau_n} & \mathbb{L}_{A1}\mathbb{A}_n(\mathbb{R}) \\
\phi_n \times \psi_n \downarrow & & \downarrow \phi_n \\
\mathbb{L}_{A1}\mathbb{A}_{n-1}(\mathbb{R}) \times \mathbb{R}_{n-1}[x] & \xrightarrow{\tau_{n-1}} & \mathbb{L}_{A1}\mathbb{A}_{n-1}(\mathbb{R})
\end{array} \tag{32}$$

is commutative, thus the mapping $F_n = \phi_n \times \psi_n$ is a homomorphism of the automaton $\mathcal{A}_n = (\mathbb{L}_{A1}\mathbb{A}_n(\mathbb{R}), \mathbb{R}_n[x], \tau_n)$ into the automaton $\mathcal{A}_{n-1} = (\mathbb{L}_{A1}\mathbb{A}_{n-1}(\mathbb{R}), \mathbb{R}_{n-1}[x], \tau_{n-1})$.

Proof. Let $[L(m, \vec{p}), f] \in \mathbb{L}_{A1}\mathbb{A}_n(\mathbb{R}) \times \mathbb{R}_n[z]$, $f(x) = \sum_{k=0}^{n} a_k x^k$. Then

$$(\phi_n \circ \tau_n)(L(m, \vec{p}), f) = \phi_n\left(\tau_n\left(L(m, p_{n-1}, \ldots, p_0), \sum_{k=0}^{n} a_k x^k\right)\right) =$$

$$= \phi_n((m, a_n + p_{n-1}, \ldots, a_1 + p_0 + a_0)) = L(m, a_{n-1} + p_{n-2}, \ldots, a_1 + p_0 + a_0) =$$

$$= \tau_{n-1}\left(L(m, p_{n-2}, \ldots, p_0), \sum_{k=0}^{n-1} a_k x^k\right) =$$

$$= \tau_{n-1}\left((\phi_n \times \psi_n)\left(L(m, p_{n-1}, \ldots, p_0), \sum_{k=0}^{n} a_k x^k\right)\right) =$$

$$= (\tau_{n-1} \circ (\phi_n \times \phi_n)))(L(m, \vec{p}), f), \tag{33}$$

Thus the diagram (32) is commutative. □

Using the above defined homomorphism of automata we obtain the sequence of automata with linking homomorphisms $F_k : \mathcal{A}_k \to \mathcal{A}_{k-1}, k \in \mathbb{N}, k \geq 2$:

$$(\mathbb{L}_{A1}\mathbb{A}_1(\mathbb{R}), \tau_1) \xleftarrow{F_2} (\mathbb{L}_{A1}\mathbb{A}_2(\mathbb{R}), \tau_2) \xleftarrow{F_3} \ldots \xleftarrow{F_{n-2}} (\mathbb{L}_{A1}\mathbb{A}_{n-2}(\mathbb{R}), \mathbb{R}_{n-2}[x], \tau_{n-2}) \xleftarrow{F_{n-1}} \ldots \xleftarrow{F_{n-1}} (\mathbb{L}_{A1}\mathbb{A}_{n-1}(\mathbb{R}), \mathbb{R}_{n-1}[x], \tau_{n-1}) \xleftarrow{F_n} (\mathbb{L}_{A1}\mathbb{A}_n(\mathbb{R}), \mathbb{R}_n[x], \tau_n) \tag{34}$$

Here, for $L(m, P-1, P-0) \in \mathbb{L}_{A1}\mathbb{A}_2(\mathbb{R})$ we have

$$L(m, p_1, p_0)y(x) = m\frac{d^2y(x)}{dx^2} + p_1(x)\frac{dy(x)}{dx} + p_0(x)y(x) \tag{35}$$

for any $y(x) \in \mathbb{C}^2(\mathbb{R})$ and any $L(m, p_0) \in \mathbb{L}_{A1}\mathbb{A}_1(\mathbb{R})$ it holds $L(m, p_0)y(x) = m\frac{dy(x)}{dx} + p_0(x)y(x)$.

The obtained sequence of automata can be transformed into a countable sequence of quasi-multiautomata. We already know that the transition function

$$\sigma_n : \mathbb{L}_{A1}\mathbb{A}_n(\mathbb{R}) \times \mathbb{R}_n[x] \to \mathbb{L}_{A1}\mathbb{A}_n(\mathbb{R})$$

satisfies GMAC. Further, suppose $f, g \in \mathbb{R}_n[x]$, $f(x) = \sum_{k=0}^m a_k x^k$, $g(x) = \sum_{k=0}^m b_k x^k$. Then

$$\psi_n(f\#g) = \psi_n(\{h; h \in \mathbb{R}_n[x], f + g \leqq h\}) =$$
$$= \{\psi_n(h); h \in \mathbb{R}_n[x], f + g \leqq h\} =$$
$$= \{u; u \in \mathbb{R}_{n-1}[x], \psi_n(f) + \psi_n(g) \leqq u\} =$$
$$= \psi_n(f)\#\psi_n(g);$$

here $\mathrm{grad}\,\psi_n(h) < \mathrm{grad}\,h$ for any polynomial $h \in f\#g$. Thus the mapping $\psi_n : (\mathbb{R}_n[x], \#) \to (\mathbb{R}_{n-1}[x], \#)$ is a good homomorphism of corresponding hypergroups.

Now we are going to construct a sequence of automata with increasing dimensions, i.e., in a certain sense sequence "dual" to the previous sequence. First of all, we need a certain "reduction" member to the definition of a transition function

$$\lambda_n^* : \mathbb{L}_{A1}\mathbb{A}_n(\mathbb{R}) \times \mathbb{R}_n[x] \to \mathbb{L}_{A1}\mathbb{A}_n(\mathbb{R}), \tag{36}$$

namely $\mathrm{red}_{n-1} : \mathbb{R}_r[x] \to \mathbb{R}_{n-1}[x]$ whenever $r > n-1$. In detail, if $f(x) = \sum_{k=0}^r a_k x^k \in \mathbb{R}_r[x]$, then

$$\mathrm{red}_{n-1}(f) = \mathrm{red}_{n-1}\left(\sum_{k=0}^r a_k x^k\right) = \mathrm{red}_{n-1}\left(\sum_{k=n}^r a_k x^k + \sum_{k=0}^{n-1} a_k x^{k-1}\right) =$$
$$= \sum_{k=0}^{n-1} a_k x^{k-1} \in \mathbb{R}_{n-1}[x]. \tag{37}$$

Further, $L(m, p_{n-2}, \ldots, p_0)_n \in \mathbb{L}_{A1}\mathbb{A}_n(\mathbb{R})$ is acting by

$$L(m, p_{n-2}, \ldots, p_0)_n y(x) = m\frac{d^n y(x)}{dx^n} + \sum_{k=0}^{n-2} p_k(x)\frac{d^k y(x)}{dx^k} \tag{38}$$

whereas $L(m, p_{n-2}, \ldots, p_0)_{n-1} \in \mathbb{L}_{A1}\mathbb{A}_{n-1}(\mathbb{R})$, i.e.,

$$L(m, p_{n-2}, \ldots, p_0)_{n-1} y(x) = m\frac{d^{n-1} y(x)}{dx^{n-1}} + \sum_{k=0}^{n-2} p_k(x)\frac{d^k y(x)}{dx^k}. \tag{39}$$

Then for any pair $(L(m, p_{n-1}, \ldots, p_0), f) \in \mathbb{L}_{A1}\mathbb{A}_n(\mathbb{R}) \times \mathbb{R}_n[x]$, where $f(x) = \sum_{k=0}^n a_k x^k$, we obtain

$$\lambda_n = \lambda_n^* \circ (\mathrm{id} \times \mathrm{red}_{n-1}) \tag{40}$$

and

$$(\lambda_n^\star \circ (\mathrm{id} \times \mathrm{red}_{n-1}))(L(m, p_{n-1}, \ldots, p_0, f) = \lambda_n^\star \left(L(m, p_{n-1}, \ldots, p_0), \sum_{k=0}^{n-1} a_k x^k \right) =$$
$$= L(m, p_{n-1} + a_{n-1}, \ldots, p_0 + a_0) \in \mathbb{L}_{A1}\mathbb{A}_n(\mathbb{R}), \tag{41}$$

thus the mapping $\lambda_n : \mathbb{L}_{A1}\mathbb{A}_n(\mathbb{R}) \times \mathbb{R}_n[x] \to \mathbb{L}_{A1}\mathbb{A}_n(\mathbb{R})$ is well defined. We should verify validity of MAC and commutativity of the square diagram determining a homomorphism between automata.

Suppose $L(m, p_{n-1}, \ldots, p_0) \in \mathbb{L}_{A1}\mathbb{A}_n(\mathbb{R})$, $f, g \in \mathbb{R}_n[x]$, $f(x) = \sum_{k=0}^n a_k x^k$, $g(x) = \sum_{k=0}^n b_k x^k$. Then

$$\lambda_n(\lambda_n(L(m, p_{n-1}, \ldots, p_0), f), g) =$$
$$= \lambda_n(L(m, p_{n-1} + + a_{n-1}, \ldots, p_0 + a_0, g) =$$
$$= L(m, p_{n-1} + a_{n-1} + b_{n-1}, \ldots, p_0 + a_0 + b_0) =$$
$$= \lambda_n \left(L(m, p_{n-1}, \ldots, p_0), \sum_{k=0}^n (a_k + b_k) x^k \right) =$$
$$= \lambda_n(L(m, p_{n-1}, \ldots, p_0), f + g), \tag{42}$$

thus MAC is satisfied.

Further, we are going to verify commutativity of this diagram

$$\begin{array}{ccc} \mathbb{L}_{A1}\mathbb{A}_{n-1}(\mathbb{R}) \times \mathbb{R}_{n-1} & \xrightarrow{\lambda_{n-1}} & \mathbb{L}_{A1}\mathbb{A}_{n-1}(\mathbb{R}) \\ {\scriptstyle \xi_{n-1} \times \eta_{n-1}}\downarrow & & \downarrow{\scriptstyle \eta_{n-1}} \\ \mathbb{L}_{A1}\mathbb{A}_n(\mathbb{R}) \times \mathbb{R}_n[x] & \xrightarrow{\lambda_n} & \mathbb{L}_{A1}\mathbb{A}_n(\mathbb{R}) \end{array} \tag{43}$$

where $\xi_{n-1}(L(m, p_{n-2}, \ldots, p_0)_{n-1}) = L(m, p_{n-1}, \ldots, p_0)_n$, i.e.,

$$m\frac{d^{n-1}}{dx^{n-1}} + p_{n-1}(x)\frac{d^{n-2}}{dx^{n-1}} + \cdots + p_0(x)\mathrm{id} \to m\frac{d^n}{dx^n} + p_{n-2}(x)\frac{d^{n-2}}{dx^{n-2}} + \cdots + p_0(x)\mathrm{id} \tag{44}$$

and $\eta_{n-1}\left(\sum_{k=0}^{n-1} a_k x^k\right) = a_{n-1} x^n + \sum_{k=0}^{n-1} a_k x^k$.

Considering $L(m, p_{n-2}, \ldots, p_0)$ and the polynomial $f(x) = \sum_{k=0}^{n-1} a_k x^k$ similarly as above, we have

$$(\eta_{n-1})(L(m, p_{n-2}, \ldots, p_0), f) = \eta_{n-1}(L(m, p_{n-2} + a_{n-2}, \ldots, p_0 + a_0)_{n-1}) =$$
$$= \eta_{n-1}(\lambda_{n-1}(L(m, p_{n-2}, \ldots, p_0), f) = (\eta_{n-1} \circ \lambda_{n-1})(L(m, p_{n-2}, \ldots, p_0), f). \tag{45}$$

Thus the above diagram is commutative. Now, denoting by T_{n-1} the pair of mappings (ξ_{n-1}, η_{n-1}), we obtain that $T_{n-1} : (\mathbb{L}_{A1}\mathbb{A}_{n-1}(\mathbb{R}), \mathbb{R}_{n-1}[x], \lambda_{n-1}) \to (\mathbb{L}_{A1}\mathbb{A}_n(\mathbb{R}), \mathbb{R}_n[x], \lambda_n)$ is a homomorphism of the given automata. Finally, using T_k as connecting homomorphism, we obtain the sequence

$$(\mathbb{L}_{A1}\mathbb{A}_1(\mathbb{R}), \mathbb{R}_1[x], \lambda_1) \xrightarrow{T_1} (\mathbb{L}_{A1}\mathbb{A}_2(\mathbb{R}), \mathbb{R}_2[x], \lambda_2) \xrightarrow{T_2} (\mathbb{L}_{A1}\mathbb{A}_3(\mathbb{R}), \mathbb{R}_3[x], \lambda_3) \xrightarrow{T_3} \ldots$$
$$\ldots \xrightarrow{T_{n-2}} (\mathbb{L}_{A1}\mathbb{A}_{n-1}(\mathbb{R}), \mathbb{R}_{n-1}[x], \lambda_{n-1}) \xrightarrow{T_{n-1}} (\mathbb{L}_{A1}\mathbb{A}_n(\mathbb{R}), \mathbb{R}_n[x], \lambda_n) \xrightarrow{T_n} \ldots \tag{46}$$

4. Practical Applications of the Sequences

In this section, we will include several examples of the above reasoning. We will apply the theoretical results in the area of artificial neurons, i.e., in a way, continue with the paper [11] which focuses on artificial neurons. For notation, recall [11]. Further on we consider a generalization of the usual concept of artificial neurons. We assume that the inputs ux_i and weight w_i are functions of an argument t, which belongs into a linearly ordered (tempus) set T with the least element 0. The index set is, in our case, the set $C(J)$ of all continuous functions defined on an open interval $J \subset \mathbb{R}$. Now, denote by W the set of all non-negative functions $w : T \to \mathbb{R}$. Obviously W is a subsemiring of the ring of all real functions of one real variable $x : \mathbb{R} \to \mathbb{R}$. Further, denote by $Ne(\vec{w}_r) = Ne(w_{r1}, \ldots, w_{rn})$ for $r \in C(J), n \in \mathbb{N}$ the mapping

$$y_r(t) = \sum_{k=1}^{n} w_{r,k}(t) x_{r,k}(t) + b_r$$

which will be called the artificial neuron with the bias $b_r \in \mathbb{R}$. By $\mathbb{AN}(T)$ we denote the collection of all such artificial neurons.

4.1. Cascades of Neurons Determined by Right Translations

Similarly as in the group of linear differential operators we will define a binary operation in the system $\mathbb{AN}(T)$ of artificial neurons $Ne(\cdot)$ and construct a non-commutative group.

Suppose $Ne(\vec{w}_r), Ne(\vec{w}_s) \in \mathbb{AN}(T)$ such that $r, s \in C(J)$ and $\vec{w}_r = (w_{r,1}, \ldots, w_{r,n})$, $\vec{w}_s = (w_{s,1}, \ldots, w_{s,n})$, where $n \in \mathbb{N}$. Let $m \in \mathbb{N}$, $1 \leq m \leq n$ be a such an integer that $w_{r,m} > 0$. We define

$$Ne(\vec{w}_r) \cdot_m Ne(\vec{w}_s) = Ne(\vec{w}_u),$$

where

$$\vec{w}_u = (w_{u,1}, \ldots, w_{u,n}) = (w_{u,1}(t), \ldots, w_{u,n}(t)),$$
$$\vec{w}_{u,k}(t) = w_{r,m}(t) w_{s,k}(t) + (1 - \delta_{m,k}) w_{r,k}(t), t \in T$$

and, of course, the neuron $Ne(\vec{w}_u)$ is defined as the mapping $y_u(t) = \sum_{k=1}^{n} w_k(t) x_k(t) + b_u$, $t \in T$, $b_u = b_r b_s$.

The algebraic structure $(\mathbb{AN}(T), \cdot_m)$ is a non-commutative group. We proceed to the construction of the cascade of neurons. Let $(\mathbb{Z}, +)$ be the additive group of all integers. Let $Ne(\vec{w}_s(t)) \in \mathbb{AN}(T)$ be an arbitrary but fixed chosen artificial neuron with the output function

$$y_s(t) = \sum_{k=1}^{n} w_{s,k}(t) x_{s,k}(t) + b_s.$$

Denote by $\rho_s : \mathbb{AN}(T) \to \mathbb{AN}(T)$ the right translation within the group of time varying neurons determined by $Ne(\vec{w}_s(t))$, i.e.,

$$\rho_s(Ne(\vec{w}_p(t))) = Ne(\vec{w}_p(t)) \cdot_m Ne(\vec{w}_s(t))$$

for any neuron $Ne(\vec{w}_p(t)) \in \mathbb{AN}(T)$. In what follows, denote by ρ_s^r the r-th iteration of ρ_s for $r \in \mathbb{Z}$. Define the projection $\pi_s : \mathbb{AN}(T) \times \mathbb{Z} \to \mathbb{AN}(T)$ by

$$\pi_s(Ne(\vec{w}_p(t)), r) = \rho_s^r(Ne(\vec{w}_p(t))).$$

One easily observes that we get a usual (discrete) transformation group, i.e., the action of $(\mathbb{Z}, +)$ (as the phase group) on the group $\mathbb{AN}(T)$. Thus the following two requirements are satisfied:

1. $\pi_s(Ne(\vec{w}_p(t)), 0) = Ne(\vec{w}_p(t))$ for any neuron $Ne(\vec{w}_p(t)) \in \mathbb{AN}(T)$.
2. $\pi_s(Ne(\vec{w}_p(t)), r + u) = \pi_s(\pi_s(Ne(\vec{w}_p(t)), r), u)$ for any integers $r, u \in \mathbb{Z}$ and any artificial neuron $Ne(\vec{w}_p(t))$. Notice that the just obtained structure is called a cascade within the framework of the dynamical system theory.

4.2. An Additive Group of Differential Neurons

As usually denote by $\mathbb{R}_n[t]$ the ring of polynomials of variable t over $\mathbb{R}$ of the grade at most $n \in \mathbb{N}_0$. Suppose $\vec{w} = (w_1(t), \ldots, w_n(t))$ be the fixed vector of continuous functions

$w_k : \mathbb{R} \to \mathbb{R}$, b_p be the bias for any polynomial $p \in \mathbb{R}_n[t]$. For any such polynomial $p \in \mathbb{R}_n[t]$ we define a differential neuron $DNe(\vec{w})$ given by the action

$$y(t) = \sum_{k=1}^{n} w_k(t) \frac{d^{k-1}p(t)}{dt^{k-1}} + b_0 \frac{d^n p(t)}{dt^n}. \tag{47}$$

Considering the additive group of $\mathbb{R}_n[t]$ we obtain an additive commutative group $\mathbb{DN}(T)$ of differential neurons which is assigned to the group of $\mathbb{R}_n[t]$. Thus for $DNe_p(\vec{w})$, $DNe_q(\vec{w}) \in \mathbb{DN}(T)$ with actions

$$y(t) = \sum_{k=1}^{n} w_k(t) \frac{d^{k-1}p(t)}{dt^{k-1}} + b_0 \frac{d^n p(t)}{dt^n}$$

and

$$z(t) = \sum_{k=1}^{n} w_k(t) \frac{d^{k-1}q(t)}{dt^{k-1}} + b_0 \frac{d^n q(t)}{dt^n}$$

we have $DNe_{p+q}(\vec{w}) = DNe_p(\vec{w}) + DNe_q(\vec{w}) \in \mathbb{DN}(T)$ with the action

$$u(t) = y(t) + z(t) = \sum_{k=1}^{n} w_k(t) \frac{d^{k-1}(p(t)+q(t))}{dt^{k-1}} + b_0 \frac{d^n(p(t)+q(t))}{dt^n}$$

Considering the chain of inclusions

$$\mathbb{R}_n[t] \subset \mathbb{R}_{n+1}[t] \subset \mathbb{R}_{n+2}[t] \dots$$

we obtain the corresponding sequence of commutative groups of differential neurons.

4.3. *A Cyclic Subgroup of the Group* $\mathbb{AN}(T)_m$ *Generated by Neuron* $Ne(\vec{w}_r) \in \mathbb{AN}(T)_m$

First of all recall that if $Ne(\vec{w}_r), Ne(\vec{w}_s) \in \mathbb{AN}(T)_m$, $r,s \in C(J)$, where $\vec{w}_r(t) = (w_{r,1}(t), \dots, w_{r,n}(t))$, $\vec{w}_s(t) = (w_{s,1}(t), \dots, w_{s,n}(t))$, are vector function of weights such that $w_{r,m}(t) \neq 0 \neq w_{s,m}(t)$, $t \in T$ with outputs $y_r(t) = \sum_{k=1}^{n} w_{r,k}(t) x_k(t) + b_r$, $y_s(t) = \sum_{k=1}^{n} w_{s,k}(t) x_k(t) + b_s$ (with inputs $x_k(t)$), then the product $Ne(\vec{w}_r) \cdot_m Ne(\vec{w}_s) = Ne(\vec{w}_u)$ has the vector of weights

$$\vec{w}_u(t) = (w_{u,1}(t), \dots, w_{u,n}(t))$$

with $w_{u,k}(t) = w_{r,m}(t) w_{s,k}(t) + (1 - \delta_{m,k}) w_{r,k}(t), t \in T$.

The binary operation "$\cdot_m$" is defined under the assumption that all values of functions which are m-th components of corresponding vectors of weights are different from zero.

Let us denote by $\mathbb{ZAN}_r(T)$ the cyclic subgroup of the group $\mathbb{AN}(T)_m$ generated by the neuron $Ne(\vec{w}_r) \in \mathbb{AN}(T)_m$. Then denoting the neutral element by $N1(\vec{e})_m$ we have $\mathbb{ZAN}_r(T) =$

$$= \{\dots, [Ne(\vec{w}_r)]^{-2}, [Ne(\vec{w}_r)]^{-1}, N1(\vec{e})_m, Ne(\vec{w}_r), [Ne(\vec{w}_r)]^2, \dots, [Ne(\vec{w}_r)]^p, \dots\}.$$

Now we describe in detail objects

$$[Ne(\vec{w}_r)]^2, [Ne(\vec{w}_r)]^p, p \in \mathbb{N}, p \geq 2, N1(\vec{e})_m \text{ and } [Ne(\vec{w}_r)]^{-1}, \tag{48}$$

i.e., the inverse element to the neuron $Ne(\vec{w}_r)$.

Let us denote $[Ne(\vec{w}_r)]^2 = Ne(\vec{w}_s)$, with $\vec{w}_s(t) = (w_{s,1}(t), \dots, w_{s,n}(t))$. then

$$w_{s,k}(t) = w_{r,m}(t) w_{r,k}(t) + (1 - \delta_{m,k}) w_{r,k}(t) = (w_{r,m}(t) + 1 - \delta_{m,k}) w_{r,k}(t).$$

Then the vector of weights of the neuron $[Ne(\vec{w}_r)]^2$ is

$$\vec{w}_s(t) = ((w_{r,m}(t)+1)w_{r,1}(t), \ldots, w_{r,m}^2(t), \ldots, (w_{r,m}(t)+1)w_{r,n}(t)),$$

the output function is of the form

$$y_s(t) = \sum_{\substack{k=1 \\ k \neq m}}^{n} (w_{r,m}(t)+1)w_{r,k}(t)x_k(t)) + w_{r,m}^2 x_n(t) + b_r^2.$$

It is easy to calculate the vector of weights of the neuron $[Ne(\vec{w}_r)]^3$:

$$((w_{r,m}^2(t)+1)w_{r,1}(t), \ldots, w_{r,m}^3(t), \ldots, (w_{r,m}^2(t)+1)w_{r,n}(t)).$$

Finally, putting $[Ne(\vec{w}_r)]^p = Ne(\vec{w}_v)$ for $p \in \mathbb{N}$, $p \geq 2$, the vector of weights of this neuron is

$$\vec{w}_v(t) = ((w_{r,m}^{p-1}(t)+1)w_{r,1}(t), \ldots, w_{r,m}^p(t), \ldots, (w_{r,m}^{p-1}(t)+1)w_{r,n}(t)).$$

Now, consider the neutral element (the unit) $N1(\vec{e})_m$ of the cyclic group $\mathbb{Z}A\mathbb{N}_r(T)$. Here the vector $\vec{e}$ of weights is $\vec{e} = (e_1, \ldots, e_m, \ldots, e_n)$, where $e_m = 1$ and $e_k = 0$ for each $k \neq m$. Moreover the bias $b = 1$.

We calculate products $Ne(\vec{w}_s) \cdot N1(\vec{e})_m$, $N1(\vec{e})_m) \cdot Ne(\vec{w}_s)$. Denote $Ne(\vec{w}_u)$, $Ne(\vec{w}_v)$ results of corresponding products, respectively–we have $\vec{w}_u(t) = (w_{u,1}(t), \ldots, w_{u,n}(t))$, where

$$w_{u,k}(t) = w_{s,m}(t)e_k(t) + (1-\delta_{m,k})w_{s,k}(t) = w_{s,k}(t)$$

if $k \neq m$ and $w_{u,k}(t) = w_{s,m}(t)(e_m(t) + 0 \cdot w_{s,m}(t)) = w_{s,m}(t)$ for $k = m$. Since the bias is $b = 1$, we obtain $y_u(t) = x_m(t) + 1$. Thus $Ne(\vec{w}_u) = Ne(\vec{w}_s)$. Similarly, denoting $\vec{w}_v(t) = (w_{v,1}(t), \ldots, w_{v,n}(t))$, we obtain $w_{v,k}(t) = e_m(t)w_{s,k}(t) + (1-\delta_{m,k})e_k(t) = w_{s,k}(t)$ for $k \neq m$ and $w_{v,k}(t) = w_{s,m}(t)$ if $k = m$, thus $\vec{w}_v(t) = (w_{s,1}(t), \ldots, w_{s,n}(t))$, consequently $Ne(\vec{w}_u) = Ne(\vec{w}_s)$ again.

Consider the inverse element $[Ne(\vec{w}_r)]^{-1}$ to the element $Ne(\vec{w}_r \in \mathbb{Z}A\mathbb{N}(T)_m$. Denote $Ne(\vec{w}_s) = [Ne(\vec{w}_r)]^{-1}$, $\vec{w}_s(t) = (w_{s,1}(t), \ldots, w_{s,n}(t))$, $t \in T$. We have $Ne(\vec{w}_r) \cdot Ne(\vec{w}_s) = Ne(\vec{w}_r) \cdot_m [Ne(\vec{w}_r)]^{-1} = N1(\vec{e})_m$. Then

$$0 = e_1 = w_{r,m}(t)w_{s,1}(t) + w_{r,1}(t),$$

$$0 = e_2 = w_{r,m}(t)w_{s,2}(t) + w_{r,2}(t),$$

$$\ldots\ldots\ldots\ldots\ldots\ldots\ldots\ldots\ldots$$

$$1 = e_m = w_{r,m}(t)w_{s,m}(t),$$

$$\ldots\ldots\ldots\ldots\ldots\ldots\ldots\ldots\ldots$$

$$0 = e_n = w_{r,m}(t)w_{s,n}(t) + w_{r,n}(t).$$

From the above equalities we obtain

$$w_{s,1}(t) = -\frac{w_{r,1}(t)}{w_{r,m}(t)}, \ldots, w_{s,m}(t) = \frac{1}{w_{r,m}(t)}, \ldots, w_{s,n}(t) = -\frac{w_{r,n}(t)}{w_{r,m}(t)}.$$

Hence, for $[Ne(\vec{w}_r)]^{-1} = Ne(\vec{w}_s)$, we get

$$\vec{w}_s(t) = \left(-\frac{w_{r,1}(t)}{w_{r,m}(t)}, \ldots, \frac{1}{w_{r,m}(t)}, \ldots, -\frac{w_{r,n}(t)}{w_{r,m}(t)}\right) =$$

$$= \frac{1}{w_{r,m}(t)}(-w_{r,1}(t), \ldots, 1, \ldots, -w_{r,n}(t)),$$

where the number 1 is on the m-th position. Of course, the bias of the neuron $[Ne(\vec{w}_r)]^{-1}$ is b_r^{-1}, where b_r is the bias of the neuron $Ne(\vec{w}_r)$.

5. Conclusions

The scientific school of O. Borůvka and F. Neuman used, in their study of ordinary differencial equations and their transformations [1,28–30], the algebraic approach with the group theory as a main tool. In our study, we extended this existing theory with the employment of hypercomposiional structures—semihypergroups and hypergroups. We constructed hypergroups of ordinary linear differential operators and certain sequences of such structures. This served as a background to investigate systems of artificial neurons and neural networks.

Author Contributions: Investigation, J.C., M.N., B.S.; writing—original draft preparation, J.C., M.N., D.S.; writing—review and editing, M.N., B.S.; supervision, J.C. All authors have read and agreed to the published version of the manuscript.

Funding: This research received no external funding.

Institutional Review Board Statement: Not applicable.

Informed Consent Statement: Not applicable.

Data Availability Statement: Not applicable.

Conflicts of Interest: The authors declare no conflict of interest.

References

1. Neuman, F. *Global Properties of Linear Ordinary Differential Equations*; Academia Praha-Kluwer Academic Publishers: Dordrecht, The Netherlands; Boston, MA, USA; London, UK, 1991.
2. Chvalina, J.; Chvalinová, L. Modelling of join spaces by n-th order linear ordinary differential operators. In Proceedings of the Fourth International Conference APLIMAT 2005, Bratislava, Slovakia, 1–4 February 2005; Slovak University of Technology: Bratislava, Slovakia, 2005; pp. 279–284.
3. Chvalina, J.; Moučka, J. Actions of join spaces of continuous functions on hypergroups of second-order linear differential operators. In *Proc. 6th Math. Workshop with International Attendance*, FCI Univ. of Tech. Brno 2007. [CD-ROM]. Available online: https://math.fce.vutbr.cz/pribyl/workshop_2007/prispevky/ChvalinaMoucka.pdf (accessed on 4 February 2021).
4. Chvalina, J.; Novák, M. Laplace-type transformation of a centralizer semihypergroup of Volterra integral operators with translation kernel. In *XXIV International Colloquium on the Acquisition Process Management*; University of Defense: Brno, Czech Republic, 2006.
5. Chvalina, J.; Křehlík, Š.; Novák, M. Cartesian composition and the problem of generalising the MAC condition to quasi-multiautomata. *Analele Universitatii "Ovidius" Constanta-Seria Matematica* **2016**, *24*, 79–100. [CrossRef]
6. Křehlík, Š.; Vyroubalová, J. The Symmetry of Lower and Upper Approximations, Determined by a Cyclic Hypergroup, Applicable in Control Theory. *Symmetry* **2020**, *12*, 54. [CrossRef]
7. Novák, M.; Křehlík, Š. EL–hyperstructures revisited. *Soft Comput.* **2018**, *22*, 7269–7280. [CrossRef]
8. Novák, M.; Křehlík, Š.; Staněk, D. n-ary Cartesian composition of automata. *Soft Comput.* **2020**, *24*, 1837–1849.
9. Cesarano, C. A Note on Bi-orthogonal Polynomials and Functions. *Fluids* **2020**, *5*, 105. [CrossRef]
10. Cesarano, C. Multi-dimensional Chebyshev polynomials: A non-conventional approach. *Commun. Appl. Ind. Math.* **2019**, *10*, 1–19. [CrossRef]
11. Chvalina, J.; Smetana, B. Series of Semihypergroups of Time-Varying Artificial Neurons and Related Hyperstructures. *Symmetry* **2019**, *11*, 927. [CrossRef]
12. Jan, J. *Digital Signal Filtering, Analysis and Restoration*; IEEE Publishing: London, UK, 2000.
13. Koudelka, V.; Raida, Z.; Tobola, P. Simple electromagnetic modeling of small airplanes: neural network approach. *Radioengineering* **2009**, *18*, 38–41.
14. Krenker, A.; Bešter, J.; Kos, A. Introduction to the artificial neural networks. In *Artificial Neural Networks: Methodological Advances and Biomedical Applications*; Suzuki, K., Ed.; InTech: Rijeka, Croatia, 2011; pp. 3–18.
15. Raida, Z.; Lukeš, Z.; Otevřel, V. Modeling broadband microwave structures by artificial neural networks. *Radioengineering* **2004**, *13*, 3–11.
16. Srivastava, N.; Hinton, G.; Krizhevsky, A.; Sutskever, I.; Salakhutdinov, R. Dropout: a simple way to prevent neural networks from overfitting. *J. Mach. Learn. Res.* **2014**, *15*, 1929–1958.
17. Neuman, F. Global theory of ordinary linear homogeneous differential equations in the ral domain. *Aequationes Math.* **1987**, *34*, 1–22. [CrossRef]

18. Křehlík, Š.; Novák, M. From lattices to H_v–matrices. *An. Şt. Univ. Ovidius Constanţa* **2016**, *24*, 209–222. [CrossRef]
19. Novák, M. On EL-semihypergroups. *Eur. J. Comb.* **2015**, *44*, 274–286; ISSN 0195-6698. [CrossRef]
20. Novák, M. Some basic properties of *EL*-hyperstructures. *Eur. J. Comb.* **2013**, *34*, 446–459. [CrossRef]
21. Bavel, Z. The source as a tool in automata. *Inf. Control* **1971**, *18*, 140–155. [CrossRef]
22. Dörfler, W. The cartesian composition of automata. *Math. Syst. Theory* **1978**, *11*, 239–257. [CrossRef]
23. Gécseg, F.; Peák, I. *Algebraic Theory of Automata*; Akadémia Kiadó: Budapest, Hungary, 1972.
24. Massouros, G.G. Hypercompositional structures in the theory of languages and automata. *An. Şt. Univ. A.I. Çuza Iaşi, Sect. Inform.* **1994**, *III*, 65–73.
25. Massouros, G.G.; Mittas, J.D. Languages, Automata and Hypercompositional Structures. In Proceedings of the 4th International Congress on Algebraic Hyperstructures and Applications, Xanthi, Greece, 27–30 June 1990.
26. Massouros, C.; Massouros, G. Hypercompositional Algebra, Computer Science and Geometry. *Mathematics* **2020**, *8*, 138. [CrossRef]
27. Křehlík, Š. *n*-Ary Cartesian Composition of Multiautomata with Internal Link for Autonomous Control of Lane Shifting. *Mathematics* **2020**, *8*, 835. [CrossRef]
28. Borůvka, O. *Lineare Differentialtransformationen 2. Ordnung*; VEB Deutscher Verlager der Wissenschaften: Berlin, Germany, 1967.
29. Borůvka, O. *Linear Differential Transformations of the Second Order*; The English University Press: London, UK, 1971.
30. Borůvka, O. *Foundations of the Theory of Groupoids and Groups*; VEB Deutscher Verlager der Wissenschaften: Berlin, Germany, 1974.

Article

On Factorizable Semihypergroups

Dariush Heidari [1] and Irina Cristea [2],*

[1] Faculty of Science, Mahallat Institute of Higher Education, Mahallat 37811-51958, Iran; dheidari@mahallat.ac.ir or dheidari82@gmail.com

[2] Centre for Information Technologies and Applied Mathematics, University of Nova Gorica, Vipavska cesta 13, 5000 Nova Gorica, Slovenia

* Correspondence: irina.cristea@ung.si or irinacri@yahoo.co.uk; Tel.: +386-0533-15-395

Received: 29 May 2020; Accepted: 17 June 2020; Published: 1 July 2020

Abstract: In this paper, we define and study the concept of the factorizable semihypergroup, i.e., a semihypergroup that can be written as a hyperproduct of two proper sub-semihypergroups. We consider some classes of semihypergroups such as regular semihypergroups, hypergroups, regular hypergroups, and polygroups and investigate their factorization property.

Keywords: semihypergroup; factorization; regular semihypergroup; polygroup

1. Introduction

The decomposition property of a set appears as a key element in many topics in algebra, being connected, for example, with equivalence relations (and called partition), factorizable semigroups [1,2], factorization of groups [3], breakable semigroups [4], or recently introduced breakable semihypergroups [5]. A group G is called factorized if it can be written as a product of two subgroups A and B. This means that any element g in G has the form $g = ab$ for some $a \in A$ and $b \in B$ [3]. One of the most famous results about factorization of groups, proven in 1955 by Ito [6], concerns the product of two abelian subgroups. More precisely, it is proven that any group $G = AB$ written as the product of two abelian subgroups is metabelian. A survey on some topics related to factorization of groups was published in 2009 by Amberg and Kazarin [3]. The problem of the factorization of groups has been immediately extended to the theory of semigroups. We mention here the systematic studies of Tolo [2], Catino [1], or Tirasupa [7]. A semigroup is called factorizable if it can be written as the product of two proper subgroups [1]. Besides the concept of left univocal factorization, (A, B) of a semigroup $S = AB$ is defined by the supplementary condition: for any $a, a' \in A$ and $b, b' \in B$, the equality $ab = a'b'$ implies that $a = a'$ (and similarly for the right univocal factorization). In the same article [1], after constructing all the semigroups that have a univocal factorization with factors isomorphic with a pair of prescribed semigroups, the author presented necessary and sufficient conditions that a factorization of a semigroup by right simple semigroups be univocal and characterized the semigroups with this factorization property.

Motivated by the above-mentioned studies, in this paper we introduce the notion of the factorizable semihypergroup, and we investigate the factorization property for special semihypergroups, as regular semihypergroups, hypergroups, regular hypergroups, and polygroups. Then, we also present some properties connected with the left (right) univocal factorization. It is worth mentioning here that the problem of decomposition or factorization in hypercompositional algebra has been previously studied by the authors in [5] related with breakable semihypergroups, or by Massouros in [8,9]. In [8], the separation aspect was studied for hypergroups, join spaces, and convexity hypergroups, while in [9], it was proven that from the general decomposition theorems, which are valid in hypergroups, well known decomposition theorems for convex sets were derived as corollaries, like Kakutani's lemma, Stone and Helly's theorem, etc.

2. Preliminaries

For the basic concepts and terminology of semihypergroups or hypergroups, the reader is refereed to the fundamental books [10–12]. In the following, we will recall those related to the identity element, invertible element, or zero element, as well as the basic properties of regular semihypergroups and regular hypergroups, or polygroups. See also [13,14] for more details regarding small polygroups.

Let S be a semihypergroup, i.e., S is a nonempty set endowed with a hyperoperation $\circ : S \times S \longrightarrow \mathcal{P}^*(S)$, so a function from the Cartesian product $S \times S$ to the family of nonempty subsets of S, which is associative: any three elements $a, b, c \in S$ satisfies the property $(a \circ b) \circ c = a \circ (b \circ c)$, where the left side member of the equality means the union of all hyperproducts $u \circ c$, with $u \in a \circ b$. A semihypergroup is a hypergroup if the reproduction axiom is verified, as well: for any $a \in S$, there is $a \circ S = S = S \circ a$. Sometimes, especially in the case of semihypergroups, when there is no risk of confusion, the hyperoperation is omitted, and the hyperproduct between two elements is simply denoted by ab, as we will do also throughout this paper. Besides, we denote the cardinality of a set S by $|S|$. An element $a \in S$ is called left (right) scalar if $|ax| = 1$ (respectively, $|xa| = 1$), for all $x \in S$. If a is both a left and right scalar, then it is called a two-sided scalar, or simply a scalar. An element e in a semihypergroup S is called the left (right) identity if $x \in ex$ (respectively, $x \in xe$), for all $x \in S$. If e is both a left and an identity, then it is called a two-sided identity, or simply an identity. An element e in a semihypergroup S is called a left (right) scalar identity if $ex = \{x\}$ (respectively, $xe = \{x\}$), for all $x \in S$. If e is both a left and right scalar identity, then it is called a two-sided scalar identity, or simply a scalar identity. An element $a' \in S$ is called a left (right) inverse of the element $a \in S$ if there exists a left (right) identity $e \in S$ such that $e \in a'a$ (respectively, $e \in aa'$). If a' is both a left and right inverse of a, then it is called a two-sided inverse, or simply an inverse of a. An element zero in a semihypergroup S is called a left (right) zero element if $0x = \{0\}$ (respectively, $x0 = \{0\}$), for all $x \in S$. If zero is both a left and a right zero, then it is called a two-sided zero, or simply a zero.

A regular hypergroup is a hypergroup that has at least one identity and every element has at least one inverse. The regularity property for semihypergroups is not the same, in the sense that a semihypergroup S is called regular if every element x of S is regular, i.e., if there exists $s \in S$ such that $x \in xsx$. Notice that any regular semigroup is a regular semihypergroup, and moreover, based on the reproduction axiom, every hypergroup is a regular semihypergroup (but not necessarily a regular hypergroup).

Definition 1 ([12]). *A polygroup is a system $\langle P, \circ, 1, ^{-1} \rangle$, where $1 \in P$, $^{-1}$ is a unitary operation on P, $\circ$ maps $P \times P$ into the family of non-empty subsets of P, and the following axioms hold for all $x, y, z \in P$:*

(P_1) $x \circ (y \circ z) = (x \circ y) \circ z$,
(P_2) $1 \circ x = x = x \circ 1$,
(P_3) $x \in y \circ z$ implies $y \in x \circ z^{-1}$ and $z \in y^{-1} \circ x$.

Theorem 1. *Let S be a semihypergroup with the scalar identity element one such that every element $x \in S$ has a unique inverse, denoted by x^{-1}. Then, S is a polygroup if and only if:*

(P_3') $\forall x, y \in S, (xy)^{-1} = y^{-1}x^{-1}$,

where $A^{-1} = \{a^{-1} \mid a \in A\}$ for $A \subseteq S$.

Proof. We can restrict our attention to the "only if" part, since the "if" part is obvious. Let $x, y, z \in S$ such that $x \in yz$. Then, $1 \in xx^{-1} \subseteq y(zx^{-1})$, and thus, $y^{-1} \in zx^{-1}$. Hence, the property (P_3') implies $y \in xz^{-1}$. Similarly, $z \in y^{-1}x$. Therefore, S is a polygroup. □

3. Factorizable Semihypergroups

Based on the definition of factorizable semigroups, in this section, we first introduce and study the main properties of factorizable semihypergroups.

Definition 2. *A semihypergroup S is said to be factorizable if there exist some proper sub-semihypergroups A and B of S such that $S = AB$ and $|ab| = 1$ for every $a \in A$ and $b \in B$. The pair (A, B) is called a factorization of S, with factors A and B.*

It is clear that every factorizable semigroup can be considered as a factorizable semihypergroup. In the following, we give some examples of factorizable proper semihypergroups (i.e., they are not semigroups).

Example 1. *Let $S = \{1, 2, 3\}$ be a semihypergroup represented by the Cayley table:*

∘	1	2	3
1	1	1	1
2	1	2	3
3	{1,2,3}	3	{2,3}

Then, the pair (A, B), where $A = \{1, 2\}$ and $B = \{2, 3\}$, is a factorization of S.

Example 2. *Let $S = \{1, 2, 3, 4, 5\}$ be a semihypergroup with the following Cayley table:*

∘	1	2	3	4	5
1	1	2	3	4	5
2	2	1	3	4	5
3	3	3	{1,2,3}	5	{4,5}
4	4	4	5	{1,2,4}	{3,5}
5	5	5	{4,5}	{3,5}	{1,2,3,4,5}

Then, the pair (A, B), where $A = \{1, 2, 3\}$ and $B = \{1, 2, 4\}$, is a factorization of S.

Lemma 1. *If a semihypergroup S is factorizable as $S = AB$, where A has a left identity and B has a right identity, then S has a two-sided identity.*

Proof. Suppose that e is a left identity of the sub-semihypergroup A and e' is a right identity of the sub-semihypergroup B. Since $e, e' \in S$, we have $e = ab$ and $e' = a'b'$ for some $a, a' \in A$ and $b, b' \in B$. Thus, $e = ab \in a(be') = (ab)e' = ee'$; thus, $e = ee'$. Similarly, one obtains $e' = ee'$. Therefore, $e = e' \in A \cap B$ is a two-sided identity element of S. □

Similarly, we can prove the following lemma.

Lemma 2. *If a semihypergroup S is factorizable as $S = AB$, where A has a left scalar identity and B has a right scalar identity, then A, B, and S have a two-sided scalar identity.*

In the next theorem, we present a sufficient condition such that a factorizable semihypergroup is a hypergroup.

Theorem 2. *If a semihypergroup S is factorizable as $S = AB$, where A and B are hypergroups such that $A \cap B \neq \emptyset$, then S is a hypergroup.*

Proof. Let S be a semihypergroup and (A, B) a factorization of S, where A and B are hypergroups such that $e \in A \cap B$. Then, for every $x \in S$, there exist $a \in A$ and $b \in B$ such that $x = ab$. By the reproducibility property of A and B, there exist $a', a'', a^* \in A$ and $b', b'', b^* \in B$ such that $a \in a'a''$, $a'' \in ea^*$, $b \in b'b''$, and $e \in b'b^*$. Thus, we can write:

$$x = ab \in (a'a'')b \subseteq a'(ea^*)b \subseteq a'(b'b^*)a^*b = (a'b')(b^*a^*b).$$

Hence, $x \in (a'b')z = yz$, with $z \in b^*a^*b$ and $y = a'b'$. Similarly, one proves that $x \in ty$, with $t \in S$, meaning that the reproducibility property in S holds, as well. Therefore, S is a hypergroup. □

The following example shows that the converse of the above theorem does not hold necessarily.

Example 3. *Let $S = \{1,2,3,4,5,6\}$ be a hypergroup with the Cayley table:*

∘	1	2	3	4	5	6
1	1	2	3	4	5	6
2	2	1	4	3	6	5
3	1	2	3	4	5	6
4	2	1	4	3	6	5
5	{1,3,5}	{2,4,6}	{1,3,5}	{2,4,6}	{1,3,5}	{2,4,6}
6	{2,4,6}	{1,3,5}	{2,4,6}	{1,2,5}	{2,4,6}	{1,3,5}

Then, the pair (A, B), where $A = \{1,2,3,4\}$ and $B = \{1,3,5\}$, is a factorization of S, but A is not a hypergroup, while B is a hypergroup.

Theorem 3. *Let the semihypergroup S be factorizable as $S = AB$, where A is a group. If B is contained in the semihypergroup class $C_i, (i = 1, 2, 3)$, then so, S is, where:*

$C_1 = $ *the class of regular semihypergroups;*
$C_2 = $ *the class of hypergroups;*
$C_3 = $ *the class of regular hypergroups.*

Proof. We denote by e the identity of the group A, which is, based on Lemma 1, a left scalar identity of S.

(1) Assume that B is a regular semihypergroup, and let $x = ab$ be an arbitrary element in $S = AB$. Then, there exits $b' \in B$ such that $b \in bb'b$. Thus, $x = ab \in abb'a^{-1}ab$ and, thereby, $x \in xx'x$, for some $x' \in b'a^{-1}$. This means that any element in S is regular, so the semihypergroup S is regular, as well.

(2) Assume that B is a hypergroup and $S = AB$. Then there exist $a \in A$ and $b \in B$ such that $e = ab$. Thus $a^{-1} = eb = b \in A \cap B$. So, by Theorem 2, we conclude that S is a hypergroup.

(3) Assume that B is a regular hypergroup. Then, by Lemma 1, the identity elements of S, A, and B are the same. Let $x = ab$, with $a \in A$ and $b \in B$, be an arbitrary element of S and $y \in b^{-1}a^{-1}$. Then, $ya = b^{-1}$; hence, $1 \in b^{-1}b = yab$, so $y = (ab)^{-1}$. Thus, $(ab)^{-1} = b^{-1}a^{-1}$. Therefore, S is a regular hypergroup.
□

Definition 3. *A factorization (A, B) of a semihypergroup S is called left univocal (respectively right univocal) if the following condition holds:*

$$(\forall a, a' \in A, b, b' \in B) \ ab = a'b' \implies a = a' (\text{ respectively, } b = b').$$

A factorization (A, B) of a semihypergroup S is called univocal if it is both left and right univocal.

Example 4. *Let $S = \{1,2,3,4\}$ be a semihypergroup with the following Cayley table:*

∘	1	2	3	4
1	1	2	3	4
2	2	{1,2}	4	{3,4}
3	3	4	{1,3}	{2,4}
4	4	{3,4}	{2,4}	{1,2,3,4}

Then, the pair (A, B), where $A = \{1, 2\}$ and $B = \{1, 3\}$, is a left univocal factorization of S.

Example 5. *Let $S = \{1, 2, 3, 4\}$ be a semihypergroup with the following Cayley table:*

∘	1	2	3	4	5	6
1	1	2	3	4	5	6
2	2	1	3	5	4	6
3	3	3	{1,2}	6	6	{4,5}
4	4	5	6	1	2	3
5	5	4	6	2	1	3
6	6	6	{4,5}	3	3	{1,2}

Then, the pair (A, B), where $A = \{1, 4\}$ and $B = \{1, 2, 3\}$, is a left univocal factorization of S.

Lemma 3. *Let (A, B) be a left univocal factorization of a semihypergroup S. Then, the following assertions hold:*

(i) *If $a = a'b$ for $a, a' \in A$ and $b \in B$, then $a = a'$.*
(ii) *Let $a \in A$, $A_1 \subseteq A$, and $B_1 \subseteq B$ be such that $a \in A_1 B_1$. Then, $a \in A_1$.*
(iii) *Let $a \in A$, $A_1 \subseteq A$, $b \in B$, and $B_1 \subseteq B$ be such that $ab \in A_1 B_1$. Then, $a \in A_1$.*
(iv) *If $a \in A_1 b$, with $A_1 \subseteq A$, then $a \in A_1$ and $a = ab$.*
(v) *For every $a \in A$, there exists $b \in B$ such that $a = ab$.*

Proof.

(i) Suppose that $a = a'b$ for $a, a' \in A$ and $b \in B$. Then, $ab \in a'(bb)$, and thus, $ab = a'b'$ for $b' \in bb$, which implies that $a = a'$.
(ii) Let $a \in A$, $A_1 \subseteq A$ and $B_1 \subseteq B$ be such that $a \in A_1 B_1$. Then, we have $a = a_1 b_1$, with $a_1 \in A_1$ and $b_1 \in B_1$. Hence, Assertion (i) implies that $a = a_1 \in A_1$.
(iii) Let $a \in A$, $A_1 \subseteq A$, $b \in B$, and $B_1 \subseteq B$ be such that $ab \in A_1 B_1$. Then, we have $ab = a_1 b_1$, with $a_1 \in A_1$ and $b_1 \in B$, so $a = a_1 \in A_1$.
(iv) This follows immediately from Assertion (i).
(v) For every $a \in A$, there exist $a' \in A$ and $b \in B$ such that $a = a'b$. Then, Assertion (i) implies $a = a'$, as required. □

Lemma 4. *Let (A, B) be a left univocal factorization of a semihypergroup S. Then, $A \cap B$ is a left zero semigroup formed by right scalar identities of A.*

Proof. First we prove that the intersection $A \cap B$ is nonempty. Let b be an element of B such that $b = eb'$, with $e \in A$ and $b' \in B$. Then, for every $a \in A$, it follows that $ab \in (ae)b'$; thus, Lemma 3 (iii) implies $a \in ae$. Again, by Lemma 3 (v), there exists $b'' \in B$ such that $e = eb''$, while $b'' = a^*b^*$, with $a^* \in A$ and $b^* \in B$. Thus, $a^*b'' \in (a^*a^*)b^*$. It follows that $a^* \in a^*a^*$ by Lemma 3 (iii). Hence, $a^* \in a^*e = a^*eb''$, so $a^* = a^*b''$ by Lemma 3 (iv). Thereby, $a^* = a^*b'' = (a^*a^*)b^*$, and thus, $a^* = a^*b^* = b''$, which shows that $A \cap B \neq \emptyset$.

Now, let e be an arbitrary element in $A \cap B$. Then, $ee \in A \cap B$, with $|ee| = 1$ and $e(ee) = (ee)e$; thus, $e = ee$, because of the univocal factorization. Moreover, since $ae = (ae)e$, it follows that $a = ae$; thus, e is a scalar right identity of A. Therefore, $A \cap B$ is a left zero sub-semigroup of S formed by right scalar identities of A. □

Remark 1. *Note that the property "to be a scalar" is essential in Lemma 4, i.e., if (A, B) is a left univocal factorization of a semihypergroup S and e is a right identity, but not a scalar one, then $e \notin A \cap B$, as we can see in the following example.*

Example 6. Let $S = \{1, 2, 3, 4\}$ be a semihypergroup with the following Cayley table:

∘	1	2	3	4
1	1	{1,2}	1	1
2	1	{1,2}	1	1
3	1	2	{3,4}	3
4	1	2	{3,4}	4

Then, the pair (A, B), where $A = \{3, 4\}$ and $B = \{1, 2, 4\}$, is a factorization of S. We observe that the element 3 is a right, but not scalar, identity of A, which does not belong to $A \cap B$.

Theorem 4. Let (A, B) be a left univocal factorization of a polygroup P. Then, the following assertions hold:

(i) A and B are sub-polygroups of P.
(ii) $A \cap B = \{1\}$, where one is the identity of the polygroup P.

Proof.

(i) Let P be a polygroup with the identity one and (A, B) be a left univocal factorization of P. First, we prove that both A and B contain one. Since there exists $a \in A$ such that $1 = aa^{-1}$, it follows that $a \in a^2 a^{-1}$, and hence, $a = a'a^{-1}$ for some $a' \in a^2$. Thus, $a \in aa^{-1} = 1$; hence, $1 = a = a^{-1} \in A \cap B$.

Now, for every $a \in A$, there exist $a' \in A$ and $b' \in B$ such that $a^{-1} = a'b'$. Thus, $1 \in aa^{-1} = aa'b'$; and hence, $1 \in aa'$, by Lemma 3 (ii). This means that A is a polygroup.

Furthermore, let $b \in B$. Then, $b^{-1} = a'b'$ with $a' \in A$ and $b' \in B$. Thus, $1 \in b^{-1}b = a'b'b$, so $a' = 1$, by Lemma 3 (iv). Hence, $b^{-1} = b' \in B$, meaning that B is a polygroup.

(ii) This follows from Assertion (i) and Lemma 4. □

Example 7. Let $P = \{1, 2, 3, 4, 5, 6\}$ be a polygroup with the following Cayley table:

∘	1	2	3	4	5	6
1	1	2	3	4	5	6
2	2	1	3	5	4	6
3	3	3	{1,2}	6	6	{4,5}
4	4	5	6	1	2	3
5	5	4	6	2	1	3
6	6	6	{4,5}	3	3	{1,2}

Then, the pair (A, B) where $A = \{1, 2, 3\}$ and $B = \{1, 4\}$ is a factorization of P.

Lemma 5. Let (A, B) be a left univocal factorization of a polygroup P. Then, $|P| = |A| \cdot |B|$.

Proof. Let (A, B) be a left univocal factorization of a polygroup P. Since $P = AB$, it is sufficient to prove that for every $x \in P$, there exist unique $a \in A$ and $b \in B$ such that $x = ab$. If $x = ab = a'b'$, then, by definition, we get $a = a'$, and hence, $ab = ab'$. Therefore, $b \in aa^{-1}b'$. Thus, there exists $a'' \in A$ such that $b \in a''b'$. Hence, $a'' \in bb'^{-1}$. It follows that $a'' \in A \cap B$, meaning that $a'' = 1$, by Theorem 4. We conclude that $b = b'$. □

Proposition 1. The minimum order of a non-commutative left univocal factorizable polygroup is six.

Proof. First, we show that every left univocal factorizable polygroup of order less than six is commutative. Let (A, B) be a left univocal factorization of a polygroup P and $|P| < 6$. Then, Lemma 5

implies $|P| = 4$ and $|A| = |B| = 2$. Consider $P = \{1, a, b, c\}$, $A = \{1, a\}$, and $B = \{1, b\}$. Since A and B are sub-polygroups of P, by Theorem 2, we have $a = a^{-1}$ and $b = b^{-1}$, and hence, $c = c^{-1}$. Now, for every $x, y \in P$ we have:
$$xy = (xy)^{-1} = y^{-1}x^{-1} = yx.$$

Therefore, P is a commutative polygroup. Now, one can see that the symmetric group on the set $\{1, 2, 3\}$, denoted $S_3 = \{(), (1\ 2), (1\ 3), (2\ 3), (1\ 2\ 3), (1\ 3\ 2)\}$, as a polygroup has a left univocal factorization $S_3 = \{(), (1\ 2)\}\{(), (1\ 2\ 3), (1\ 3\ 2)\}$, which completes the proof. □

4. Conclusions

Many properties from semigroup theory have been extended to semihypergroup theory, showing their similarities but also differences, and the factorization property, discussed within this note, is one of them. Following the classical paper of Catino [1], we defined a factorizable semihypergroup as a semihypergroup that can be written as a hyperproduct of its two proper sub-semihypergroups. One of the main results presented here showed that if the semihypergroup S is factorized as $S = AB$ with A a group and B a regular semihypergroup, a hypergroup, or a regular hypergroup, then also S has the same algebraic hypercompositional structure as B. Regarding the polygroups, it was proven that for a left univocal factorizable polygroup $P = AB$, both factors are sub-polygroups of P, with the intersection containing only the identity of P. Moreover, we determined that the minimum order of a non-commutative left univocal factorizable polygroup is six.

Author Contributions: Conceptualization, D.H. and I.C.; funding acquisition, I.C.; investigation, D.H. and I.C.; methodology, D.H. and I.C.; writing, original draft, D.H.; writing, review and editing, I.C. All authors read and agreed to the published version of the manuscript.

Funding: The second author acknowledges the financial support from the Slovenian Research Agency (Research Core Funding No. P1-0285).

Conflicts of Interest: The authors declare no conflict of interest.

References

1. Catino, F. Factorizable semigroups. *Semigroup Forum* **1987**, *36*, 167–174.
2. Tolo, K. Factorizable semigroups. *Pac. J. Math.* **1969**, *31*, 523–535.
3. Amberg, B.; Kazarin, L.S. Factorizations of groups and related topics. *Sci. China Ser. A Math.* **2009**, *52*, 217–230.
4. Redei, L. *Algebra I*; Pergamon Press: Oxford, UK, 1967.
5. Heidari, D.; Cristea, I. Breakable semihypergroups. *Symmetry* **2019**, *11*, 100.
6. Ito, N. Uber das Produkt von zwei abelschen Gruppen. *Math. Z.* **1955**, *62*, 400–401.
7. Tirasupa, Y. Factorizable transformation semigroups. *Semigroup Forum* **1979**, *18*, 15–19.
8. Massouros, C.G. Separation and Relevant Properties in Hypergroups. *AIP Conf. Proc.* **2016**, *1738*, 480051-1–480051-4.
9. Massouros, C.G. On connections between vector spaces and hypercompositional structures. *Int. J. Pure Appl. Math.* **2015**, *34*, 133–150.
10. Corsini, P. *Prolegomena of Hypergroup Theory*; Aviani Editore: Tricesimo, Italy, 1993.
11. Davvaz, B. *Semihypergroup Theory*; Elsevier/Academic Press: London, UK, 2016.
12. Davvaz, B. *Polygroup Theory and Related Systems*; World Sciences Publishing Co. Pte. Ltd.: Hackensack, NJ, USA, 2013.
13. Heidari, D.; Davvaz, B. Characterization of small polygroups by their fundamental groups. *Discret. Math. Algorithms Appl.* **2019**, *11*, 1950047.
14. Heidari, D.; Amooshahi, M.; Davvaz, B. Generalized Cayley graphs over polygroups. *Commun. Algebra* **2019**, *47*, 2209–2219.

© 2020 by the authors. Licensee MDPI, Basel, Switzerland. This article is an open access article distributed under the terms and conditions of the Creative Commons Attribution (CC BY) license (http://creativecommons.org/licenses/by/4.0/).

Article

1-Hypergroups of Small Sizes

Mario De Salvo [1,*], **Dario Fasino** [2], **Domenico Freni** [2] **and Giovanni Lo Faro** [1]

1. Dipartimento di Scienze Matematiche e Informatiche, Scienze Fisiche e Scienze della Terra, Università di Messina, 98122 Messina, Italy; lofaro@unime.it
2. Dipartimento di Scienze Matematiche, Informatiche e Fisiche, Università di Udine, 33100 Udine, Italy; dario.fasino@uniud.it (D.F.); domenico.freni@uniud.it (D.F.)
* Correspondence: desalvo@unime.it

Abstract: In this paper, we show a new construction of hypergroups that, under appropriate conditions, are complete hypergroups or non-complete 1-hypergroups. Furthermore, we classify the 1-hypergroups of size 5 and 6 based on the partition induced by the fundamental relation β. Many of these hypergroups can be obtained using the aforesaid hypergroup construction.

Keywords: hypergroups; complete hypergroups; fundamental relations

Citation: De Salvo, M.; Fasino, D.; Freni, D.; Lo Faro, G. 1-Hypergroups of Small Sizes. *Mathematics* **2021**, *9*, 108. https://doi.org/10.3390/math9020108

Received: 7 December 2020
Accepted: 2 January 2021
Published: 6 January 2021

Publisher's Note: MDPI stays neutral with regard to jurisdictional claims in published maps and institutional affiliations.

Copyright: © 2021 by the authors. Licensee MDPI, Basel, Switzerland. This article is an open access article distributed under the terms and conditions of the Creative Commons Attribution (CC BY) license (https://creativecommons.org/licenses/by/4.0/).

1. Introduction

Hypercompositional algebra is a branch of Algebra experiencing a surge of activity nowadays that concerns the study of hyperstructures, that is, algebraic structures where the composition of two elements is a set rather than a single element [1]. The subjects, methods, and goals of the hypercompositional algebra are very different from those of classic algebra. However, the two fields are connected by certain equivalence relations, called fundamental relations [2,3]. Through fundamental relations, the analysis of algebraic hyperstructures can make use of the wealth of tools typical of classical algebra. Indeed, fundamental relations are peculiar equivalence relations defined on hyperstructures, in such a way that the associated quotient set is one of the classical algebraic structures.

More precisely, a fundamental relation is the smallest equivalence relation defined on the support of a hyperstructure such that the corresponding quotient set is a classical structure having operational properties analogous to those of the hyperstructure [4–7]. For example, the quotient structure modulo the equivalence β^* defined on a semihypergroup (or a hypergroup) is a semigroup (or a group, respectively) [2,8–10]. Analogous definitions and results are also known in hyperstructures endowed with more than one operation, see e.g., [11]. Moreover, hypergroups can be classified according to the height of a β^*-class, that is, the least number of order-2 hyperproducts that can cover that class, see [12].

If $(H, \circ)$ is a hypergroup and $\varphi : H \mapsto H/\beta^*$ is the canonical projection then the kernel $\omega_H = \varphi^{-1}(1_{H/\beta^*})$ is the heart of $(H, \circ)$. The heart of a hypergroup plays a very important role in hypergroup theory. Indeed, if we know the structure of ω_H then we have detailed information on the partition determined by relation β^* since $\beta^*(x) = \omega_H \circ x = x \circ \omega_H$, for all $x \in H$. When the heart of a hypergroup $(H, \circ)$ has only one element ε, this element is also the identity of $(H, \circ)$, since $x \in \beta^*(x) = x \circ \varepsilon = \varepsilon \circ x$. According to a definition introduced by Corsini in [4], the hypergroups whose heart has size 1 are called 1-hypergroups. In ([12] Theorem 2), we characterized the 1-hypergroups in terms of the height of their heart, and in [13] Sadrabadi and Davvaz investigated sequences of join spaces associated with non-complete 1-hypergroups.

In this paper, we deepen the knowledge of 1-hypergroups. In particular, we classify the 1-hypergroups of cardinalities up to 6 on the basis of the partition of H induced by β^*. This technique allows us to explicitly construct all 1-hypergroups of order 5, and enumerate those of order 6 by means of scientific computing software. We recall that the study of small-size algebraic hyperstructures is both a practical tool to analyze more elaborate structures and a well-established research topic in itself. In fact, the enumeration and classification of hyperstructures having small cardinality have made it possible to solve various relevant existence issues in hyperstructure theory, see e.g., [14–17].

The plan of this paper is the following: In the forthcoming Section 2, we introduce the basic definitions, notations, and fundamental facts to be used throughout the paper. In Section 3, we present a new construction of hypergroups that, under appropriate hypotheses, are complete hypergroups or non-complete 1-hypergroups. Moreover, we prove a few results concerning the β-classes of 1-hypergroups and sufficient conditions for 1-hypergroups to be complete, which are relevant in subsequent sections. In Section 4, we determine the 1-hypergroups of size 5, up to isomorphisms. In Section 5 we classify the 1-hypergroups of size 6, up to isomorphisms. The 1-hypergroups of size 4, and many 1-hypergroups of size 5 and 6, can be determined by the construction defined in Section 3. The paper ends with some conclusions and directions for future research in Section 6.

2. Basic Definitions and Results

Let H be a non-empty set and let $\mathcal{P}^*(H)$ be the set of all non-empty subsets of H. A hyperproduct $\circ$ on H is a map from $H \times H$ to $\mathcal{P}^*(H)$. For all $x, y \in H$, the subset $x \circ y$ is the hyperproduct of x and y. If A, B are non-empty subsets of H then $A \circ B = \bigcup_{x \in A, y \in B} x \circ y$.

A semihypergroup is a non-empty set H endowed with an associative hyperproduct $\circ$, that is, $(x \circ y) \circ z = x \circ (y \circ z)$ for all $x, y, z \in H$. We say that a semihypergroup $(H, \circ)$ is a hypergroup if for all $x \in H$, we have $x \circ H = H \circ x = H$, the so-called reproducibility property.

A non-empty subset K of a semihypergroup $(H, \circ)$ is called a subsemihypergroup of $(H, \circ)$ if it is closed with respect to the hyperproduct $\circ$, that is, $x \circ y \subseteq K$ for all $x, y \in K$. A non-empty subset K of a hypergroup $(H, \circ)$ is called a subhypergroup of $(H, \circ)$ if $x \circ K = K \circ x = K$, for all $x \in K$. If a subhypergroup is isomorphic to a group, then we say that it is a subgroup of $(H, \circ)$.

Given a semihypergroup $(H, \circ)$, the relation β^* in H is the transitive closure of the relation $\beta = \bigcup_{n \geq 1} \beta_n$ where β_1 is the diagonal relation in H and, for every integer $n > 1$, β_n is defined recursively as follows:

$$x\beta_n y \iff \exists (z_1, \ldots, z_n) \in H^n : \{x, y\} \subseteq z_1 \circ z_2 \circ \cdots \circ z_n.$$

We let $\beta^*(x)$ denote the β^*-class of $x \in H$. The relations β and β^* are among the best known fundamental relations [3]. Their relevance in hyperstructure theory stems from the following facts [2]: If $(H, \circ)$ is a semihypergroup (respectively, a hypergroup) then the quotient set H/β^* equipped with the operation $\beta^*(x) \otimes \beta^*(y) = \beta^*(z)$ for all $x, y \in H$ and $z \in x \circ y$ is a semigroup (respectively, a group). Moreover, the relation β^* is the smallest strongly regular equivalence on H such that the quotient H/β^* is a semigroup (resp., a group). The canonical projection $\varphi : H \mapsto H/\beta^*$ is a good homomorphism, that is, $\varphi(x \circ y) = \varphi(x) \otimes \varphi(y)$ for all $x, y \in H$. The relations β and β^* are also useful to introduce notable families of semihypergroups and hypergroups, including the fully simple semihypergroups [18–20] and the 0-simple semihypergroups [14,21–23], having interesting connections with partially ordered sets and integer sequences. Furthermore, we recall from [8,10] that if $(H, \circ)$ is a hypergroup then β is transitive, so that $\beta = \beta^*$ in every hypergroup.

If $(H, \circ)$ is a hypergroup then H/β^* is a group and the kernel $\omega_H = \varphi^{-1}(1_{H/\beta^*})$ of φ is the heart of $(H, \circ)$. Furthermore, if $|\omega_H| = 1$ then $(H, \circ)$ is a 1-hypergroup. For later reference, we collect in the following theorem a couple of classic results concerning the heart of a hypergroup, see [2,4].

Theorem 1. *Let $(H, \circ)$ be a hypergroup. Then,*
1. *$\beta(x) = x \circ \omega_H = \omega_H \circ x$, for all $x \in H$;*
2. *$(x \circ y) \cap \omega_H \neq \varnothing \iff (y \circ x) \cap \omega_H \neq \varnothing$, for all $x, y \in H$.*

If A is a non-empty set of a semihypergroup $(H, \circ)$ then we say that A is a complete part if it fulfills the following condition: for every $n \in \mathbb{N} - \{0\}$ and $(x_1, x_2, \ldots, x_n) \in H^n$,

$$(x_1 \circ \cdots \circ x_n) \cap A \neq \varnothing \implies (x_1 \circ \cdots \circ x_n) \subseteq A.$$

For every non-empty set X of H, the intersection of all the complete parts containing X is called the complete closure of X and is denoted with $\mathfrak{C}(X)$. Clearly, X is a complete part of $(H, \circ)$ if and only if $\mathfrak{C}(X) = X$. If $(H, \circ)$ is a semihypergroup and $\varphi : H \mapsto H/\beta^*$ is the canonical projection then, for all non-empty set $A \subseteq H$, we have $\mathfrak{C}(A) = \varphi^{-1}(\varphi(A))$. Moreover, if $(H, \circ)$ is a hypergroup then

$$\mathfrak{C}(A) = \varphi^{-1}(\varphi(A)) = A \circ \omega_H = \omega_H \circ A.$$

A semihypergroup or hypergroup $(H, \circ)$ is complete if $x \circ y = \mathfrak{C}(x \circ y)$ for all $x, y \in H$. If $(H, \circ)$ is a complete (semi-)hypergroup then

$$x \circ y = \mathfrak{C}(a) = \beta^*(a),$$

for every $x, y \in H$ and $a \in x \circ y$. Recently, Sonea and Cristea analyzed in [24] the commutativity degree of complete hypergroups, stressing their similarities and differences with respect to group theory. The interested reader can find all relevant definitions, properties and applications of hyperstructures and fundamental relations, even in more abstract contexts, also in [4,25–30].

In what follows, if $(H, \circ)$ is a finite hypergroup and $|H| = n$ then we set $H = \{1, 2, \ldots, n\}$. Moreover, if $(H, \circ)$ is a (possibly infinite) 1-hypergroup then we adopt the convention $\omega_H = \{1\}$.

3. Main Results

In this section, we prove some results which will be used to classify the 1-hypergroups of sizes 4, 5 and 6. To this aim, we now give a construction of hypergroups which, under certain conditions, allows us to determine non-complete 1-hypergroups, starting from complete 1-hypergroups.

3.1. A New Construction

Let $(G, \cdot)$ be a group with $|G| \geq 2$ and let $\mathfrak{F} = \{A_k\}_{k \in G}$ be a family of non-empty and pairwise disjoint sets indexed by G. Let $i, j \in G - \{1_G\}$ be not necessarily distinct elements and let $\varphi : A_i \times A_j \mapsto \mathcal{P}^*(A_{ij})$ be any function such that for all $a \in A_i$ and $b \in A_j$

$$\bigcup_{x \in A_j} \varphi(a, x) = \bigcup_{x \in A_i} \varphi(x, b) = A_{ij}. \tag{1}$$

As a shorthand, introduce the infix notation $\star : A_i \times A_j \mapsto A_{ij}$ defined by $a \star b = \varphi(a, b)$ for every $a \in A_i$ and $b \in A_j$. This operation is naturally extended to sets as follows: for $X \in \mathcal{P}^*(A_i)$ and $Y \in \mathcal{P}^*(A_j)$ let

$$a \star Y = \bigcup_{y \in Y} a \star y, \quad X \star b = \bigcup_{x \in X} = x \star b, \quad X \star Y = \bigcup_{x \in X, y \in Y} x \star y.$$

Hence, the condition (1) can be reformulated as $A_i \star b = a \star A_j = A_{ij}$. Now, let $H = \bigcup_{k \in G} A_k$ and consider the hyperproduct $\circ : H \times H \mapsto \mathcal{P}^*(H)$ defined as follows: for all $x, y \in H$ let

$$x \circ y = \begin{cases} A_{rs} & \text{if } x \in A_r, y \in A_s \text{ and } (r,s) \neq (i,j), \\ x \star y & \text{if } x \in A_i \text{ and } y \in A_j. \end{cases}$$

The following result shows the usefulness of this construction.

Proposition 1. *In the previous notation,*

1. *for every $r, s \in G$ and $x \in A_s$ we have $A_r \circ x = A_{rs}$ and $x \circ A_r = A_{sr}$;*
2. *the hyperproduct $\circ$ is associative: for every $r, s, t \in G$, $x \in A_r$, $y \in A_s$ and $z \in A_t$, we have*

$$(x \circ y) \circ z = A_{(rs)t} = A_{r(st)} = x \circ (y \circ z);$$

3. *for every $z_1, z_2, \ldots, z_n \in H$ with $n \geq 3$ there exists $r \in G$ such that $z_1 \circ z_2 \circ \cdots \circ z_n = A_r$;*
4. *$(H, \circ)$ is a hypergroup such that $\beta = \beta_2$;*
5. *for every $x \in H$, $x \in A_k \iff \beta(x) = A_k$;*
6. *$H/\beta \cong G$ and $\omega_H = A_{1_G}$;*
7. *if $|A_{1_G}| = 1$ then $(H, \circ)$ is a 1-hypergroup;*
8. *$(H, \circ)$ is complete if and only if $a \star b = A_{ij}$ for every $a \in A_i$ and $b \in A_j$.*

Proof. In the stated hypothesis we have:

1. Let $r, s \in G$ and $x \in A_s$. If $r \neq i$ or $s \neq j$ then $A_r \circ x = \bigcup_{y \in A_r}(y \circ x) = A_{rs}$. Otherwise, if $r = i$ and $s = j$ then $A_r \circ x = A_i \circ x = A_i \star x = A_{ij}$ by Equation (1). The identity $x \circ A_r = A_{sr}$ can be derived by similar arguments.
2. For every $r, s, t \in G$ and $x \in A_r$, $y \in A_s$ and $z \in A_t$, we have

$$(r, s) \neq (i, j) \implies (x \circ y) \circ z = A_{rs} \circ z = A_{(rs)t}.$$

Moreover, since $j \neq 1_G$ and the sets of the family $\mathfrak{F}$ are pairwise disjoint, if $(r, s) = (i, j)$ then $A_{ij} \neq A_i$ and $a \circ z = A_{(ij)t} = A_{(rs)t}$, for every $a \in x \star y \subseteq A_{ij}$. Therefore,

$$(x \circ y) \circ z = (x \star y) \circ z = \bigcup_{a \in x \star y} a \circ z = A_{(rs)t}.$$

The identity $x \circ (y \circ z) = A_{(rs)t}$ follows analogously.

3. It suffices to apply points 1. and 2. above and proceed by induction on n.
4. By 2., $(H, \circ)$ is a semihypergroup. To prove that it is a hypergroup it remains to prove that the hyperproduct $\circ$ is reproducible. Let $x \in H$. If $x \in A_i$ then

$$x \circ H = \bigcup_{y \in H} x \circ y = \left(\bigcup_{y \in A_j} x \circ y\right) \cup \left(\bigcup_{y \in H - A_j} x \circ y\right)$$

$$= (x \circ A_j) \cup \left(\bigcup_{r \in G - \{j\}} A_{ir}\right) = A_{ij} \cup (H - A_{ij}) = H.$$

If $x \in A_h$ with $h \neq i$ then $x \circ H = \bigcup_{y \in H} x \circ y = \bigcup_{r \in G} A_{hr} = H$ because $hG = G$. Therefore $x \circ H = H$. The identity $H \circ x = H$ can be shown analogously, by considering separately the cases $x \in A_j$ and $x \in H - A_j$. Therefore $\circ$ is reproducible and $(H, \circ)$ is a hypergroup. Consequently, we have the chain of inclusions

$$\beta_1 \subseteq \beta_2 \subseteq \cdots \subseteq \beta_n \subseteq \cdots$$

Now, let $x, y \in H$ be such that $x\beta y$. Hence, there exists $n \geq 3$ such that $x\beta_n y$. By point 3., there exists $r \in G$ such that $\{x, y\} \subseteq A_r$. For every $a \in A_1$ we have $\{x, y\} \subseteq A_r = x \circ a$ and we obtain $x\beta_2 y$.

5. Let $x \in A_k$. If $a \in A_1$ then $A_k = x \circ a$, and so $y \in A_k$ implies $y\beta_2 x$. Conversely, if $y\beta_2 x$ then there exist $a, b \in H$ such that $\{x, y\} \subseteq a \circ b$. From the definition of the hyperproduct $\circ$ it follows that there exists $r \in G$ such that $a \circ b \subseteq A_r$. Therefore, since $x \in A_k \cap A_r$ and the sets of the family $\mathfrak{F}$ are pairwise disjoint, we obtain $y \in a \circ b \subseteq A_r = A_k$. Finally, $A_k = \beta(x)$ because $\beta_2 = \beta$.
6. The application $f : G \mapsto H/\beta$ such that $f(k) = A_k$ is a group isomorphism. Moreover, since $1_{H/\beta} = f(1_G) = A_{1_G}$, we conclude $\omega_H = A_{1_G}$.

7. The claim follows immediately from points 4. and 6.
8. Trivial. □

We stress the fact that the hypothesis $i, j \neq 1_G$ placed in the above construction is essential for the validity of Proposition 1. In fact, if that hypothesis is not fulfilled then the hyperproduct ∘ defined by our construction may not be associative, as shown by the following example.

Example 1. *Let* $G \cong \mathbb{Z}_2$, $(i, j) = (2, 1)$, $A_1 = \{a, b\}$, *and* $A_2 = \{c, d\}$. *Consider the function* $\varphi : A_2 \times A_1 \mapsto \mathcal{P}^*(A_2)$ *represented by the following table:*

⋆	a	b
c	c	d
d	d	c

In this case, the previous construction determines the following hyperproduct table:

∘	a	b	c	d
a	A_1	A_1	A_2	A_2
b	A_1	A_1	A_2	A_2
c	c	d	A_1	A_1
d	d	c	A_1	A_1

We have $c \star A_1 = d \star A_1 = A_2$ *and* $A_2 \star a = A_2 \star b = A_2$, *hence the hyperproduct* ∘ *is not associative because*

$$(c \circ a) \circ a = \{c\} \subset A_2 \qquad c \circ (a \circ a) = c \circ A_1 = A_2.$$

Remark 1. *The complete hypergroups have been characterized by Corsini in [4] by means of a construction very similar to ours. In fact, the above construction reduces to the one in [4] if the condition in Equation (1) is replaced by* $\varphi(a, b) = A_{ij}$ *for every* $a \in A_i$ *and* $b \in A_j$. *In that case, the hypergroup thus produced is complete.*

3.2. Auxiliary Results

Now, we prove two results that are valid in every hypergroup. Recall that in every hypergroup the relation β is an equivalence coinciding with β^* [8,10].

Proposition 2. *Let* $(H, \circ)$ *be a hypergroup. For all* $x, y \in H$, $x \circ \beta(y) = \beta(x) \circ \beta(y) = \beta(x) \circ y$.

Proof. By Theorem 1(1) we have $x \circ \beta(y) = x \circ (\omega_H \circ \omega_H \circ y) = (x \circ \omega_H) \circ (\omega_H \circ y) = \beta(x) \circ \beta(y) = \beta(x) \circ (\omega_H \circ y) = (\beta(x) \circ \omega_H) \circ y = \beta(x) \circ y$. □

Proposition 3. *Let* $(H, \circ)$ *be a hypergroup. If a is an elements of H such that* $\beta(a) = \{a\}$ *then both* $a \circ b$ *and* $b \circ a$ *are* β-*classes, for all* $b \in H$.

Proof. By Proposition 2, $a \circ b = \beta(a) \circ b = \beta(a) \circ \beta(b)$. The identity $b \circ a = \beta(b) \circ \beta(a)$ is obtained analogously. □

The next results concern the properties of 1-hypergroups.

Corollary 1. *Let* $(H, \circ)$ *be a 1-hypergroup. If there exists only one* β-*class of size greater than 1 then H is complete.*

Proof. Let $\beta(x)$ be the only β-class with $|\beta(x)| > 1$. By Proposition 3, we only have to prove that if $a \in \beta(x)$ then both $a \circ b$ and $b \circ a$ are β-classes, for all $b \in H$. Let $\varphi: H \mapsto H/\beta$ be the canonical projection and $c \in a \circ b$. We prove that $a \circ b = \beta(c)$. If $|\beta(c)| = 1$ then $a \circ b = \beta(c)$. If $|\beta(c)| > 1$ then $\beta(c) = \beta(x) = \beta(a)$. Consequently,

$$\varphi(x) = \varphi(c) = \varphi(a) \otimes \varphi(b) = \varphi(x) \otimes \varphi(b),$$

hence $\varphi(b) = 1_{H/\beta}$ and we have $b \in \omega_H = \{1\}$. Finally, $a \circ b = a \circ 1 = \beta(a) = \beta(c)$. Analogous arguments can prove that also $b \circ a$ is a β-class. □

Remark 2. *If H is not a complete 1-hypergroup and H owns exactly two β-classes, $\beta(a)$ and $\beta(b)$, of size greater than 1, then $\beta(a) \circ \beta(a) = \beta(b)$ or $\beta(b) \circ \beta(b) = \beta(a)$.*

From Corollary 1 we get the following results.

Proposition 4. *Let $(H, \circ)$ be a finite 1-hypergroup. If $|H/\beta| = p$ and there exists a β-class of size $|H| - p + 1$, then H is a complete hypergroup.*

The previous proposition allows us to find a simple proof to a result shown in [4] providing a taxonomy of all 1-hypergroups of size up to 4.

Theorem 2. *If $(H, \circ)$ is 1-hypergroup and $|H| \leq 4$ then $(H, \circ)$ is a complete hypergroup. Moreover, $(H, \circ)$ is either a group or is one of the hypergroups described by the following three hyperproduct tables, up to isomorphisms:*

∘	1	2	3
1	1	2,3	2,3
2	2,3	1	1
3	2,3	1	1

∘	1	2	3	4
1	1	2,3,4	2,3,4	2,3,4
2	2,3,4	1	1	1
3	2,3,4	1	1	1
4	2,3,4	1	1	1

∘	1	2	3	4
1	1	2,3	2,3	4
2	2,3	4	4	1
3	2,3	4	4	1
4	4	1	1	2,3

Proof. Let $(H, \circ)$ be a 1-hypergroup of size ≤ 4 that is not a group. Two cases are possible: (i) $|H| = 3$ and $|H/\beta| = 2$; (ii) $|H| = 4$ and $|H/\beta| \in \{2, 3\}$. In both cases $(H, \circ)$ is a complete 1-hypergroup by Proposition 4. The corresponding hyperproduct tables are derived from Remark 1. □

Proposition 5. *Let $(H, \circ)$ be a 1-hypergroup and let a, b be elements of H such that $\beta(a) \circ \beta(b) = \{1\}$ and $\beta(a) \circ \beta(a) = \beta(b)$. Then,*

1. *$\beta(b) \circ \beta(a) = \{1\}$ and $\beta(b) \circ \beta(b) = \beta(a)$;*
2. *if $a', a'' \in \beta(a)$ and $a' \circ a'' = A$ then*
    (a) $A \circ x = x \circ A = \beta(a)$ for all $x \in \beta(b)$;
    (b) if there exist $b', b'' \in \beta(b)$ such that $b' \circ b'' = \{a'\}$ or $b' \circ b'' = \{a''\}$ then $A = \beta(b)$.

Proof. 1. The claim follows immediately from Theorem 1.
2. (a) $\beta(a) = \beta(a') = a' \circ 1 = a' \circ (a'' \circ x) = (a' \circ a'') \circ x = A \circ x$ and $\beta(a) = \beta(a'') = 1 \circ a'' = (x \circ a') \circ a'' = x \circ (a' \circ a'') = x \circ A$.
(b) If $b' \circ b'' = \{a'\}$, then $A = a' \circ a'' = (b' \circ b'') \circ a'' = b' \circ (b'' \circ a'') = b' \circ 1 = \beta(b)$. In the same way, if $b' \circ b'' = \{a''\}$ then $A = a' \circ a'' = a' \circ (b' \circ b'') = (a' \circ b') \circ b'' = 1 \circ b'' = \beta(b)$.

□

In the forthcoming sections, we will determine the hyperproduct tables of 1-hypergroups of sizes 5 and 6, up to isomorphisms. Since β is an equivalence, the β-classes of a hypergroup $(H, \circ)$ determine a partition of H in disjoint subsets. By Theorem 1(1), if $(H, \circ)$ is a finite 1-hypergroup such that $H = \{1, 2, \ldots, n\}$ and $\omega_H = \{1\}$ then the first row and the first column of the hyperproduct table exhibits the sets of the partition. In order to find the 1-hypergroups of size n with $|H/\beta| = r$, we will consider all the non-increasing partitions of the integer $(n-1)$ in exactly $(r-1)$ positive integers.

4. 1-Hypergroups of Size 5

In this section we determine the hyperproduct tables of 1-hypergroups of size 5, apart of isomorphisms. Hence, we put $H = \{1, 2, 3, 4, 5\}$ and proceed with the analysis by considering the following cases, corresponding to the non-increasing partitions of 4:

1. $|H/\beta| = 2$, $\beta(2) = \{2, 3, 4, 5\}$;
2. $|H/\beta| = 3$, $\beta(2) = \{2, 3, 4\}$, $\beta(5) = \{5\}$;
3. $|H/\beta| = 3$, $\beta(2) = \{2, 3\}$, $\beta(4) = \{4, 5\}$;
4. $|H/\beta| = 4$, $\beta(2) = \{2, 3\}$, $\beta(4) = \{4\}$, $\beta(5) = \{5\}$;
5. $|H/\beta| = 5$ and $\beta(x) = \{x\}$ for all $x \in H$.

Case 1. In the first case $H/\beta \cong \mathbb{Z}_2$, so we only have the following complete hypergroup:

$\circ_1$	1	2	3	4	5
1	1	2,3,4,5	2,3,4,5	2,3,4,5	2,3,4,5
2	2,3,4,5	1	1	1	1
3	2,3,4,5	1	1	1	1
4	2,3,4,5	1	1	1	1
5	2,3,4,5	1	1	1	1

Case 2. By Proposition 4(2), $(H, \circ)$ is a complete hypergroup and so its hyperproduct table is the following, apart of isomorphisms:

$\circ_2$	1	2	3	4	5
1	1	2,3,4	2,3,4	2,3,4	5
2	2,3,4	5	5	5	1
3	2,3,4	5	5	5	1
4	2,3,4	5	5	5	1
5	5	1	1	1	2,3,4

Case 3. Here $H/\beta \cong \mathbb{Z}_3$ and, setting $\beta(2) = \{2,3\}$ and $\beta(4) = \{4,5\}$, we derive the following partial hyperproduct table:

∘	1	2	3	4	5
1	1	2,3	2,3	4,5	4,5
2	2,3			1	1
3	2,3			1	1
4	4,5	1	1		
5	4,5	1	1		

By Proposition 5,

- if a,b,a',b' are elements in $\beta(2)$ then

$$|a \circ b| = |a' \circ b'| = 1 \implies 4 \circ 4 = 4 \circ 5 = 5 \circ 4 = 5 \circ 5 = \{2,3\};$$

- if a,b,a',b' are elements in $\beta(4)$ then

$$|a \circ b| = |a' \circ b'| = 1 \implies 2 \circ 2 = 2 \circ 3 = 3 \circ 2 = 3 \circ 3 = \{4,5\}.$$

Therefore, if we denote

$$P: \begin{array}{c|cc} \circ & 2 & 3 \\ \hline 2 & & \\ 3 & & \end{array} \qquad Q: \begin{array}{c|cc} \circ & 4 & 5 \\ \hline 4 & & \\ 5 & & \end{array}$$

then we can restrict ourselves to the following three sub-cases:

- The tables P and Q do not contain any singleton entry. Here, one complete hypergroup arises,

$\circ_3$	1	2	3	4	5
1	1	2,3	2,3	4,5	4,5
2	2,3	4,5	4,5	1	1
3	2,3	4,5	4,5	1	1
4	4,5	1	1	2,3	2,3
5	4,5	1	1	2,3	2,3

- The table P contains (one or more) singleton entries in the main diagonal only. Without loss of generality, we can set $2 \circ 2 = \{4\}$ and obtain

$$P: \begin{array}{c|cc} \circ & 2 & 3 \\ \hline 2 & 4 & 4,5 \\ 3 & 4,5 & R \end{array} \qquad Q: \begin{array}{c|cc} \circ & 4 & 5 \\ \hline 4 & 2,3 & 2,3 \\ 5 & 2,3 & S \end{array}$$

where $R \in \{\{4\},\{5\},\{4,5\}\}$ and $S \in \{\{3\},\{2,3\}\}$, that is to say there are 6 tables to examine. Rejecting the hyperproduct tables that are not reproducible and the isomorphic copies, we are left with the following 4 hypergroups:

$\circ_4$	1	2	3	4	5
1	1	2,3	2,3	4,5	4,5
2	2,3	4	4,5	1	1
3	2,3	4,5	4,5	1	1
4	4,5	1	1	2,3	2,3
5	4,5	1	1	2,3	3

$\circ_5$	1	2	3	4	5
1	1	2,3	2,3	4,5	4,5
2	2,3	4	4,5	1	1
3	2,3	4,5	4,5	1	1
4	4,5	1	1	2,3	2,3
5	4,5	1	1	2,3	2,3

$\circ_6$	1	2	3	4	5
1	1	2,3	2,3	4,5	4,5
2	2,3	4	4,5	1	1
3	2,3	4,5	5	1	1
4	4,5	1	1	2,3	2,3
5	4,5	1	1	2,3	2,3

$\circ_7$	1	2	3	4	5
1	1	2,3	2,3	4,5	4,5
2	2,3	4	4,5	1	1
3	2,3	4,5	4	1	1
4	4,5	1	1	2,3	2,3
5	4,5	1	1	2,3	2,3

- The table P contains at least one singleton entry off the main diagonal, for instance $2 \circ 3 = \{4\}$. Consequently, from Proposition 5 we have

$$P: \begin{array}{c|cc} \circ & 2 & 3 \\ \hline 2 & & 4 \\ 3 & & \end{array} \qquad Q: \begin{array}{c|cc} \circ & 4 & 5 \\ \hline 4 & 2,3 & 2,3 \\ 5 & 2,3 & 2,3 \end{array}$$

where every empty cell can be filled with $\{4\}$ or $\{5\}$ or $\{4,5\}$, giving rise to 27 more tables. After checking reproducibility and isomorphisms, we find the following 8 hypergroups:

$\circ_8$	1	2	3	4	5
1	1	2,3	2,3	4,5	4,5
2	2,3	4,5	4	1	1
3	2,3	4,5	4,5	1	1
4	4,5	1	1	2,3	2,3
5	4,5	1	1	2,3	2,3

$\circ_9$	1	2	3	4	5
1	1	2,3	2,3	4,5	4,5
2	2,3	5	4	1	1
3	2,3	4,5	4,5	1	1
4	4,5	1	1	2,3	2,3
5	4,5	1	1	2,3	2,3

$\circ_{10}$	1	2	3	4	5
1	1	2,3	2,3	4,5	4,5
2	2,3	4,5	4	1	1
3	2,3	4,5	5	1	1
4	4,5	1	1	2,3	2,3
5	4,5	1	1	2,3	2,3

$\circ_{11}$	1	2	3	4	5
1	1	2,3	2,3	4,5	4,5
2	2,3	4,5	4	1	1
3	2,3	5	4,5	1	1
4	4,5	1	1	2,3	2,3
5	4,5	1	1	2,3	2,3

$\circ_{12}$	1	2	3	4	5
1	1	2,3	2,3	4,5	4,5
2	2,3	4,5	4	1	1
3	2,3	4	4,5	1	1
4	4,5	1	1	2,3	2,3
5	4,5	1	1	2,3	2,3

$\circ_{13}$	1	2	3	4	5
1	1	2,3	2,3	4,5	4,5
2	2,3	4,5	4	1	1
3	2,3	4	5	1	1
4	4,5	1	1	2,3	2,3
5	4,5	1	1	2,3	2,3

$\circ_{14}$	1	2	3	4	5
1	1	2,3	2,3	4,5	4,5
2	2,3	5	4	1	1
3	2,3	4,5	5	1	1
4	4,5	1	1	2,3	2,3
5	4,5	1	1	2,3	2,3

$\circ_{15}$	1	2	3	4	5
1	1	2,3	2,3	4,5	4,5
2	2,3	5	4	1	1
3	2,3	4	5	1	1
4	4,5	1	1	2,3	2,3
5	4,5	1	1	2,3	2,3

Case 4. Here, being $|H/\beta| = 4$, three more 1-hypergroups are obtained by considering that the quotient group H/β is isomorphic to either the group $\mathbb{Z}_4$ or the group $\mathbb{Z}_2 \times \mathbb{Z}_2$.

- If $H/\beta \cong \mathbb{Z}_4$ and the β-class $\beta(2)$ is associated with a generator of $\mathbb{Z}_4$ then we have

$\circ_{16}$	1	2	3	4	5
1	1	2,3,	2,3	4	5
2	2,3	4	4	5	1
3	2,3	4	4	5	1
4	4	5	5	1	2,3
5	5	1	1	2,3	4

- If $H/\beta \cong \mathbb{Z}_4$ and the β-class $\beta(2)$ is not associated with a generator of $\mathbb{Z}_4$ then we have

$\circ_{17}$	1	2	3	4	5
1	1	2,3	2,3	4	5
2	2,3	1	1	5	4
3	2,3	1	1	5	4
4	4	5	5	2,3	1
5	5	4	4	1	2,3

- If $H/\beta \cong \mathbb{Z}_2 \times \mathbb{Z}_2$ then we have

$\circ_{18}$	1	2	3	4	5
1	1	2,3	2,3	4	5
2	2,3	1	1	5	4
3	2,3	1	1	5	4
4	4	5	5	1	2,3
5	5	4	4	2,3	1

Case 5. Lastly, in this case we have trivially $H \cong \mathbb{Z}_5$ as $|H/\beta| = 5$.

Therefore we have obtained the following result.

Theorem 3. *Apart of isomorphisms, there are 19 1-hypergroups of size 5. Of these hypergroups, exactly 7 are complete.*

Remark 3. With the only exception of the hypergroup $(H, \circ_4)$ in case 3, the 1-hypergroups of size 5 can be determined by the construction defined in Section 3.1. In fact, the hypergroups $(H, \circ_k)$ with $k \in \{1,2,3\}$ are also complete. The hypergroups $(H, \circ_k)$ with $k \in \{5, 6, \cdots, 15\}$ are obtained by considering $G \cong \mathbb{Z}_3$, $A_1 = \{1\}$, $A_2 = \{2,3\}$, $A_3 = \{4,5\}$ and the functions $\varphi_k : A_2 \times A_2 \mapsto \mathcal{P}^*(A_3)$ defined as $\varphi_k(a,b) = a \circ_k b$ for $a, b \in A_2$ and $k \in \{5, 6, \cdots, 15\}$.

5. 1-Hypergroups of Size 6

In this section we classify the product tables of 1-hypergroups of size 6, apart of isomorphisms. Hence, we assume $H = \{1,2,3,4,5,6\}$, $\omega_H = \{1\}$ and distinguish the following nine cases:

1. $|H/\beta| = 2$, $\beta(2) = \{2,3,4,5,6\}$;
2. $|H/\beta| = 3$, $\beta(2) = \{2,3,4,5\}$, $\beta(6) = \{6\}$;
3. $|H/\beta| = 3$, $\beta(2) = \{2,3,4\}$, $\beta(5) = \{5,6\}$;
4. $|H/\beta| = 4$, $H/\beta \cong \mathbb{Z}_4$, $\beta(2) = \{2,3,4\}$, $\beta(5) = \{5\}$, $\beta(6) = \{6\}$;
5. $|H/\beta| = 4$, $H/\beta \cong \mathbb{Z}_4$, $\beta(2) = \{2,3\}$, $\beta(4) = \{4,5\}$, $\beta(6) = \{6\}$;
6. $|H/\beta| = 4$, $H/\beta \cong \mathbb{Z}_2 \times \mathbb{Z}_2$, $\beta(2) = \{2,3,4\}$, $\beta(5) = \{5\}$, $\beta(6) = \{6\}$;
7. $|H/\beta| = 4$, $H/\beta \cong \mathbb{Z}_2 \times \mathbb{Z}_2$, $\beta(2) = \{2,3\}$, $\beta(4) = \{4,5\}$, $\beta(6) = \{6\}$;
8. $|H/\beta| = 5$, $\beta(2) = \{2,3\}$, $\beta(4) = \{4\}$, $\beta(5) = \{5\}$, $\beta(6) = \{6\}$;
9. $|H/\beta| = 6$.

In all aforesaid cases, except case 3, we can give the hyperproduct tables of the 1-hypergroups, apart of isomorphisms. To achieve this goal, we use the partition of H into β-classes, the involved quotient group and the reproducibility condition that the hyperproduct tables must satisfy. In case 3, we obtain a too high number of tables and it is impossible to list them. Nevertheless, with the help of a computer algebra system, we are able to perform an exhaustive search of all possible hyperproduct tables and to determine their number, apart from isomorphisms. To improve readability, we postpone the discussion of case 3 at the end of this chapter.

Case 1. The quotient group H/β is isomorphic to $\mathbb{Z}_2$.

∘	1	2	3	4	5	6
1	1	2,3,4,5,6	2,3,4,5,6	2,3,4,5,6	2,3,4,5,6	2,3,4,5,6
2	2,3,4,5,6	1	1	1	1	1
3	2,3,4,5,6	1	1	1	1	1
4	2,3,4,5,6	1	1	1	1	1
5	2,3,4,5,6	1	1	1	1	1
6	2,3,4,5,6	1	1	1	1	1

Case 2. The quotient group H/β is isomorphic to $\mathbb{Z}_3$.

∘	1	2	3	4	5	6
1	1	2,3,4,5	2,3,4,5	2,3,4,5	2,3,4,5	6
2	2,3,4,5	6	6	6	6	1
3	2,3,4,5	6	6	6	6	1
4	2,3,4,5	6	6	6	6	1
5	2,3,4,5	6	6	6	6	1
6	6	1	1	1	1	2,3,4,5

Case 4. By Corollary 1, we obtain two complete non-isomorphic hypergroups. In particular, where the only β-class of size larger than 1 is associated to a generator of $\mathbb{Z}_4$, we have the following hyperproduct table:

∘	1	2	3	4	5	6
1	1	2,3,4	2,3,4	2,3,4	5	6
2	2,3,4	5	5	5	6	1
3	2,3,4	5	5	5	6	1
4	2,3,4	5	5	5	6	1
5	5	6	6	6	1	2,3,4
6	6	1	1	1	2,3,4	5

Instead, if the only β-class of size larger than 1 is associated to a non-generator of $\mathbb{Z}_4$, we obtain the following table:

∘	1	2	3	4	5	6
1	1	2,3,4	2,3,4	2,3,4	5	6
2	2,3,4	1	1	1	6	5
3	2,3,4	1	1	1	6	5
4	2,3,4	1	1	1	6	5
5	5	6	6	6	2,3,4	1
6	6	5	5	5	1	2,3,4

Case 5. Considering that the group $\mathbb{Z}_4$ has only one element x of order 2 and that $\beta(6)$ is the only β-class of size 1, we have to examine two sub-cases, depending on whether the class $\beta(6)$ is associated to the element x or not.

1. $|H/\beta| = 4$, $H/\beta \cong \mathbb{Z}_4$, $\beta(2) = \{2,3\}$, $\beta(4) = \{4,5\}$, $\beta(6) = \{6\}$ and $\beta(6)$ associated to the only element of $\mathbb{Z}_4$ having order two;
2. $|H/\beta| = 4$, $H/\beta \cong \mathbb{Z}_4$, $\beta(2) = \{2,3\}$, $\beta(4) = \{4,5\}$, $\beta(6) = \{6\}$ and $\beta(6)$ associated to a generator of $\mathbb{Z}_4$.

In the first case we obtain a complete hypergroup,

$\circ$	1	2	3	4	5	6
1	1	2,3	2,3	4,5	4,5	6
2	2,3	6	6	1	1	4,5
3	2,3	6	6	1	1	4,5
4	4,5	1	1	6	6	2,3
5	4,5	1	1	6	6	2,3
6	6	4,5	4,5	2,3	2,3	1

In the second case, by using the multiplicative table of $\mathbb{Z}_4$ and the reproducibility of H, we obtain the following partial table:

$\circ$	1	2	3	4	5	6
1	1	2,3	2,3	4,5	4,5	6
2	2,3	X	Y	6	6	1
3	2,3	Z	T	6	6	1
4	4,5	6	6	1	1	2,3
5	4,5	6	6	1	1	2,3
6	6	1	1	2,3	2,3	4,5

with $X \cup Y = Z \cup T = X \cup Z = Y \cup T = \{4,5\}$. If we suppose that $X \in \{\{4\}, \{4,5\}\}$, up to isomorphisms, we obtain 12 hyperproduct tables corresponding to the following values of the sets X, Y, Z, T:

($\star_1$) $X = \{4\}, Y = \{5\}, Z = \{5\}, T = \{4\}$;
($\star_2$) $X = \{4\}, Y = \{5\}, Z = \{5\}, T = \{4,5\}$;
($\star_3$) $X = \{4\}, Y = \{5\}, Z = \{4,5\}, T = \{4\}$;
($\star_4$) $X = \{4\}, Y = \{5\}, Z = \{4,5\}, T = \{4,5\}$;
($\star_5$) $X = \{4\}, Y = \{4,5\}, Z = \{5\}, T = \{4,5\}$;
($\star_6$) $X = \{4\}, Y = \{4,5\}, Z = \{4,5\}, T = \{4\}$;
($\star_7$) $X = \{4\}, Y = \{4,5\}, Z = \{4,5\}, T = \{5\}$;
($\star_8$) $X = \{4\}, Y = \{4,5\}, Z = \{4,5\}, T = \{4,5\}$;
($\star_9$) $X = \{4,5\}, Y = \{4\}, Z = \{4\}, T = \{4,5\}$;
($\star_{10}$) $X = \{4,5\}, Y = \{4\}, Z = \{5\}, T = \{4,5\}$;
($\star_{11}$) $X = \{4,5\}, Y = \{4\}, Z = \{4,5\}, T = \{4,5\}$;
($\star_{12}$) $X = \{4,5\}, Y = \{4,5\}, Z = \{4,5\}, T = \{4,5\}$.

Remark 4. *The previous 12 hypergroups can be derived from the construction shown in Section 3.1, where we let $G \cong \mathbb{Z}_4$, $A_1 = \{1\}$, $A_2 = \{2,3\}$, $A_3 = \{4,5\}$, $A_4 = \{6\}$, and $\varphi : A_2 \times A_2 \mapsto \mathcal{P}^*(A_3)$ is the function defined as $\varphi(a,b) = a \star_k b$ for $a, b \in A_2$ and $k \in \{1, 2, \ldots, 12\}$. Incidentally, we note that the hypergroup arising from $\star_{12}$ is also complete.*

Case 6. In this case we obtain only one 1-hypergroup, which is also complete:

∘	1	2	3	4	5	6
1	1	2,3,4	2,3,4	2,3,4	5	6
2	2,3,4	1	1	1	6	5
3	2,3,4	1	1	1	6	5
4	2,3,4	1	1	1	6	5
5	5	6	6	6	1	2,3,4
6	6	5	5	5	2,3,4	1

Case 7. In this case, we also obtain only one 1-hypergroup, which is also complete:

∘	1	2	3	4	5	6
1	1	2,3	2,3	4,5	4,5	6
2	2,3	1	1	6	6	4,5
3	2,3	1	1	6	6	4,5
4	4,5	6	6	1	1	2,3
5	4,5	6	6	1	1	2,3
6	6	4,5	4,5	2,3	2,3	1

Case 8. Here the quotient group is isomorphic to $\mathbb{Z}_5$ and we deduce one complete hypergroup:

∘	1	2	3	4	5	6
1	1	2,3	2,3	4	5	6
2	2,3	4	4	5	6	1
3	2,3	4	4	5	6	1
4	4	5	5	6	1	2,3
5	5	6	6	1	2,3	4
6	6	1	1	2,3	4	5

Case 9. Here $\beta(x) = \{x\}, \forall x \in \{2,3,4,5,6\}$, and so H is a group of order 6, that is $H \cong \mathbb{Z}_6$ or $H \cong S_3$.

To conclude the review of 1-hypergroups of size 6, hereafter we consider the most challenging case, where a very high number of tables arises.

Case 3. Here the quotient group H/β is isomorphic to $\mathbb{Z}_3$, $\beta(2) = \{2,3,4\}$ and $\beta(5) = \{5,6\}$. In this case there is only one complete 1-hypergroup; its multiplicative table is the following:

∘	1	2	3	4	5	6
1	1	2,3,4	2,3,4	2,3,4	5,6	5,6
2	2,3,4	5,6	5,6	5,6	1	1
3	2,3,4	5,6	5,6	5,6	1	1
4	2,3,4	5,6	5,6	5,6	1	1
5	5,6	1	1	1	2,3,4	2,3,4
6	5,6	1	1	1	2,3,4	2,3,4

In order to find the other 1-hypergroups, we make sure that the sub-cases we are dealing with are disjoint from each other, which means that a hypergroup of a sub-case can not be isomorphic to a hypergroup of another sub-case.

If $(H, \circ)$ is not a complete hypergroup then we can start from the partial table

$\circ$	1	2	3	4	5	6
1	1	2,3,4	2,3,4	2,3,4	5,6	5,6
2	2,3,4				1	1
3	2,3,4				1	1
4	2,3,4				1	1
5	5,6	1	1	1		
6	5,6	1	1	1		

and the partial sub-tables

$P:$
$\circ$	2	3	4
2			
3			
4			

$Q:$
$\circ$	5	6
5		
6		

Taking into account Proposition 5, there are three options:

1. In the partial table Q there is at least one hyperproduct which is a singleton, for instance $\{2\}$, and for all $a, a' \in \{2, 3, 4\}$ we have $a \circ a' = \{5, 6\}$. We consider two sub-cases:

    (1a) the singleton can appear only in the main diagonal:

    $Q:$
$\circ$	5	6
5	2	R
6	S	T

    By reproducibility, we have $R, S \in \{\{3, 4\}, \{2, 3, 4\}\}$ and $T \in \mathcal{P}^*(\{2, 3, 4\})$. This yields $2^2 \cdot 7 = 28$ tables to examine.

    (1b) The singleton must appear off the main diagonal,

    $Q:$
$\circ$	5	6
5	R	2
6	T	S

    with $R, S \in \{\{3, 4\}, \{2, 3, 4\}\}$ and $T \in \mathcal{P}^*(\{2, 3, 4\})$. Thus other $2^2 \cdot 7 = 28$ tables arise.

2. The partial table Q contains at least one hyperproduct of size two, for instance $\{2, 3\}$, but there are no singletons inside Q. Moreover, for all $a, a' \in \{2, 3, 4\}$, we have $a \circ a' = \{5, 6\}$. We obtain two subcases, again:

    (2a) the hyperproduct $\{2, 3\}$ can appear only in the main diagonal,

    $Q:$
$\circ$	5	6
5	2,3	2,3,4
6	2,3,4	

    and $6 \circ 6 \in \{\{2, 3\}, \{2, 4\}, \{3, 4\}, \{2, 3, 4\}\}$. Hence, 4 cases tables arise.

(2b) the hyperproduct $\{2,3\}$ must appear out of the main diagonal,

$$Q: \begin{array}{c|ccc} \circ & 5 & 6 \\ \hline 5 & R & 2,3 \\ 6 & S & T \end{array}$$

the hyperproducts R and T belong to the set $\{\{2,4\},\{3,4\},\{2,3,4\}\}$ and $S \in \{\{2,3\},\{2,4\},\{3,4\},\{2,3,4\}\}$. Therefore $3^2 \cdot 4 = 36$ cases arise.

3. The partial table P contains at least one singleton. Without loss in generality, we can suppose that $\{5\}$ is among them. From Proposition 5 we deduce $5 \circ 5 = 5 \circ 6 = 6 \circ 5 = \{2,3,4\}$. The following two possibilities arise:

   (3a) Singletons can appear only in the main diagonal of P. Therefore we put $2 \circ 2 = \{5\}$ and obtain

   $$P: \begin{array}{c|ccc} \circ & 2 & 3 & 4 \\ \hline 2 & 5 & 5,6 & 5,6 \\ 3 & 5,6 & R & 5,6 \\ 4 & 5,6 & 5,6 & S \end{array} \qquad Q: \begin{array}{c|cc} \circ & 5 & 6 \\ \hline 5 & 2,3,4 & 2,3,4 \\ 6 & 2,3,4 & T \end{array}$$

   where $R, S \in \{\{5\}, \{6\}, \{5,6\}\}$. Moreover, from Proposition 5, we deduce that $T \neq \{2\}$, that is $T \in \mathcal{P}^*(\{2,3,4\}) - \{\{2\}\}$, and $3^2 \cdot 6 = 54$ cases arise.

   (3b) There is a singleton cell off the main diagonal of P, for instance, $2 \circ 3 = \{5\}$. We obtain

   $$P: \begin{array}{c|ccc} \circ & 2 & 3 & 4 \\ \hline 2 & & 5 & \\ 3 & & & \\ 4 & & & \end{array} \qquad Q: \begin{array}{c|cc} \circ & 5 & 6 \\ \hline 5 & 2,3,4 & 2,3,4 \\ 6 & 2,3,4 & R \end{array}$$

   We consider two sub-cases:

   i. $R = \{2,3,4\}$: the 8 empty cells in table P can be filled with either $\{5\}$, or $\{6\}$, or $\{5,6\}$. Hence, 3^8 cases arise.

   ii. $|R| < 3$: from Proposition 5, $R \neq \{2\}$, $R \neq \{3\}$, and so $R \in \{\{4\}, \{2,3\}, \{2,4\}, \{3,4\}\}$. Moreover the table P can not contain the hyperproduct $\{6\}$, that is every cell in P has to be filled with $\{5\}$ or $\{5,6\}$. Thus, $2^8 \cdot 4 = 512$ cases arise.

All the previous sub-cases have been examined with the help of a computer algebra system based on MATLAB R2018a running on an iMac 2009 with an Intel Core 2 processor (3.06 GHz, 4 GB RAM). The complete enumeration of all 1-hypergroups in case 3 took about 2 min utilizing the subdivision into sub-cases described above, while without that subdivision the running time for solving case 3 exceeded 90 min. We report in Table 1 the number of 1-hypergroups found in each sub-case considered above, up to isomorphisms.

Table 1. Number of non-isomorphic, non-complete 1-hypergroups found in case 3, $|H| = 6$.

Case	(1a)	(1b)	(2a)	(2b)	(3a)	(3b)	Total
Hypergroups	13	13	3	12	12	1180	1233

Remark 5. *The 1-hypergroups in sub-cases (1a), (1b), (2a) and (2b) can be derived from the construction shown in Section 3.1, where $G \cong \mathbb{Z}_3$, $A_1 = \{1\}$, $A_2 = \{5,6\}$, $A_3 = \{2,3,4\}$ and $\varphi: A_2 \times A_2 \mapsto \mathcal{P}^*(A_3)$ is the function defined by the corresponding partial tables Q.*

In Table 2 we summarize the results obtained in our case-by-case review of 1-hypergroups of order 6. In that table, we report the number of 1-hypergroups found in each case and the number of complete hypergroups among them. Theorem 4 states the conclusion.

Table 2. Number of non-isomorphic 1-hypergroups, $|H| = 6$.

Case	1	2	3	4	5	6	7	8	9	Total
Hypergroups	1	1	1234	2	13	1	1	1	2	1256
Complete	1	1	1	2	2	1	1	1	2	11

Theorem 4. *Up to isomorphisms, there are* 1256 *1-hypergroups of size 6, of which* 11 *are complete.*

6. Conclusions and Directions for Further Research

A 1-hypergroup is a hypergroup $(H, \circ)$ where the kernel of the canonical projection $\varphi : H \mapsto H/\beta$ is a singleton. In this paper, we enumerate the 1-hypergroups of size 5 and 6. The main results are given in Theorem 3 for $|H| = 5$ and Theorem 4 for $|H| = 6$. In particular, in Section 4 we show a representation of the 19 1-hypergroups of size 5. To achieve this goal, we exploit the partition of H induced by β. In this way, we reduce the analysis of a tough problem to that of a few sub-problems that can be solved explicitly or by means of scientific computing software on an ordinary desktop computer. Moreover, in Section 3.1 we give a construction of hypergroups which, under certain conditions, are 1-hypergroups. That construction is very flexible and many 1-hypergroups of size 5 and 6 can be determined in that way.

To highlight a direction for possible further research, we point out that many hypergroups found in the present work are also join spaces or transposition hypergroups. To be precise, let $(H, \circ)$ be a hypergroup and, for every $a, b \in H$, let a/b and $b \backslash a$ denote the sets $\{x \in H \mid a \in x \circ b\}$ and $\{x \in H \mid a \in b \circ x\}$, respectively. The commutative hypergroups fulfilling the transposition axiom, that is

$$a/b \cap c/d \neq \emptyset \implies a \circ d \cap b \circ c \neq \emptyset$$

for all $a, b, c, d \in H$ are called join spaces. These hypergroups have been widely used in Geometry [31,32]. In [33] Jantosciak generalized the transposition axiom to arbitrary hypergroups as follows:

$$b \backslash a \cap c/d \neq \emptyset \implies a \circ d \cap b \circ c \neq \emptyset,$$

for all $a, b, c, d \in H$. These particular hypergroups are called transposition hypergroups. A number of results on transposition hypergroups can be found in, e.g., [33–35]. For example, it is known that the complete hypergroups are also transposition hypergroups. The construction shown in Section 3.1 produces transposition hypergroups when $a \star d \cap b \star c \neq \emptyset$, for all $a, b \in A_i$ and $c, d \in A_j$. Indeed, if $x \in b \backslash a \cap c/d$ then $a \in b \circ x$ and $c \in x \circ d$. Thus, we have $a \circ d \cup b \circ c \subseteq b \circ x \circ d$. By point 3. of Proposition 1, there exists $k \in G$ such that $b \circ x \circ d = A_k$. By definition of $\circ$, if $k \neq ij$ then $a \circ d = b \circ c = A_k$. Otherwise, if $k = ij$ then we have $a, b \in A_i, c, d \in A_j, a \circ d = a \star d$ and $b \circ c = b \star c$. Hence, by hypotesis, $a \circ d \cap b \circ c \neq \emptyset$.

Based on the preceding comment, we plan to characterize and enumerate the 1-hypergroups of small size that also are join spaces or transposition hypergroups in further works.

Author Contributions: Conceptualization, investigation, writing—original draft: M.D.S., D.F. (Domenico Freni), and G.L.F.; software, writing—review and editing: D.F. (Dario Fasino). All authors have read and agreed to the published version of the manuscript.

Funding: The research work of Mario De Salvo was funded by Università di Messina, Italy, grant FFABR Unime 2019. Giovanni Lo Faro was supported by INdAM-GNSAGA, Italy, and by Università di Messina, Italy, grant FFABR Unime 2020. The work of Dario Fasino was partially supported by INdAM-GNCS, Italy.

Conflicts of Interest: The authors declare no conflict of interest.

References

1. Massouros, G.; Massouros, C. Hypercompositional algebra, computer science and geometry. *Mathematics* **2020**, *8*, 1338. [CrossRef]
2. Koskas, H. Groupoïdes, demi-hypergroupes et hypergroupes. *J. Math. Pures Appl.* **1970**, *49*, 155–192.
3. Vougiouklis, T. Fundamental relations in hyperstructures. *Bull. Greek Math. Soc.* **1999**, *42*, 113–118.
4. Corsini, P. *Prolegomena of Hypergroup Theory*; Aviani Editore: Tricesimo, Italy, 1993.
5. Davvaz, B.; Salasi, A. A realization of hyperrings. *Commun. Algebra* **2006**, *34*, 4389–4400. [CrossRef]
6. De Salvo, M.; Lo Faro, G. On the $n*$-complete hypergroups. *Discret. Math.* **1999**, *208/209*, 177–188. [CrossRef]
7. De Salvo, M.; Fasino, D.; Freni, D.; Lo Faro, G. Semihypergroups obtained by merging of 0-semigroups with groups. *Filomat* **2018**, *32*, 4177–4194.
8. Freni, D. Une note sur le cœur d'un hypergroup et sur la clôture transitive $β^*$ de $β$. *Riv. Mat. Pura Appl.* **1991**, *8*, 153–156.
9. Freni, D. Strongly transitive geometric spaces: Applications to hypergroups and semigroups theory. *Commun. Algebra* **2004**, *32*, 969–988. [CrossRef]
10. Gutan, M. On the transitivity of the relation $β$ in semihypergroups. *Rend. Circ. Mat. Palermo* **1996**, *45*, 189–200. [CrossRef]
11. Norouzi, M.; Cristea, I. Fundamental relation on m-idempotent hyperrings. *Open Math.* **2017**, *15*, 1558–1567. [CrossRef]
12. De Salvo, M.; Fasino, D.; Freni, D.; Lo Faro, G. On hypergroups with a $β$-class of finite height. *Symmetry* **2020**, *12*, 168. [CrossRef]
13. Sadrabadi, E.; Davvaz, B. Atanassov's intuitionistic fuzzy grade of a class of non-complete 1-hypergroups. *J. Intell. Fuzzy Syst.* **2014**, *26*, 2427–2436. [CrossRef]
14. Fasino, D.; Freni, D. Fundamental relations in simple and 0-simple semi-hypergroups of small size. *Arab. J. Math.* **2012**, *1*, 175–190. [CrossRef]
15. Freni, D. Minimal order semi-hypergroups of type U on the right. II. *J. Algebra* **2011**, *340*, 77–89. [CrossRef]
16. Heidari, D.; Freni, D. On further properties of minimal size in hypergroups of type U on the right. *Comm. Algebra* **2020**, *48*, 4132–4141. [CrossRef]
17. Fasino, D.; Freni, D. Existence of proper semi-hypergroups of type U on the right. *Discrete Math.* **2007**, *307*, 2826–2836. [CrossRef]
18. De Salvo, M.; Fasino, D.; Freni, D.; Lo Faro, G. Fully simple semihypergroups, transitive digraphs, and Sequence A000712. *J. Algebra* **2014**, *415*, 65–87. [CrossRef]
19. De Salvo, M.; Freni, D.; Lo Faro, G. Fully simple semihypergroups. *J. Algebra* **2014**, *399*, 358–377. [CrossRef]
20. De Salvo, M.; Freni, D.; Lo Faro, G. Hypercyclic subhypergroups of finite fully simple semihypergroups. *J. Mult. Valued Log. Soft Comput.* **2017**, *29*, 595–617.
21. De Salvo, M.; Fasino, D.; Freni, D.; Lo Faro, G. A family of 0-simple semihypergroups related to sequence A00070. *J. Mult. Valued Log. Soft Comput.* **2016**, *27*, 553–572.
22. De Salvo, M.; Freni, D.; Lo Faro, G. On hypercyclic fully zero-simple semihypergroups. *Turk. J. Math.* **2019**, *43*, 1905–1918. [CrossRef]
23. De Salvo, M.; Freni, D.; Lo Faro, G. On further properties of fully zero-simple semihypergroups. *Mediterr. J. Math.* **2019**, *16*, 48. [CrossRef]
24. Sonea, A.; Cristea, I. The class equation and the commutativity degree for complete hypergroups. *Mathematics* **2020**, *8*, 2253. [CrossRef]
25. Antampoufis, N.; Spartalis, S.; Vougiouklis, T. Fundamental relations in special extensions. In *Algebraic Hyperstructures and Applications*; Vougiouklis, T., Ed.; Spanidis Press: Xanthi, Greece, 2003; pp. 81–89.
26. Davvaz, B. *Semihypergroup Theory*; Academic Press: London, UK, 2016.
27. Davvaz, B.; Leoreanu-Fotea, V. *Hyperring Theory and Applications*; International Academic Press: Palm Harbor, FL, USA, 2007.
28. De Salvo, M.; Lo Faro, G. A new class of hypergroupoids associated with binary relations. *J. Mult. Valued Log. Soft Comput.* **2003**, *9*, 361–375.
29. Massouros, G.G. The subhypergroups of the fortified join hypergroup. *Ital. J. Pure Appl. Math.* **1997**, *2*, 51–63.
30. Naz, S.; Shabir, M. On soft semihypergroups. *J. Intell. Fuzzy Syst.* **2014**, *26*, 2203–2213. [CrossRef]
31. Prenowitz, W. A contemporary approach to classical geometry. *Am. Math. Month.* **1961**, *68*, 1–67. [CrossRef]
32. Prenowitz, W.; Jantosciak, J. *Join Geometries. A Theory of Convex Sets and Linear Geometry*; Springer: New York, NY, USA, 1979.
33. Jantosciak, J. Transposition hypergroups: Noncommutative join spaces. *J. Algebra* **1997**, *187*, 97–119. [CrossRef]
34. Jantosciak, J.; Massouros, C.G. Strong identities and fortification in transposition hypergroups. *J. Discrete Math. Sci. Cryptogr.* **2003**, *6*, 169–193. [CrossRef]
35. Massouros, C.G.; Massouros, G.G. The transposition axiom in hypercompositional structures. *Ratio Math.* **2011**, *21*, 75–90.

Article

n-Ary Cartesian Composition of Multiautomata with Internal Link for Autonomous Control of Lane Shifting

Štěpán Křehlík

CDV—Transport Research Centre, Líšeňská 33a, 636 00 Brno, Czech Republic; stepan.krehlik@cdv.cz;
Tel.: +420-724-863-410

Received: 27 April 2020; Accepted: 19 May 2020; Published: 21 May 2020

Abstract: In this paper, which is based on a real-life motivation, we present an algebraic theory of automata and multi-automata. We combine these (multi-)automata using the products introduced by W. Dörfler, where we work with the cartesian composition and we define the internal links among multiautomata by means of the internal links' matrix. We used the obtained product of n-ary multi-automata as a system that models and controls certain traffic situations (lane shifting) for autonomous vehicles.

Keywords: hypergroup; automata theory; cartesian composition; autonomous vehicles; cooperative intelligent transport systems

1. Introduction

1.1. Motivation

Every physical object in real time and space is defined by its specific properties such as its position in spacetime, temperature, shape, or dimension. This set of properties may be regarded as a state in which the physical object evinces. While focusing on transport infrastructure, there are a lot of elements of characteristic properties. These characteristics are considered as states of elements in the transport infrastructure, for example, a state of traffic lights, a state of velocity, a state of mileage, or a state of traffic density. A clear description of such states is not the only requirement. The possibility of a state change is an equally important requirement. In this respect, the algebraic theory of automata suggests a specific tool (automata or multi-automata) that enables to change a state using the transition function and input words. In other words, the input symbol (or input word) is applied to a state by a transition function, and consequently, a new state is obtained. In the past, these structures were considered as systems that transmitted a specific type of information. Nowadays, this concept cannot sufficiently describe or even control real-life applications as these are too complex. In such complex and difficult systems, a change of a state causes a change of another state, for example, a train that passes a railroad crossing stops vehicles on a road. In current traffic, a driver of a vehicle usually responds to the change in the state of the surrounding elements in transport—generally, this is what we consider a human factor. Traffic situations will get more complicated when autonomous vehicles or even whole autonomous systems that control the traffic are included. In the case of autonomous systems, we suggest that every vehicle communicates with other vehicles or the infrastructure, and as a result, it influences the next state of a system. The fact that autonomous vehicles can communicate and send information about their state or information about a planned change of their state is considered as an advantage that makes the traffic infrastructure more effective. In this paper, we construct a cartesian composition of multi-automata, i.e., we combine some multi-automata (or automata) and we add internal links between their particular states. This new approach to a system of multi-automata with an internal link (SMA*il*) is defined by matrices and enables to describe and control the above-mentioned problems.

1.2. Basic Terminology of the Hyperstructure Theory

Before we introduce a theory of automata, recall some basic notions of the algebraic hyperstructure theory (or theory of algebraic hypercompositional structures). For further reference see, for example, books [1,2]. A hypergroupoid is a pair $(H, *)$, where H is a nonempty set and the mapping $* : H \times H \longrightarrow \mathcal{P}^*(H)$ is a binary hyperoperation (or hypercomposition) on H (here $\mathcal{P}^*(H)$ denotes the system of all nonempty subsets of H). If $a * (b * c) = (a * b) * c$ holds for all $a, b, c \in H$, then $(H, *)$ is called a semihypergroup. If moreover the reproduction axiom, i.e., relation $a * H = H = H * a$ for all $a \in H$, is satisfied, then the semihypergroup $(H, *)$ is called hypergroup. By extensive hypergroup in the sense of hypercompositional structures we mean a hypergroup that $\{a, b\} \subseteq a * b$ for all $a, b \in H$, i.e., that the elements are included in its "resalt of hyperoperation". For the same application on this theory in other science or real-life problems see, for example, [3–7].

1.3. Some Notions of the (Multi-)Automata Theory

In the field of *automata theory*, several types of automata are considered. In this work, we focus on *automata without outputs*, i.e., a structure composed from a triad of an *input alphabet*, a *set of states*, and a *transition function*. For completeness, let us note that the automaton is a quintuple consisting of the above-mentioned triad plus an *output alphabet*, and an *output function*. For more details, see [8–11]. For details regarding terminology and some minor deviations from standard usage, see Novák, Křehlík, and Staněk [12]. Further on, we recall the following definition (notice that condition 1 is sometimes omitted if we regard a semigroup instead of a monoid).

Definition 1 ([13]). *By a quasi–automaton we mean a structure* $\mathbb{A} = (I, S, \delta)$ *such that* $I \neq \emptyset$ *is a monoid,* $S \neq \emptyset$ *and* $\delta : I \times S \to S$ *satisfies the following condition:*

1. *There exists an element* $e \in I$ *such that* $\delta(e, s) = s$ *for any state* $s \in S$.
2. $\delta(y, \delta(x, s)) = \delta(xy, s)$ *for any pair* $x, y \in I$ *and any state* $s \in S$.

The set I is called the input set or input alphabet, the set S is called the state set and the mapping δ is called next-state or transition function. Condition 1 is called the unit condition (UC) while condition 2 is called the Mixed Associativity Condition (MAC).

Next, we are going to work with the idea of a *quasi-multiautomaton* as a hyperstructure generalization of a quasi-automaton, see [14–18]. When adjusting the conditions imposed on the transition function δ, it must be defined cautiously because we get a state on the left-hand side of Definition 1, condition 2, whereas we get a *set of states* on the right-hand side.

Definition 2 ([13]). *A quasi–multiautomaton is a triad* $\mathbb{MA} = (I, S, \delta)$, *where* $(I, *)$ *is a semihypergroup, S is a non-empty set and* $\delta : I \times S \to S$ *is a transition function satisfying the condition:*

$$\delta(b, \delta(a, s)) \in \delta(a * b, s) \text{ for all } a, b \in I, \ s \in S. \tag{1}$$

The semihypergroup $(I, *)$ is called the input semihypergroup of the quasi–multiautomaton $\mathbb{A}$ (I alone is called the input set or input alphabet), the set S is called the state set of the quasi–multiautomaton $\mathbb{A}$, and δ is called next-state or transition function. Elements of the set S are called states, elements of the set I are called input symbols or letters. Condition (1) is called Generalized Mixed Associativity Condition (GMAC).

In the theory of automata, ways to combine automata into one structure have been described. We will recall *homogeneous, heterogeneous products*, and *cartesian composition*, which was introduced by [9]. These products were constructed and investigated primarily on classical automata without output in [19]. In the following definition, we recall all three types of products/compositions introduced by W. Dörfler

Definition 3 ([20]). *Let $\mathbb{A}_1 = (I, S, \delta)$, $\mathbb{A}_2 = (I, R, \tau)$ and $\mathbb{B} = (J, T, \sigma)$ be quasi–automata. By the homogeneous product $\mathbb{A}_1 \times \mathbb{A}_2$ we mean the quasi–automaton $(I, S \times R, \delta \times \tau)$, where $\delta \times \tau : I \times (S \times R) \to S \times R$ is a mapping satisfying, for all $s \in S, r \in R, a \in H$, $(\delta \times \tau)(a,(s,r)) = (\delta_1(a,s), \tau(a,r))$, while the heterogeneous product $\mathbb{A}_1 \otimes \mathbb{B}$ is the quasi-automaton $(I \times J, S \times T, \delta \otimes \sigma)$, where $\delta \otimes \tau : (I \times J) \times (S \times T) \to S \times T$ is a mapping satisfying, for all $a \in I, b \in J, s \in S, t \in T$, $\delta \otimes \sigma((a,b),(s,t)) = (\delta(a,s), \sigma(b,t))$. For I, J disjoint we, by $\mathbb{A} \cdot \mathbb{B}$, denote the cartesian composition of $\mathbb{A}$ and $\mathbb{B}$, i.e., the quasi–automaton $(I \cup J, S \times T, \delta \cdot \sigma)$, where $\delta \cdot \sigma : (I \cup J) \times (S \times T) \to S \times T$ is defined, for all $x \in I \cup J, s \in S$ and $t \in T$, by*

$$(\delta \cdot \sigma)(x,(s,t)) = \begin{cases} (\delta(x,s), t) & \text{if } x \in I, \\ (s, \sigma(x,t)) & \text{if } x \in J. \end{cases}$$

One can see that in the *homogeneous product*, we have the same input set, which operates on every component of the state set. In the *heterogeneous product*, the input is a pair of symbols such that each input symbol from the pair is operated on the respective component of the state pair. In the cartesian composition, the input is directed to the corresponding state and operates on one component of the state pair only.

Notice that generalizing the homogeneous or heterogeneous product of automata to the case of quasi-multiautomata is rather straightforward; see Chvalina, Novák, and Křehlík [20]. However, applying the GMAC condition, which distinguishes the transition function of a quasi-multiautomaton from a transition function of a quasi-automaton, on cartesian composition of quasi-multiautomata is not straightforward. In Chvalina, Novák, and Křehlík [13] two extensions of this condition:

- Extension Generalized Mixed Associativity Condition (E-GMAC),
- Small Extension Generalized Mixed Associativity Condition (SE-GMAC)

and their applications were suggested.

Definition 4 ([13]). *Let $\mathbb{A} = (I, S, \delta)$, $\mathbb{B} = (J, T, \sigma)$ be e-quasi-multiautomata with input semihypergroups $(I \circ), (J, *)$ and transition maps $\delta : I \times S \to S, \sigma : J \times T \to T$ satisfying conditions*

$$\delta(y, \delta(x,s)) \in \delta(x \cdot y, s) \cup \delta(I, s) \text{ for all } x, y \in I, s \in S, \tag{2}$$

$$\sigma(y, \sigma(x,t)) \in \sigma(x \cdot y, I) \cup \sigma(J, t) \text{ for all } x, y \subset J, t \subset T. \tag{3}$$

By the cartesian composition of e-quasi-multiautomata $\mathbb{A}$ and $\mathbb{B}$, denoted as $\mathbb{A} \cdot_E \mathbb{B}$, we mean the e-quasi-multiautomaton $\mathbb{A} \cdot_E \mathbb{B} = (((I \cup J), \diamond), S \times T, \delta \cdot \sigma)$, where $\delta \cdot \sigma$ is for all $x \in I \cup J, s \in S$ and $t \in T$ defined by

$$(\delta \cdot \sigma)(x,(s,t)) = \begin{cases} (\delta(x,s), t) & \text{if } x \in I, \\ (s, \sigma(x,t)) & \text{if } x \in J, \end{cases} \tag{4}$$

and $\diamond : (I \cup J) \times (I \cup J) \to \mathcal{P}^*(I \cup J)$ is for all $x, y \in I \cup J$ defined by

$$x \diamond y = \begin{cases} x \circ y \subseteq I & \text{if } x, y \in I, \\ x * y \subseteq J & \text{if } x, y \in J, \\ \{x, y\} & \text{if } x \in I, y \in J \text{ or } x \in J, y \in I \end{cases}$$

and $\delta \cdot \sigma : (I \cup J) \times (S \times T) \to (S \times T)$

When modelling the process of data aggregation from underwater wireless sensor networks, Křehlík and Novák [21] suggested that using an internal link between respective automata in their cartesian composition describes the real-life context better because the internal link influences not only the given individual automaton but to a certain extent the whole system. In this paper we make use of [21]. Therefore, we include the following definition. For an example explaining the concept (together with calculations) see [21], Example 1.

Definition 5 ([21]). *Let $\mathbb{A} = (I, S, \delta_\mathbb{A})$ and $\mathbb{B} = (J, T, \delta_\mathbb{B})$ be two quasi–automata $\delta_\mathbb{A} : I \times S \to S$, $\delta_\mathbb{B} : J \times T \to T$ with disjoint input sets I, J. By $\mathbb{A} \cdot \mathbb{B}$ we denote the automaton $\mathbb{A} \cdot \mathbb{B} = (I \cup J, S \times T, \delta_\mathbb{A} \cdot \delta_\mathbb{B})$, where $\delta_\mathbb{A} \cdot \delta_\mathbb{B} : (I \cup J) \times (S \times T) \to S \times T$ and $\varphi : S \to J$, $\varrho : T \to I$ is defined by*

$$(\delta_\mathbb{A} \cdot \delta_\mathbb{B})(x,(s,t)) = \begin{cases} (\delta_\mathbb{A}(x,s); \delta_\mathbb{B}(\varphi(\delta_\mathbb{A}(x,s),t))) & \text{if } x \in I, \\ (\delta_\mathbb{A}(\varrho(\delta_\mathbb{B}(x,t),s)); \delta_\mathbb{B}(x,t)) & \text{if } x \in J, \end{cases} \quad (5)$$

for all $x \in I \cup J$, $s \in S$ and $t \in T$. The quasi–automaton $\mathbb{A} \cdot \mathbb{B}$ is called the cartesian composition of quasi-automata $\mathbb{A}$ and $\mathbb{B}$ with an internal link. If in Definition 4 we replace Condition (4) with (5), we call the resulting quasi-multiautomaton the cartesian composition of quasi–multiautomata $\mathbb{A}$ and $\mathbb{B}$ with an internal link.

2. New Theoretical Model

In the classical theory, automata were considered as systems for transfering information of specific types. However, given complicated systems used nowadays, their benefits may not seem sufficient. In [9,13,19,20] we presented some real-life applications. In this section, we introduce an extension of ideas first included in [21] regarding internal links in the cartesian composition of quasi-multiautomata. We discuss systems in which there is a whole set of internal links. For our theoretical purposes we will organize them in a matrix considered by, for example, Golestan et al. [22]. In Section 3, we show the application of our theoretical results in the context of autonomous cars and their navigation.

Notation 1. *We are going to use the following notation:*

$$\text{Cartesian product of state sets}: \quad \bigotimes_{i=1}^{n} S_i = S_1 \times S_2 \times S_3 \times \ldots \times S_n$$

$$\text{Cartesian product of transiton functions}: \quad \prod_{i=1}^{n} \delta_i = \delta_1 \cdot \delta_2 \cdot \delta_3 \cdot \ldots \cdot \delta_n$$

Definition 6. *Let $\mathbb{MA}_i = (I_i, S_i, \delta_i)$ be e-quasi-multiautomata with input semihypergroups $(I_i, *_i)$ and $\delta_i : I_i \times S_i \to S_i$ satisfying condition*

$$\delta_i(y, (\delta_i(x,s)) \in \delta_i(y *_i x, s) \cup \delta_i(I_i, s) \text{ for all } x, y \in I_i, s \in S_i, i \in \{1, 2 \ldots, n\}. \quad (6)$$

By n-ary cartesian composition of the e-quasi-multiautomata we mean the following system of e-quasi-multiautomata with internal link (SMAil):

$$\mathbb{A}^{ncc} = \left(\left(\bigcup_{i=1}^{n} I_i, \diamond \right), \bigotimes_{i=1}^{n} S_i, \prod_{i=1}^{n} \delta_i, \mathbf{M}_{nn}(\varphi) \right),$$

where $\prod_{i=1}^{n} \delta_i : \bigcup_{i=1}^{n} I_i \times \bigotimes_{i=1}^{n} S_i \to \bigotimes_{i=1}^{n} S_i$ and matrix $\mathbf{M}_{nn}(\varphi)$, called matrix of internal links, where $\varphi_{ij} : S_i \to I_j$ for $i, j \in \{1, 2 \ldots, n\}$, i.e.,

$$\mathbf{M}_{nn}(\varphi) = \begin{pmatrix} \varphi_{11} & \varphi_{12} & \cdots & \varphi_{1n} \\ \varphi_{21} & \varphi_{22} & \cdots & \varphi_{2n} \\ \vdots & \cdots & \ddots & \vdots \\ \varphi_{n1} & \varphi_{n2} & \cdots & \varphi_{nn} \end{pmatrix} \text{ is defined by}$$

$$\left(\prod_{i=1}^{n}\delta_i\right)(x,(s_1;s_2;\ldots;s_n)) = (\delta_1 \cdot \delta_2 \cdot \ldots \cdot \delta_n)(x,(s_1;s_2;\ldots;s_n)) =$$

$$= \begin{cases} (\delta_1(\varphi_{11}(\delta_1(x,s_1)),s_1);\delta_2(\varphi_{12}(\delta_1(x,s_1)),s_2);\ldots;\delta_n(\varphi_{1n}(\delta_1(x,s_1)),s_n)) & \text{if } x \in I_1, \\ (\delta_1(\varphi_{21}(\delta_2(x,s_2)),s_1);\delta_2(\varphi_{22}(\delta_2(x,s_2)),s_2);\ldots;\delta_n(\varphi_{2n}(\delta_2(x,s_2)),s_n)) & \text{if } x \in I_2, \\ \vdots & \vdots \\ (\delta_1(\varphi_{n1}(\delta_n(x,s_n)),s_1);\delta_2(\varphi_{n2}(\delta_n(x,s_n)),s_2);\ldots;\delta_n(\varphi_{nn}(\delta_n(x,s_n)),s_n)) & \text{if } x \in I_n, \end{cases} \quad (7)$$

and $\diamond : \bigcup_{i=1}^{n} I_i \times \bigcup_{i=1}^{n} I_i \to \mathcal{P}^*\left(\bigcup_{i=1}^{n} I_i\right)$ is for all $x,y \in \bigcup_{i=1}^{n} I_i$ defined by

$$x \diamond y = \begin{cases} x *_i y \subseteq I_i & \text{if } x,y \in I_i, \\ \{x,y\} & \text{if } x \in I_i, y \in I_k \text{ where } i \neq k. \end{cases}$$

satisfies the condition:

$$\left(\prod_{i=1}^{n}\delta_i\right)\left(y,\left(\prod_{i=1}^{n}\delta_i\right)(x,(s_1;s_2;\ldots;s_n))\right) \in \left(\prod_{i=1}^{n}\delta_i\right)(x \diamond y,(s_1;s_2;\ldots;s_n)) \cup \bigotimes_{i=1}^{n}\delta_i(I_i,s_i). \quad (8)$$

One can easily explain what is meant by Definition 6 using Figure 1. When an arbitrary input $x \in \bigcup_{i=1}^{n} I_i$ is applied, the system, i.e., the cartesian composition, has to find out in which input set x belongs. Therefore, we determine the correct line in Equation (7), which through using we obtain a new state of the system. If, for example, $x \in I_1$, the new state $s'_1 \in S_1$ will be computed. Then, it will be adjusted to $s''_1 \in S_1$ using φ_{11}, i.e., by the self-mapping internal link. This new state $s''_1 \in S_1$ will be mapped to remaining input sets $I_2, I_3, \ldots$ as defined by the matrix of internal links. In other words, the matrix determines which other states can be influenced by the change of the primary state. We apply inputs on states using respective transition functions. If there is 0 instead of φ_{ij} in the matrix of internal links, then the link between the state set S_{ij} and input set I_j does not exist. Notice that in Equation (7) e.g., $\delta_1(\varphi_{n1}(\delta_n(x,s_n)),s_1) = \delta_1(\varphi_{n1}(s'),s_1) = \delta_1(i_1,s_1) = s''$. Since we obviously have to regard directions (see the oriented arrows), the matrix of internal links is not symmetric.

In Section 3 we apply the matrix of internal links in the context of modelling the navigation of autonomous vehicles. In this context, each vehicle obviously adjusts its behaviour based on the behaviour of other vehicles. Also obviously, not all vehicles need to be autonomous, which explains the use of zeros in the matrix (only autonomous vehicles communicate).

The above reasoning can be seen applied in the following example. Notice that we will use it also for the construction of the state set in Theorems 2 and 4 of Section 3.

Example 1. *Consider a section of a road in front of an autonomously controlled intersection in Figure 2. All vehicles are autonomous with compatible self-driving control parameters. Every car is in a certain state s, for example, car A_3 is in state $s = (3;45;250;1;0;0)$, which means, for example, that car number 3 going at 45 km/h is in the position 250 m from the intersection in the 1st lane from the right. The first 0 means no change in speed while the second 0 means that the car is about to turn left.*

In our system it is obvious that vehicle A_3 does not need to communicate with vehicle A_4 because it is not an obstacle in its intended driving operation. Thus, introducing an internal link between A_3 and A_4 is not necessary. However, vehicles A_1 and A_2 are problematics for A_3 because it intends to turn left. Therefore, A_1 should increase its speed while A_2 should slow down.

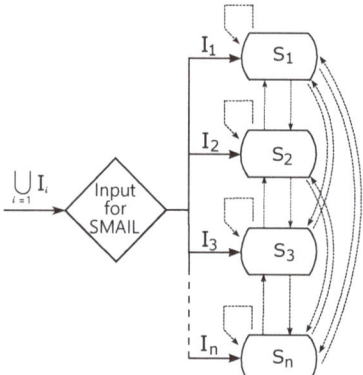

Figure 1. $\mathcal{SMA}il$—A system of multiautomata with internal links.

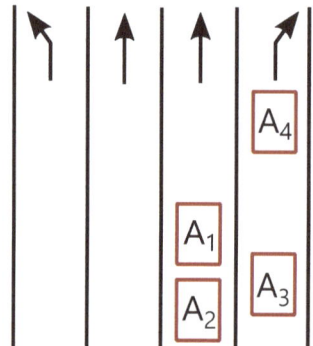

Figure 2. Lane shifting.

Remark 1. *Notice that some theoretical requests in the above example are not possible or even absurd, i.e., we have to keep in mind their feasibility. We already faced this problem in [12,23], where it was eliminated by using a suitable state set, input set, or operations on them. In this paper, we will choose special input and state sets as well.*

3. $\mathcal{SMA}il$—Application of the Theoretical Model for Autonomous Cooperation

We are going to construct our system, called $\mathcal{SMA}il$, for the above context of Example 1. Naturally, Example 1 is an example of a possible usage of $\mathcal{SMA}il$ only.

First, we need to define suitable sets $\mathbb{S}_{\mathcal{RM}}^p$ and $I_{\mathcal{VM}}^p$, where $p \in \mathbb{N}$ is the index. The elements of the state set are ordered pairs—the first component being an ordered sextuple of numbers while the second component is a matrix. The state set is

$$\mathbb{S}_{\mathcal{RM}}^p = \left\{ [s; \mathbf{A}], |s \in \mathcal{R}_6^p, \mathbf{A} \in \mathbf{M}_{m,n} \right\}, \tag{9}$$

where

$$\mathcal{R}_6^p = \left\{ s = (s_1, s_2, \ldots s_6) | s_1 = p; s_2, s_3, s_5 \in \mathbb{R}_0^+; s_4, \in \mathbb{N}; s_6 \in \langle -\frac{\pi}{4}, \frac{\pi}{4} \rangle \right\} \tag{10}$$

and

$$\mathbf{M}_{m,n} = \left\{ \begin{bmatrix} a_{11} & \cdots & a_{1n} \\ \vdots & \ddots & \vdots \\ a_{m1} & \cdots & a_{mn} \end{bmatrix} \middle| a_{ij} \in \left\{ \begin{array}{ll} \{0\} & \text{for } i = \frac{m+1}{2}, j = \frac{n+1}{2}; \\ \{-1,0,1\} & \text{for } otherwise \end{array} \right. \right\} \quad (11)$$

where m, n are a odd numbers. The state set describes the complete position of every vehicle in all lanes. Notice that this is the same approach as in [24,25]. Now, the elements of $\mathcal{R}_6^p$ defined by Equation (10) correspond to states mentioned in Example 1, i.e., s_1 stands for the vehicle number, s_2 for its speed, s_3 for distance from the intersection, s_4 for the respective lane (calculated from the right using odd numbers only, even number are reserved for positions between two lanes), s_5 stands for changing velocity (i.e., an interval $(0,1)$ is deceleration, an interval $(1,\infty)$ is acceleration, 1 stands for constant speed), s_6 stands for changing direction (i.e., an interval $[-\frac{\pi}{4}, 0)$ means manoeuvring left, interval $(0, \frac{\pi}{4}]$ means manoeuvring right, 0 means straight direction). Of course, more parameters can be used; for an example see [26]. Matrices $\mathbf{A}$ used in Equation (9) are taken from the set $\mathbf{M}_{m,n}$ defined by Equation (11). Notice that we use these matrices in the form suggested in [25,27,28] where, for the purpose of its control policy, the intersection is divided into a grid of reservation tiles. To be more precise,

$$\mathbf{A} = \begin{bmatrix} -1 & -1 & -1 & -1 & -1 & -1 & 1 & 1 & 1 \\ 1 & 1 & 1 & -1 & -1 & -1 & 1 & 1 & 1 \\ 1 & 1 & 1 & -1 & -1 & -1 & -1 & -1 & -1 \\ 1 & 1 & 1 & -1 & -1 & -1 & -1 & -1 & -1 \\ -1 & -1 & -1 & 0 & 0 & 0 & -1 & -1 & -1 \\ -1 & -1 & -1 & 0 & 0 & 0 & 1 & 1 & 1 \\ -1 & -1 & -1 & 0 & 0 & 0 & 1 & 1 & 1 \\ -1 & -1 & -1 & -1 & -1 & -1 & 1 & 1 & 1 \\ -1 & -1 & -1 & -1 & -1 & -1 & 1 & 1 & 1 \\ -1 & -1 & -1 & -1 & -1 & -1 & -1 & -1 & -1 \end{bmatrix} \in \mathbf{M}_{m,n} \quad (12)$$

where 0 are parts of the grid (or tiles) occupied by the given vehicle, 1 stands for tiles occupied by some other vehicles and -1 are free tiles. The above matrix $\mathbf{A}$ describes the situation depicted in Figure 3 from the point of view of A_3.

Next, for construction of the i-th quasi-multiautomaton and consequently $\mathcal{SMA}il$, we will need an input alphabet, i.e., input sets (since we will be working with quasi-multiautomata, these will be algebraic hyperstructures). As mentioned above, we will define $I_{\mathcal{VM}}^p$, where $p \in \mathbb{N}$ is the index of the input set.

$$I_{\mathcal{VM}}^p = \left\{ [\vec{i}, \mathbf{C}_{m,n}] = \left[(i_1, i_2, \ldots, i_6), \begin{bmatrix} c_{11} & \cdots & c_{1n} \\ \vdots & \ddots & \vdots \\ c_{m1} & \cdots & c_{mn} \end{bmatrix} \right] \middle| i_1 = p, i_2 \in \mathbb{R}_0^+, i_3 \in \langle 0, 1 \rangle, i_4 \in \mathbb{N}_0, i_5 \in \mathbb{R}^+, i_5 \in \left\langle -\frac{\pi}{4}, \frac{\pi}{4} \right\rangle, \\ c_{ij} \in \{-1, 1\}, i \in \{1, 2, \ldots, n\}, j \in \{1, 2, \ldots, m\} \right\} \quad (13)$$

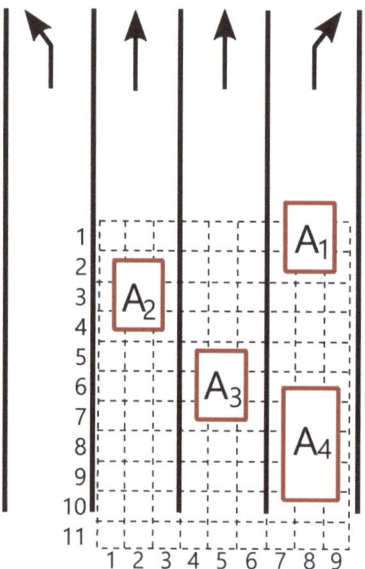

Figure 3. Lane shifting; grid view of vehicle A_3.

It is obvious that input sets $I_{V,M}^p$ and $I_{V,M}^q$ are disjunct for $p \neq q$, because the first component of every vector used in $[\vec{v}, \mathbf{A}_{n,m}] \in I_{V,M}^p$ is p while the first component of every vector used in $[\vec{u}, \mathbf{B}_{n,m}] \in I_{V,M}^q$ is q, i.e.,

$$I_{V,M}^p \cap I_{V,M}^q = \varnothing, \text{ for every } p, q \in \mathbb{N}, \text{ where } p \neq q.$$

Since vector components are real (or even natural) numbers, we can suppose that their sets are ordered; we write $\leq$ for this ordering. For all $p \in \mathbb{N}$ we define hyperoperation $*_p : I_{V,M}^p \times I_{V,M}^p \to \mathcal{P}^*(I_{V,M}^p)$ by:

$$[\vec{u}, \mathbf{A}_{m,n}] *_p [\vec{v}, \mathbf{B}_{m,n}] = \quad (14)$$
$$\{[\vec{w}, \mathbf{C}_{m,n}] \mid \min\{u_i, v_i\} \leq w_i \leq \max\{u_i, v_i\}, \text{ for every } i \in \mathbb{N}, \mathbf{C}_{m,n} \in \{\mathbf{A}_{m,n}, \mathbf{B}_{m,n}\}\}.$$

In the following example we show what we mean by the hyperoperation. Notice that the hyperoperation will be later on used to construct a hypergroup in Theorem 1, which will fulfil the assumption regarding input alphabet stated in Definition 2.

Example 2. Consider two elements $[\vec{a}, A_{2,3}], [\vec{c}, C_{2,3}] \in I_{V,M}^2$, where $[\vec{a}, A_{2,3}] = \left[(2; 6.2; 0; 1; 4; -\frac{\pi}{8}), \begin{bmatrix} 1 & -1 & 1 \\ 1 & -1 & -1 \end{bmatrix}\right]$ and $[\vec{c}, C_{2,3}] = \left[(2; 8.1; 1; 2; 5; \frac{\pi}{6}), \begin{bmatrix} -1 & -1 & -1 \\ 1 & -1 & -1 \end{bmatrix}\right]$. Then the hyperoperarion

$$[\vec{a}, A_{2,3}] *_p [\vec{c}, C_{2,3}] = \left[(2; 6.2; 0; 1; 4; -\frac{\pi}{8}), \begin{bmatrix} 1 & -1 & 1 \\ 1 & -1 & -1 \end{bmatrix}\right] *_p \left[(2; 8.1; 1; 2; 5; \frac{\pi}{6}), \begin{bmatrix} -1 & -1 & -1 \\ 1 & -1 & -1 \end{bmatrix}\right] =$$
$$\left\{[(2; r_2; r_3; r_4; r_5; r_6), D_{2,3}] \mid 6.2 \leq r_2 \leq 8.1; 0 \leq r_3 \leq 1; 1 \leq r_4 \leq 2; 4 \leq r_5 \leq 5; -\frac{\pi}{8} \leq r_6 \leq \frac{\pi}{6}\right\},$$

where $[\vec{r}, D_{2,3}]$ is defined by Equation (14), i.e., $D_{2,3} \in \left\{\begin{bmatrix} 1 & -1 & 1 \\ 1 & -1 & -1 \end{bmatrix}, \begin{bmatrix} -1 & -1 & -1 \\ 1 & -1 & -1 \end{bmatrix}\right\}$.

Theorem 1. *For every index $p \in \mathbb{N}$, the pairs $(I_{\mathcal{VM}}^p, *_p)$ are hypergroups.*

Proof. First, we have to show that the associativity axiom holds, i.e., for all $[\vec{u}, \mathbf{A}_{m,n}]; [\vec{v}, \mathbf{B}_{m,n}]; [\vec{w}, \mathbf{C}_{m,n}] \in I_{\mathcal{VM}}^p$ there is $[\vec{u}, \mathbf{A}_{m,n}] *_p ([\vec{v}, \mathbf{B}_{m,n}] *_p [\vec{w}, \mathbf{C}_{m,n}]) = ([\vec{u}, \mathbf{A}_{m,n}] *_p [\vec{v}, \mathbf{B}_{m,n}]) *_p [\vec{w}, \mathbf{C}_{m,n}]$. Without losing generality, we can show that associativity axiom holds for both matrices and vectors. First, we consider vectors:

$$\vec{u} *_p (\vec{v} *_p \vec{w}) = \vec{u} *_p \{\vec{r}, \min\{v_i, w_i\} \le r_i \le \max\{v_i, w_i\}\} =$$
$$\bigcup_{\vec{s} \in \{\vec{r}, \min\{v_i, w_i\} \le r_i \le \max\{v_i, w_i\}\}} \vec{u} *_p \vec{s} = \bigcup_{\vec{s} \in \{\vec{r}, \min\{v_i, w_i\} \le r_i \le \max\{v_i, w_i\}\}} \{\vec{t}, \min\{u_i, s_i\} \le t_i \le \max\{u_i, s_i\}\} =$$
$$\{\vec{t}, \min\{u_i, \min\{v_i, w_i\}\} \le t_i \le \max\{u_i, \max\{v_i, w_i\}\}\} = \{\vec{t}, \min\{u_i, v_i, w_i\} \le t_i \le \max\{u_i, v_i, w_i\}\} =$$
$$\{\vec{t}, \min\{\min\{u_i, v_i\}, w_i\} \le t_i \le \max\{\min\{u_i, v_i\}, w_i\}\} =$$
$$\bigcup_{\vec{a} \in \{\vec{b}, \min\{u_i, v_i\} \le b_i \le \max\{u_i, v_i\}\}} \{\vec{c}, \min\{w_i, a_i\} \le c_i \le \max\{w_i, a_i\}\} = \bigcup_{\vec{a} \in \{\vec{b}, \min\{u_i, v_i\} \le b_i \le \max\{u_i, v_i\}\}} \vec{a} *_p \vec{w} =$$
$$\{\vec{b}, \min\{u_i, v_i\} \le b_i \le \max\{u_i, v_i\}\} *_p \vec{w} = (\vec{u} *_p \vec{v}) *_p \vec{w}.$$

In the case of matrices, the proof is straightforward:

$$\mathbf{A} *_p (\mathbf{B} *_p \mathbf{C}) = \mathbf{A} *_p \{\mathbf{B}, \mathbf{C}\} = \mathbf{A} *_p \mathbf{B} \cup \mathbf{A} *_p \mathbf{C} = \{\mathbf{A}, \mathbf{B}\} \cup \{\mathbf{A}, \mathbf{C}\} = \{\mathbf{A}, \mathbf{B}, \mathbf{C}\}$$
$$= \{\mathbf{A}, \mathbf{C}\} \cup \{\mathbf{B}, \mathbf{C}\} = \mathbf{A} *_p \mathbf{C} \cup \mathbf{B} *_p \mathbf{C} = \{\mathbf{A}, \mathbf{B}\} *_p \mathbf{C} = (\mathbf{A} *_p \mathbf{B}) *_p \mathbf{C}.$$

The reproductivity axiom holds automatically because the hyperoperation defined above is obviously extensive, i.e., for all $\mathbf{A}, \mathbf{B} \in I_{\mathcal{VM}}^p$ there is $\{\mathbf{A}, \mathbf{B}\} \subseteq \mathbf{A} *_p \mathbf{B}$. So for arbitrary $[\vec{v}, \mathbf{B}_{m,n}]$ we have $[\vec{v}, \mathbf{B}_{m,n}] * I_{\mathcal{VM}}^p = \bigcup_{[\vec{u}, \mathbf{C}_{m,n}] \in I_{\mathcal{VM}}^p} [\vec{v}, \mathbf{B}_{m,n}] * [\vec{u}, \mathbf{C}_{m,n}] = I_{\mathcal{VM}}^p$. Thus, for all indices $p \in \mathbb{N}$ the structures $(I_{\mathcal{VM}}^p, *_p)$ are hypergroups. □

At this point, everything is ready for the construction of an e-quasi-multiautomaton with the state set $\mathbb{S}_{\mathcal{RM}}^p$ and input set $I_{\mathcal{VM}}^p$ for p-th quasi-automaton, where $p \in \mathbb{N}$.

Remark 2. *We consider ordered pairs of vectors and matrices as elements of the input set and also of the state set. As a result, we will denote elements with index ι for input and with index s for the state, i.e., $\left[\vec{a}^\iota, \;_{m,n}^\iota\right] A$ is an input word and $[u^s, \mathbf{M}_{m,n}^s]$ is a state.*

Theorem 2. *For every index $p \in \mathbb{N}$ the triple $\mathbb{MA}_p = (I_{\mathcal{VM}}^p, \mathbb{S}_{\mathcal{RM}}^p, \delta_p)$ is an e-quasi-multiautomtaton with input hypergroup $(I_{\mathcal{VM}}^p, *_p)$, where transition function $\delta_p : I_{\mathcal{VM}}^p \times \mathbb{S}_{\mathcal{RM}}^p \to \mathbb{S}_{\mathcal{RM}}^p$ is defined by*

$$\delta_p \left(\left[\vec{a}^\iota, \;_{m,n}^\iota\right] C, [v^s, \mathbf{K}_{m,n}^s] \right) = \tag{15}$$

$$\left[(\sqrt{a_1 \cdot v_1}; \; a_2 \cdot v_2 + a_2; \; a_3 \cdot v_3; \; a_4 + v_4 \equiv \mod 2l - 1; \; a_5 \cdot v_5; \; a_6 + 0 \cdot v_6)_s, \begin{bmatrix} c_{11}k_{11} & \dots & c_{1n}k_{1n} \\ \dots & \dots & \dots \\ c_{m1}k_{m1} & \dots & c_{mn}k_{mn} \end{bmatrix}^s \right].$$

Proof. This proof is constructed as follows: first we calculate left hand side condition E-GMAC, second we show that left hand side is included in right hand side.

The left hand side:

$$\delta_p\left(\left[\vec{a}^i, \,_{m,n}^l\right] A, \delta_p\left(\left[\vec{b}^i, \mathbf{B}_{m,n}^l\right], [u^s, \mathbf{U}_{m,n}^s]\right)\right) =$$

$$\delta_p\left(\left[\vec{a}^i, \,_{m,n}^l\right] A, \left[\left(\sqrt{b_1 u_1}; b_2 u_2 + b_2; b_3 u_3; b_4 + u_4 \equiv \mod 2l - 1; b_5 u_5; b_6 + 0 \cdot u_6\right)_s, (b_{ij} u_{ij})^s\right]\right) =$$

$$\delta_p\left(\left[\vec{a}^i, \,_{m,n}^l\right] A, \left[\left(\sqrt{p^2}; b_2 u_2 + b_2; b_3 u_3; b_4 + u_4 \equiv \mod 2l - 1; b_5 u_5; b_6\right)_s, (b_{ij} u_{ij})^s\right]\right) =$$

$$\left[(\sqrt{a_1 p}; a_2(b_2 u_2 + b_2) + a_2; a_3 b_3 u_3; a_4 + b_4 + u_4 \equiv \mod 2l - 1; a_5 b_5 u_5; a_6 + (0 \cdot b_6))_s, (a_{ij} b_{ij} u_{ij})^s\right] =$$

$$\left[\left(\sqrt{p^2}; a_2(b_2 u_2 + b_2) + a_2; a_3 b_3 u_3; a_4 + b_4 + u_4 \equiv \mod 2l - 1; a_5 b_5 u_5; a_6\right)_s, (a_{ij} b_{ij} u_{ij})^s\right] =$$

$$\left[(p; a_2(b_2 u_2 + b_2) + a_2; a_3 b_3 u_3; a_4 + b_4 + u_4 \equiv \mod 2l - 1; a_5 b_5 u_5; a_6)_s, (a_{ij} b_{ij} u_{ij})^s\right] =$$

$$[v^s, \mathbf{V}_{m,n}^s].$$

The right hand side:

$$\delta_p\left(\left[\vec{a}^i, \,_{m,n}^l\right] A, *_p\left[\vec{b}^i, \mathbf{B}_{m,n}^l\right], [u^s, \mathbf{U}_{m,n}^s]\right) \cup \delta_p\left(I_{\mathcal{VM}}^p, [u^s, \mathbf{U}_{m,n}^s]\right).$$

From the definition of hyperoproduct $*_p$ in Equation (14), we simply observe that the left hand side is not included in the right part of the right hand side union. Therefore, we have to proof, that the left hand side belongs to the left part of the right hand side union. Therefore we calculate:

$$\delta_p\left(I_{\mathcal{VM}}^p, [u^s, \mathbf{U}_{m,n}^s]\right) = \bigcup_{[\vec{p}^i, \mathbf{P}_{m,n}^l] \in I_{\mathcal{VM}}^p} \delta_p\left(\left[\vec{p}^i, \mathbf{P}_{m,n}^l\right], [u^s, \mathbf{U}_{m,n}^s]\right)$$

There exist an input word $\left[\vec{c}^i, \mathbf{C}_{m,n}^l\right] \in I_{\mathcal{VM}}^p$ such that the vector is $\vec{c} = \left(p; \frac{a_2 i_2 s + a_2 i_2 + a}{s+1}; a_3 b_3; a_4 + b_4; a_5 b_5; a_6\right)$ and the matrix is $\mathbf{C} = \begin{bmatrix} a_{11} b_{11} & \cdots & a_{1n} b_{1n} \\ \cdots & \cdots & \cdots \\ a_{m1} b_{m1} & \cdots & a_{mn} b_{mn} \end{bmatrix}^i$. Thus,

$$[v^s, \mathbf{V}_{m,n}^s] = \delta_p\left(\left[\vec{c}^i, \mathbf{C}_{m,n}^l\right], [u^s, \mathbf{U}_{m,n}^s]\right) \in \bigcup_{[\vec{p}^i, \mathbf{P}_{m,n}^l] \in I_{\mathcal{VM}}^p} \delta_p\left(\left[\vec{p}^i, \mathbf{P}_{m,n}^l\right] \cdot [u^s, \mathbf{U}_{m,n}^s]\right),$$

As a result, E-GMAC holds and the structure $\mathbb{MA}_p = (I_{\mathcal{VM}}^p, S_{V,M}^p, \delta_p)$ is an e-quasi-multiautomaton. □

Once Theorem 2 is proved we will show a practical application af e-quasi-multiautomata in intelligent transport systems. In the following example, the e-quasi-multiautomaton represents an autonomous vehicle. Its state is described by parameters organized into a sextuple while the matrix is used to detect its environment. This example is also linked to Example 1.

Example 3. *We consider matrix* $\mathbf{B}$ *equivalent to matrix* $\mathbf{A}$ *of Equation* (12), *which describes the situation in Figure* 3. *The vehicle* A_3—*in Figure* 4 *depicted in green (while other, non-autonomous, vehicles are red)—detects its surroundings and saves the data to matrix* $\mathbf{B}$. *This model is suitable for a situation where only one vehicle, such as* A_3, *is autonomous and it is not possible to establish communication with other vehicles (because they are not autonomous). We consider a state* $[a^s, \mathbf{B}_{m,n}^s] \in S_{V,M}^3$, *where* $\vec{a}_s = (3; 30; 200; 2; 0; 0)$ *and values in matrix* $\mathbf{B}$ *corresponding to values in the first picture of Figure* 4. *If we apply the input* $\left[\vec{s}^i, \mathbf{S}_{m,n}^l\right] \in I_{\mathcal{VM}}^3$ *by the transition* function δ_3, where $\vec{s^i} = (3;1;0,99;1;0;-\frac{\pi}{4})$ and $\mathbf{S}^i_{m,n} = \begin{bmatrix} 1 & 1 & 1 & 1 & 1 & 1 & -1 & 1 & 1 \\ -1 & 1 & 1 & -1 & 1 & 1 & -1 & 1 & 1 \\ -1 & 1 & 1 & -1 & 1 & 1 & 1 & 1 & 1 \\ -1 & 1 & 1 & -1 & 1 & 1 & 1 & 1 & 1 \\ 1 & 1 & 1 & 1 & 1 & 1 & 1 & 1 & 1 \\ 1 & 1 & 1 & 1 & 1 & 1 & -1 & 1 & 1 \\ 1 & 1 & 1 & 1 & 1 & 1 & -1 & 1 & 1 \\ 1 & 1 & 1 & 1 & 1 & 1 & -1 & 1 & 1 \\ 1 & 1 & 1 & 1 & 1 & 1 & -1 & 1 & 1 \\ 1 & 1 & 1 & 1 & 1 & 1 & -1 & 1 & 1 \end{bmatrix}$, then the first component of the input word is used to control the direction, i.e., it changes the state of the sextuple. The second component of the input word changes the matrix that detects the surroundings of a vehicle. This way we obtain a new state $[r^s, \mathbf{R}^s_{m,n}]$, where $\vec{s}_i = (3;30;198;3;0;-\frac{\pi}{4})$ and the entries of the matrix $\mathbf{R}^s_{m,n}$ are -1 for $a_{21}, a_{24}, a_{31}, a_{34}, a_{41}, a_{44}, a_{17}, a_{27}, a_{67}, a_{77}, a_{87}, a_{97}, a_{10,7}$ and 1 for other cases.

We know from the new state $[r^s, \mathbf{R}^s_{m,n}]$ that the vehicle A_3 does not change its velocity or its acceleration. We also know that the vehicle A_3 is positioned about $2m$ far from the intersection border and is between lane 2 and 3, and it turns to the left. We can see the position surroundings in Figure 4. Consequently, this is what we use in the next step as input for full inclusion in lane 3.

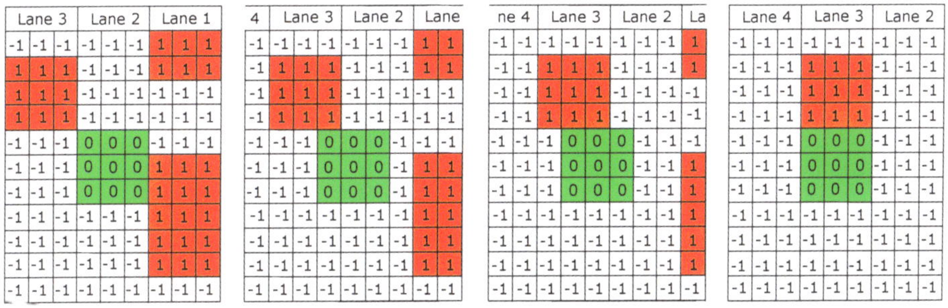

Figure 4. Lane change.

We will use the e-quasi-multiautomata from Theorem 2 for the construction of a system called $\mathcal{SMA}il$. Necessarily, a change on any component of the state of the system must trigger a change on another component of the state. However, there may be situations where the change of other components are not necessarily required. Therefore, we include the following theorem in which we assert that there are such inputs for which the state of the multiautomata do not change.

Theorem 3. *For every e-quasi-mulltiautomaton* $\mathbb{MA}_p = (I^p_{\mathcal{VM}}, \mathbb{S}^p_{\mathcal{RM}}, \delta_p)$ *there is an input* $\left[\vec{e^i}, E^i_{m,n}\right]$ *for which holds*

$$\delta_p\left(\left[\vec{e^i}, E^i_{m,n}\right], [s^s, S^s_{m,n}]\right) = [s^s, S^s_{m,n}]. \tag{16}$$

Proof. Consider an arbitrary state $[s^s, \mathbf{S}^s_{m,n}] \in \mathbb{S}^p_{\mathcal{RM}}$ of the e-quasi-mulltiautomaton $\mathbb{MA}_p$. For input words $[\vec{e}^t, \mathbf{E}^t_{m,n}] = \left[(e_1; e_2; e_3; e_4; e_5; e_6), \begin{bmatrix} 1_{11} & \cdots & 1_{1n} \\ \cdots & \cdots & \cdots \\ 1_{m1} & \cdots & 1_{mn} \end{bmatrix}\right]^i$, where $e_1 = p$; $e_2 = \frac{s_2}{s_2+1}$; $e_3 = 1$; $e_4 = 0$; $e_5 = 1$; $e_6 = s_6$, there holds Equation (16). Indeed,

$$\delta_p\left([\vec{e}^t, \mathbf{E}^t_{m,n}], [s^s, \mathbf{S}^s_{m,n}]\right) =$$

$$\delta_p\left(\left[(p; \frac{s_2}{s_2+1}; 1; 0; 1; s_6)_i, \begin{bmatrix} 1_{11} & \cdots & 1_{1n} \\ \cdots & \cdots & \cdots \\ 1_{m1} & \cdots & 1_{mn} \end{bmatrix}^i\right], \left[(s_1; s_2; s_3; s_4; s_5; s_6)_s, \begin{bmatrix} a_{11} & \cdots & a_{1n} \\ \cdots & \cdots & \cdots \\ a_{m1} & \cdots & a_{mn} \end{bmatrix}^s\right]\right) =$$

$$\left[\left(\sqrt{p^2}, \frac{s_2}{s_2+1}s_2 + \frac{s_2}{s_2+1}, 1 \cdot s_3, 0 + s_4 \equiv \mod k - 1, 1 \cdot s_5, s_6 + 0 \cdot s_6\right), \begin{bmatrix} 1_{11}a_{11} & \cdots & 1_{1n}a_{1n} \\ \cdots & \cdots & \cdots \\ 1_{m1}a_{m1} & \cdots & 1_{mn}a_{mn} \end{bmatrix}^s\right] =$$

$$\left[\left(p, \frac{s_2(s_2+1)}{s_2+1}s_3, s_4, s_5, s_6\right), \begin{bmatrix} a_{11} & \cdots & a_{1n} \\ \cdots & \cdots & \cdots \\ a_{m1} & \cdots & a_{mn} \end{bmatrix}^s\right] = [s^s, \mathbf{S}^s_{m,n}]$$

□

In Example 4 we consider different e-quasi-multiautomata as parts of a cooperative intelligent transport system. Some e-quasi-multiautomata represent autonomous vehicles while some are non-autonomous vehicles. Obviously, non-autonomous vehicles cannot detect their surroundings. In such cases we will use e-quasi-multiautomata, in which all entries of state matrices are 0. This explains the inclusion of the following lemma.

Lemma 1. Let $\overline{\mathbb{S}^p}_{\mathcal{RM}} = \left\{[s; A] | s \in \mathcal{R}^p_6, a_{ij} = 0\right\}$, then a structure $sub\mathbb{MA}_p = (I^p_{\mathcal{VM}}, \overline{\mathbb{S}^p}_{\mathcal{RM}}, \delta_p)$ is a sub-e-quasi-multiautomaton of e-quai-multiautomaton $\mathbb{MA}_p = (I^p_{\mathcal{VM}}, \mathbb{S}^p_{\mathcal{RM}}, \delta_p)$.

Proof. It is obvious that $\overline{\mathbb{S}^p}_{\mathcal{RM}} \subset \mathbb{S}^p_{\mathcal{RM}}$ and $\delta_p : I^p_{\mathcal{VM}} \times \overline{\mathbb{S}^p}_{\mathcal{RM}} \to \overline{\mathbb{S}^p}_{\mathcal{RM}}$. Next, we can see that E-GMAC holds because all entries in the state matrix are zero. Thus, the second component of the state can not change for any input. Therefore, the proof for the first component is the same as the proof of the Theorem 2. □

In what follows, we will note that every state $[_ps^s, _p\mathbf{S}^s_{m,n}]$ is equal to $\left[(_ps^s_1, \ldots, _ps^s_6), _pb^s_{ij}\right] \in \mathbb{S}^p_{\mathcal{RM}}$ and every input $[_p\vec{a}^t, _p\mathbf{A}^t_{m,n}]$ is equal to $\left[(_pa^t_1, \ldots, _pa^t_6), _pa^t_{ij}\right] \in I^p_{\mathcal{VM}}$ from e-quasi-multiautomaton $\mathbb{MA}_p = (I^p_{\mathcal{VM}}, \mathbb{S}^p_{\mathcal{RM}}, \delta_p)$, where $p \in \mathbb{N}$ is an index. In the other words, such as $_ps^s_1$, the upper index s denotes state element, lower index 1 denotes position in the sextuple and the lower index p denotes the $p - th$ e-quasi-multiautomaton. The indices in matrices have the same meaning. As far as input words are regarded, we use upper index ι; other indices have the same meaning.

Now we need to define the matrix of internal links. First, we define $\varphi_{pp} : \mathbb{S}^p_{\mathcal{RM}} \to I^p_{\mathcal{VM}}$ for all $p \in \{1, 2, \ldots, n\}$ by

$$\varphi_{pp}\left(\left[(_ps^s_1, _ps^s_2, _ps^s_3, _ps^s_4, _ps^s_5, _ps^s_6), _p\mathbf{A}^s_{m,n}\right]\right) = \left[(_pv^t_1, _pv^t_2, _pv^t_3, _pv^t_4, _pv^t_5, _pv^t_6), _p\mathbf{B}^t_{m,n}\right].$$

where $_pv^t_1 = _ps^s_1 = p$; $_pv^t_2 = \frac{_ps^s_2}{_ps^s_2 + \frac{1}{_ps^s_5}}$; $_pv^t_3 = 0.95$; $_pv^t_4 = _ps^s_4 + l + \text{sgn}\left(_ps^s_6\right)$; $_pv^t_5 = \frac{1}{_ps^s_5}$; $_pv^t_6 = 0$ and entries of the input matrix are dependent on the occupancy of the tiles.

Corollary 1. For mapping φ_{pp} added to every e-quasi-multiautomaton $\mathrm{MA}_p = (I_{\mathcal{VM}}^p, \mathbb{S}_{\mathcal{RM}}^p, \delta_p)$ from Theorem 2 there holds the condition E-GMAC.

Proof. The proof is obvious. On the left hand side we have up to four changes of the original state in the E-GMAC condition (two inputs and two applications of the internal link). However, we have a suitable input for the right hand side, similar to the proof of Theorem 2. □

Now we will define a mapping for two different e-quasi-multiautomata, i.e., from the state set of the pth e-quasi-multiautomaton to the input set of the qth e-quasi-multiautomaton. We define mapping $\varphi_{pq}: \mathbb{S}_{\mathcal{RM}}^p \to I_{\mathcal{VM}}^q$ for all $p, q \in \{1, 2, \ldots, n\}$, where $p \neq q$ between two e-quai-multiautomata $\mathrm{MA}_p = ((I_{\mathcal{VM}}^p, \mathbb{S}_{\mathcal{RM}}^p, \delta_p))$ and $\mathrm{MA}_q = ((I_{\mathcal{VM}}^q, \mathbb{S}_{\mathcal{RM}}^q, \delta_q))$ by

$$\varphi_{pq}\left([(_ps_1^s, {_p}s_2^s, {_p}s_3^s, {_p}s_4^s, {_p}s_5^s, {_p}s_6^s), {_p}\mathbf{A}_{n,m}^s]\right) = \left[(_qv_1^l, {_q}v_2^l, {_q}v_3^l, {_q}v_4^l, {_q}v_5^l, {_q}v_6^l), {_q}\mathbf{B}_{n,m}^l\right], \quad (17)$$

where

$$_qv_1^l = q \neq {_p}s_1^s;\ _qv_2^l = \frac{_ps_2^s}{_ps_2^s + 1} + \alpha\, _ps_5^s + \lambda \cdot \frac{|_ps_6^s|}{10};\ _qv_3^l = 0,99;\ _qv_4^l = 0;\ _qv_5^l = 1;\ _qv_6^l = 0.$$

Next we add a meaning of the parameters λ and α

$$\lambda = \begin{cases} 1 & \text{if } _ps_4^s >_q s_4^s \text{ and } _ps_6^s > 0 \\ 1 & \text{if } _ps_4^s <_q s_4^s \text{ and } _ps_6^s < 0 \\ 0 & \text{otherewise} \end{cases} \quad \text{and} \quad \alpha = \begin{cases} 1 & \text{if } _ps_4^s =_q s_4^s \text{ and } _qs_3^s <_p s_4^s \text{ and } _ps_5^s > 1 \\ 0 & \text{otherewise} \end{cases}$$

At this point we can give the main theorem of this section in which we are going to use all results obtained above, i.e., e-quasi-multiautomata, hypergroups, and internal links. By Definition 6 we obtain the n-ary cartesian composition of e-quasi-multiautomata with internal links.

Theorem 4. Let $\mathrm{MA}_1 = (I_{\mathcal{VM}}^1, \mathbb{S}_{\mathcal{RM}}^1, \delta_1), \mathrm{MA}_2 = ((I_{\mathcal{VM}}^2, \mathbb{S}_{\mathcal{RM}}^2, \delta_2)), \ldots; \mathrm{MA}_p = ((I_{\mathcal{VM}}^p, \mathbb{S}_{\mathcal{RM}}^p, \delta_p))$ be an e-quasi-multiautomata with disjoint input-sets $I_{\mathcal{VM}}^p$, and $p \geq 2$. Then a quadruple

$$\mathcal{SMA}il = \left(\left(\bigcup_{i=1}^{p} I_{\mathcal{VM}}^i, \diamond\right), \bigotimes_{i=1}^{p} \mathbb{S}_{\mathcal{RM}}^i, \prod_{i=1}^{p} \delta_i, \mathbf{M}_{nn}(\varphi)\right)$$

is a system of the cartesian composition of e-quasi-multiautomata with internal links.

Proof. We have to demonstrate that the condition E-GMAC holds, i.e., that there is

$$\left(\prod_{i=1}^{p}\delta_i\right)\left([\vec{a}^l, \mathbf{A}_{m,n}^l], \left(\prod_{i=1}^{p}\delta_i\right)\left([\vec{b}^l, \mathbf{B}_{m,n}^l], \bigotimes_{i=1}^{p}[_is^s, \mathbf{S}_{m,n}^s]\right)\right) \in$$

$$\left(\prod_{i=1}^{p}\delta_i\right)\left([\vec{a}^l, \mathbf{A}_{m,n}^l] *_p [\vec{b}^l, \mathbf{B}_{m,n}^l], \bigotimes_{i=1}^{p}[_is^s, \mathbf{S}_{m,n}^s]\right) \cup \bigotimes_{i=1}^{p}\delta_i\left(I_{\mathcal{VM}}^i, [_is^s, \mathbf{S}_{m,n}^s]\right). \quad (18)$$

We prove, while maintaining generality, that

$$\left(\prod_{i=1}^{t}\delta_i\right)\left([\vec{a}^l, \mathbf{A}_{m,n}^l], \left(\prod_{i=1}^{t}\delta_i\right)\left([\vec{b}^l, \mathbf{B}_{m,n}^l], \bigotimes_{i=1}^{t}[_is^s, \mathbf{S}_{m,n}^s]\right)\right) \in \bigotimes_{i=1}^{t}\delta_i\left(I_{\mathcal{VM}}^i, [_is^s, \mathbf{S}_{m,n}^s]\right), \quad (19)$$

i.e., Formula (18) without the left part of the right hand side of E-GMAC. There are two cases. The first, both input words are in the same input set, and the second, input words are from different input sets.

(a) For the first case, we consider $\left[{}_p\vec{a^i},{}_p\mathbf{A}_{m,n}^t\right], \left[{}_p\vec{b^i},{}_p\mathbf{B}_{m,n}^t\right] \in I_{\mathcal{VM}}^p$. On the left hand of Equation (19), we know that the input $\left[{}_p\vec{b^i},{}_p\mathbf{B}_{m,n}^t\right]$ works upon one component from $\left(\left[{}_1s^s, {}_1\mathbf{S}_{m,n}^s\right], \left[{}_2s^s, {}_2\mathbf{S}_{m,n}^s\right], \ldots, \left[{}_ts^s, {}_t\mathbf{S}_{m,n}^s\right]\right)$ and the state can alter by the internal link φ_{pp}. This is performed again for input $\left[{}_p\vec{a^i},{}_p\mathbf{A}_{m,n}^t\right]$, then we have an input $\left[{}_p\vec{c^i},{}_p\mathbf{C}_{m,n}^t\right] \in I_{\mathcal{VM}}^p$ for which holds following (considering the proof of Theorem 2):

$$\delta_p\left(\left[{}_p\vec{a^i},{}_p\mathbf{A}_{m,n}^t\right], \delta_p\left(\left[{}_p\vec{b^i},{}_p\mathbf{B}_{m,n}^t\right], \left[{}_ps^s, {}_p\mathbf{S}_{m,n}^s\right]\right)\right) = \delta_p\left(\left[{}_p\vec{c^i},{}_p\mathbf{C}_{m,n}^t\right], \left[{}_ps^s, {}_p\mathbf{S}_{m,n}^s\right]\right).$$

When the inputs $\left[{}_p\vec{a^i},{}_p\mathbf{A}_{m,n}^t\right]$ or $\left[{}_p\vec{b^i},{}_p\mathbf{B}_{m,n}^t\right]$ are applied on the state $\left[{}_ts^s, {}_t\mathbf{S}_{m,n}^s\right]$ other states react to this change by mapping φ_{pq}. It is obvious that mapping φ_{pr} influences other states by the input from the respective input set. Then for every component of the state on the right side there exists an element from the corresponding input set, that state on the left hand side is included on the right hand side.

(b) For the second case, we consider different inputs $\left[{}_p\vec{a^i},{}_p\mathbf{A}_{m,n}^t\right] \in I_{\mathcal{VM}}^p, \left[{}_q\vec{b^i},{}_q\mathbf{B}_{m,n}^t\right] \in I_{\mathcal{VM}}^q$ now. On the left-hand side, we have an influence on two different components in the tuple $\left(\left[{}_1s^s, {}_1\mathbf{S}_{m,n}^s\right], \left[{}_2s^s, {}_2\mathbf{S}_{m,n}^s\right], \ldots, \left[{}_ts^s, {}_t\mathbf{S}_{m,n}^s\right]\right)$ it is evident that the same inputs are the same on the right-hand side if internal links $\varphi_{pp}, \varphi_{qq}$ do not change. At the moment, $\varphi_{pp}, \varphi_{pp}$ influence corresponding components of the state, we have suitable inputs on the right-hand side, as in proof of Theorem 2. For mapping the influence of φ_{pr} and φ_{qs} on other components, we have the same situation as case a). Thus, it is obvious that $\mathcal{SMA}il$ holds condition E-GMAC. □

In the conclusion of this section, we will demonstrate the theory explained by using the example of $\mathcal{SMA}il$ to describe and model a situation with several autonomous vehicles in traffic lanes each intending to perform some action.

Example 4. *We will consider five e-quasi-multiautomata* $\mathrm{MA}_1 = (I_{\mathcal{VM}}^1, \mathbb{S}_{\mathcal{RM}}^1, \delta_1), \mathrm{MA}_2 = (I_{\mathcal{VM}}^2, \mathbb{S}_{\mathcal{RM}}^2, \delta_2), \mathrm{MA}_3 = (I_{\mathcal{VM}}^3, \mathbb{S}_{\mathcal{RM}}^3, \delta_3), \mathrm{MA}_4 = (I_{\mathcal{VM}}^4, \mathbb{S}_{\mathcal{RM}}^4, \delta_4), \mathrm{MA}_5 = (I_{\mathcal{VM}}^5, \mathbb{S}_{\mathcal{RM}}^5, \delta_5)$, *where every e-quasi-multiautomaton represents a vehicle. The e-quasi-multiautomata* $\mathrm{MA}_2, \mathrm{MA}_3, \mathrm{MA}_4$ *are autonomous vehicles and* $\mathrm{MA}_1, \mathrm{MA}_5$ *are ordinary vehicles. Figure 5 depicts the state of every vehicle, where the vehicles* $\mathrm{MA}_1, \mathrm{MA}_5$ *are denoted in red colour in the same as verge of the road and other colour are used for autonomous vehicles. The complete situation with detection field of each vehicle is presented in Figure 6. In fact, Figure 5 is state of* $\mathcal{SMA}il$, *i.e.,*

$$\left(\left[{}_1r^s, {}_1\mathbf{R}_{m,n}^s\right], \left[{}_2s^s, {}_2\mathbf{S}_{m,n}^s\right], \left[{}_3t^s, {}_3\mathbf{T}_{m,n}^s\right], \left[{}_4u^s, {}_4\mathbf{U}_{m,n}^s\right], \left[{}_5v^s, {}_5\mathbf{V}_{m,n}^s\right]\right).$$

Next, we need a matrix of the internal link

$$\begin{bmatrix} 0 & 0 & 0 & 0 & 0 \\ 0 & \varphi & \varphi & \varphi & 0 \\ 0 & \varphi & \varphi & \varphi & 0 \\ 0 & \varphi & \varphi & \varphi & 0 \\ 0 & 0 & 0 & 0 & 0 \end{bmatrix},$$

where 0 means no internal link and φ mean internal link between vehicles given by the respective indices. We are going to describe how to proceed, i.e., what input symbols to use, if the grey car, represented by e-quasi-multiautomaton MA_3, *wants to change lane and turn left to lane 5.*

We have two approaches to start changing the lane: we can correct of the state of the vehicle directly by the component on the 2nd, 3rd and 4th position in the input word, or to use the internal link and 6th component of the input. We will use an input $\left[(3 3^t, {}_3 1^t, {}_3 0.98^t, {}_3 0^t, {}_3 1^t, {}_3 \frac{\pi}{5}^t), {}_3(b_{i,j})^t = 1\right] \in I_{\mathcal{VM}}^3$, *which will operate on* the 6th component. Then, the resulting state—full correction—will be made by the internal link. The calculation will be done according to Equations (7) and (15), i.e.,

$$\delta_3\left(\left[(_33^t, {}_31^t, {}_30.99^t, {}_30^t, {}_31^t, {}_3\frac{\pi^t}{5}), {}_3(b_{i,j})^t = 1\right], \left[{}_3t^s, {}_3T^s_{m,n}\right]\right) =$$

$$\left[\left({}_33, {}_330, {}_3192.06, {}_33, {}_31, {}_3\frac{\pi}{5}\right), {}_3T^s_{n,m}\right] \quad (20)$$

We obtain a state with new elements on the $5^{th}, 6^{th}$ positions where these components have an effect on other components by means of the internal link φ_{33} between the state and input of the e-quasi-multiautomaton $\mathbb{MA}_3$. With this internal link, we obtain a new input:

$$\varphi_{33}\left(\left[\left({}_33^s, {}_330^s, {}_3192.06^s, {}_33^s, {}_31^s, {}_3\frac{\pi^s}{5}\right), {}_3T^s_{n,m}\right]\right) = \left[({}_33^t, {}_31^t, {}_31^t, {}_31^t, {}_31^t, {}_30^s), {}_3C^t_{n,m}\right],$$

where ${}_3C^t_{n,m} = \begin{bmatrix} 1 & 1 & 1 & 1 & 1 & 1 & 1 & 1 & 1 & -1 \\ -1 & 1 & 1 & -1 & 1 & 1 & 1 & 1 & 1 & -1 \\ -1 & 1 & 1 & -1 & 1 & 1 & 1 & 1 & 1 & -1 \\ -1 & 1 & 1 & -1 & 1 & 1 & 1 & 1 & 1 & 1 \\ 1 & 1 & 1 & 1 & 1 & 1 & 1 & 1 & 1 & 1 \\ 1 & 1 & 1 & 1 & 1 & 1 & 1 & 1 & 1 & 1 \\ 1 & 1 & 1 & 1 & 1 & 1 & 1 & 1 & 1 & 1 \\ 1 & 1 & 1 & 1 & 1 & 1 & 1 & 1 & 1 & 1 \\ 1 & 1 & 1 & 1 & -1 & 1 & 1 & -1 & 1 & 1 & -1 \\ 1 & 1 & 1 & 1 & -1 & 1 & 1 & -1 & 1 & 1 & -1 \\ 1 & 1 & 1 & 1 & -1 & 1 & 1 & -1 & 1 & 1 & -1 \end{bmatrix}$, i.e., matrix to correct the state matrix

detecting other vehicles or obstacles near vehicle A_3.

When we apply a new input on the state $\left[({}_33, {}_330, {}_3192.06, {}_33, {}_31, {}_3\frac{\pi}{5}), {}_3T^s_{n,m}\right]$, we obtain

$$\delta_3\left(\left[({}_33^t, {}_31^t, {}_31^t, {}_31^t, {}_31^t, {}_30^t), {}_3C^s_{n,m}\right], \left[{}_3t^s, {}_3T^s_{m,n}\right]\right) = \left[({}_33^s, {}_330^s, {}_3192.06^s, {}_34^s, {}_31^s, {}_30^s), {}_3D^s_{n,m}\right],$$

where matrix D has the same size and entries as depicted for $\mathbb{MA}_3$ in Figure 7.

Now consider internal link φ_{32}, i.e., state of the e-quasi-multiautomaton $\mathbb{MA}_2$ will be changed by the state $\left[({}_33^s, {}_330^s, {}_3192.06^s, {}_33^s, {}_31^s, {}_3\frac{\pi^s}{5}), {}_3T^s_{n,m}\right]$ of the e-quasi-multiautomaton $\mathbb{MA}_3$, which we obtained in Equation (20).

We will proceed using the definition of the link between two different e-quasi-multiautomata given as Equation (17).

$$\varphi_{32}\left(\left[\left({}_33^s, {}_330^s, {}_3192.06^s, {}_33^s, {}_31^s, {}_3\frac{\pi^s}{5}\right), {}_3T^s_{n,m}\right]\right) =$$

$$\left[\left({}_22^t, {}_2\left(\frac{30}{31} + 0.0628\right)^t, {}_20.99^t, {}_20^t, {}_21^t, {}_20^t\right), {}_2M^t_{n,m}\right] \quad (21)$$

where ${}_2M^t_{n,m}$ is a suitable matrix which enables us to obtain a state matrix with entries given in Figure 7 for the state of $\mathbb{MA}_2$. After we apply the input obtained by the internal link φ_{32} in Equation (21) with the help of transition function δ_2, we get a new state depicted in Figure 7 for e-quasi-multiautomaton $\mathbb{MA}_2$.

Next state of the (vehicle) $\mathbb{MA}_4$ will not affect velocity by internal link φ_{24}, because the input obtained by φ_{24} from state $\left[({}_33^s, {}_330^s, {}_3192.06^s, {}_33^s, {}_31^s, {}_3\frac{\pi^s}{5}), {}_3T^s_{n,m}\right]$ has parameters $\lambda = 0$ and $\alpha = 0$.

Thus, the input $\left[\left({}_44^t, {}_3\left(\frac{30}{30+1}\right)^t, {}_40.99^t, {}_40^t, {}_31^t, {}_30^s\right), {}_3N^t_{m,n}\right]$ operates as neutral input on states of $\mathbb{MA}_4$ except for the distance given by ${}_4s^s_3$. Thus, we will present a new situation on the lanes, i.e., a new state after the application of one input. See Figure 7.

Figure 5. $\mathcal{SMA}il$ for lane change.

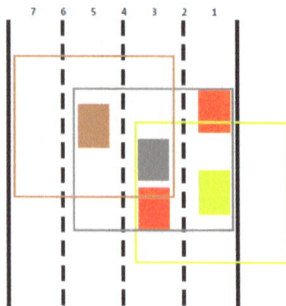

Figure 6. Complete situation in lanes with detection field of each vehicle.

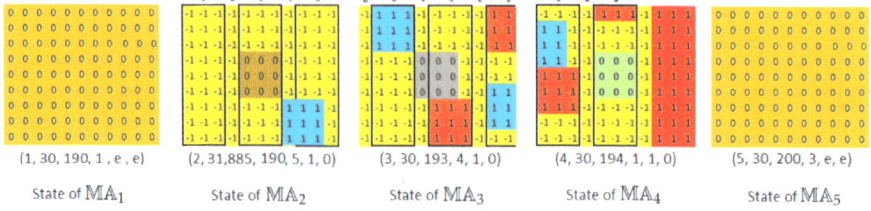

Figure 7. A new state of $\mathcal{SMA}il$ with the release position.

4. Conclusions

In this paper, we presented the concept of cartesian compositions, which was first introduced by W. Dörfler. In the past, cartesian compositions were generalized in the sense of hyperstructure theory (using complete hypergroups). Now we considered the internal link between the cartesian product of state sets. We described these internal links by matrices and decision functions. These functions determine which state will influence other components. Our modified concept of the cartesian composition is suitable to describe and control systems used for real-life applications, as was shown in examples throughout the paper. While constructing our system called $\mathcal{SMA}il$, we assumed that it was made up of e-quasi-multiautomata of a similar nature. This fact affected the proof of the E-GMAC condition (see Theorem 4), which was a substantial simplification of the procedure discussed in [13]. In our future research, we shall concentrate on answering the question of whether the internal link may not remove the necessity to use the extensions of GMAC, i.e., whether the pure GMAC could be used instead of E-GMAC.

Funding: This article was produced with the financial support of the Ministry of Education, Youth and Sports within the National Sustainability Programme I, project of Transport R&D Centre (LO1610), on the research infrastructure acquired from the Operation Programme Research and Development for Innovations (CZ.1.05/2.1.00/03.0064).

Conflicts of Interest: The author declares no conflict of interest.

References

1. Corsini, P.; Leoreanu, V. *Applications of Hyperstructure Theory*; Kluwer Academic Publishers: Dodrecht, The Netherlands; Boston, MA, USA; London, UK, 2003.
2. Davvaz, B.; Leoreanu–Fotea, V. *Applications of Hyperring Theory*; International Academic Press: Palm Harbor, FL, USA, 2007.
3. Chvalina, J.; Smetana, B. Series of Semihypergroups of Time-Varying Artificial Neurons and Related Hyperstructures. *Symmetry* **2019**, *11*, 927. [CrossRef]
4. Cristea, I.; Kocijan, J.; Novák, M. Introduction to Dependence Relations and Their Links to Algebraic Hyperstructures. *Mathematics* **2019**, *7*, 885. [CrossRef]
5. Křehlík, Š.; Vyroubalová, J. The Symmetry of Lower and Upper Approximations, Determined by a Cyclic Hypergroup, Applicable in Control Theory. *Symmetry* **2020**, *12*, 54. [CrossRef]
6. Novák, M.; Křehlík, Š. *EL*–hyperstructures revisited. *Soft. Comput.* **2018**, *22*, 7269–7280. [CrossRef]
7. Al Tahan, M.; Hoskova-Mayerova, Š.; Davvaz, B. Some Results on (Generalized) Fuzzy Multi-Hv-Ideals of Hv-Rings. *Symmetry* **2019**, *11*, 1376. [CrossRef]
8. Chvalina, J. *Functional Graphs, Quasi-Ordered Sets and Commutative Hypergroups*; Masaryk University: Brno, Czech Republic, 1995. (In Czech)
9. Dörfler, W. The cartesian composition of automata. *Math. Syst. Theory* **1978**, *11*, 239–257. [CrossRef]
10. Gécseg, F.; Peák, I. *Algebraic Theory of Automata*; Akadémia Kiadó: Budapest, Hungary, 1972.
11. Ginzburg, A. *Algebraic Theory of Automata*; Academic Press: New York, NY, USA, 1968.
12. Novák, M.; Křehlík, Š.; Staněk, D. *n*-ary Cartesian composition of automata. *Soft. Comput.* **2019**, *24*, 1837–1849. [CrossRef]
13. Chvalina, J.; Křehlík, Š.; Novák, M. Cartesian composition and the problem of generalising the MAC condition to quasi-multiautomata. *An. Univ. Ovidius Constanta-Ser. Mat.* **2016**, *24*, 79–100.
14. Ashrafi, A.R.; Madanshekaf, A. Generalized action of a hypergroup on a set. *Ital. J. Pure Appl. Math.* **1998**, *15*, 127–135.
15. Chvalina, J. Infinite multiautomata with phase hypergroups of various operators. In Proceedings of the 10th International Congress on Algebraic Hyperstructures and Applications, Brno, Czech Republic, 3 September 2008; Hošková, Š., Ed.; University of Defense: Washington, DC, USA; pp. 57–69.
16. Chvalina, J.; Chvalinová, L. State hypergroups of automata. *Acta Math. Inform. Univ. Ostrav.* **1996**, *4*, 105–120.
17. Chvalina, J.; Hošková-Mayerová, Š.; Dehghan Nezhad, A. General actions of hyperstructures and some applications. *An. Univ. Ovidius Constanta-Ser. Mat.* **2013**, *21*, 59–82. [CrossRef]
18. Massouros, G.G. Hypercompositional structures in the theory of languages and automata. *An. Şt. Univ. A.I Cuza Iaşi, Sect. Inform.* **1994**, *3*, 65–73.
19. Dörfler, W. The direct product of automata and quasi-automata. In *Mathematical Foundations of Computer Science: 5th Symposium*; Mazurkiewicz, A., Ed.; Springer: Gdansk, Poland, 1976; pp. 6–10.
20. Chvalina, J.; Novák, M.; Křehlík, Š. Hyperstructure generalizations of quasi-automata induced by modelling functions and signal processing. In Proceedings of the 16th International Conference of Numerical Analysis and Applied Mathematics, Rhodes, Greece, 13–18 September 2018.
21. Křehlík, Š.; Novák, M. Modifed Product of Automata as a Better Tool for Description of Real-Life Systems. In Proceedings of the 17th International Conference of Numerical Analysis and Applied Mathematics, Rhodes, Greece, 23–29 September, 2019.
22. Golestan, K.; Seifzadeh, S.; Kamel, M.; Karray, F.; Sattar, F. Vehicle Localization in VANETs Using Data Fusion and V2V Communication. In Proceedings of the Second ACM International Symposium on Designand Analysis of Intelligent Vehicular Networks and Applications, Paphos, Cyprus, 21 October 2012; pp. 123–130.
23. Novák, N.; Křehlík, Š.; Ovaliadis, K. Elements of hyperstructure theory in UWSN design and data aggregation. *Symmetry* **2019**, *11*, 734. [CrossRef]

24. Dresner, K.; Stone, P. A multiagent approach to autonomous intersection management, *J. Artif. Intell. Res.* **2008**, *31*, 591–656. [CrossRef]
25. Fajardo, D.; Au, T.-C.; Waller, S.T.; Stone, P.; Yang, D. Automated Intersection Control: Performance of Future Innovation Versus Current Traffic Signal Control. *Transp. Res. Rec.* **2011**, *2259*, 223–232. [CrossRef]
26. Huifu, J.; Jia, H.; Shi, A.; Meng, W.; Byungkyu, B.P. Eco approaching at an isolated signalized intersection under partially connected and automated vehicles environment. *Transp. Res. Part C Emerg. Technol.* **2017**, *79*, 290–307.
27. Latombe, J. *Robot Motion Planning*; Kluwer Academic Publishers: Boston, MA, USA, 1991.
28. Liu, F.; Naraynan, A.; Bai, Q. Effective methods for generating collision free paths for multiple robots based on collision type (demonstration). In Proceedings of the 11th International Conference on Autonomous Agents and Multiagent Systems, Valencia, Spain, 4–8 June 2012; Volume 3, pp. 1459–1460.

 © 2020 by the author. Licensee MDPI, Basel, Switzerland. This article is an open access article distributed under the terms and conditions of the Creative Commons Attribution (CC BY) license (http://creativecommons.org/licenses/by/4.0/).

Article
Derived Hyperstructures from Hyperconics

Vahid Vahedi [1], Morteza Jafarpour [1,*,†], Sarka Hoskova-Mayerova [2],
Hossein Aghabozorgi [1], Violeta Leoreanu-Fotea [3] and Svajone Bekesiene [4]

[1] Department of Mathematics, Vali-e-Asr University of Rafsanjan, Rafsanjan 7718897111, Iran; v.vahedi@vru.ac.ir (V.V.); h.aghabozorgi@vru.ac.ir (H.A.)
[2] Department of Mathematics and Physics, University of Defence in Brno, Kounicova 65, 662 10 Brno, Czech Republic; sarka.mayerova@unob.cz
[3] Department of Mathematics, Al. I. Cuza University, 6600 Iashi, Romania; violeta.fotea@uaic.ro
[4] Departament of Defence Technologies, The General Jonas Žemaitis Military Academy of Lithuania, Silo g.5, 10322 Vilnius, Lithuania; svajone.bekesiene@lka.lt
* Correspondence: m.j@vru.ac.ir
† The corresponding author in deep gratitude dedicates this work and gives special thanks to Cardiovascular Surgeon A. R. Alizadeh Ghavidel.

Received: 26 January 2020; Accepted: 11 March 2020; Published: 16 March 2020

Abstract: In this paper, we introduce generalized quadratic forms and hyperconics over quotient hyperfields as a generalization of the notion of conics on fields. Conic curves utilized in cryptosystems; in fact the public key cryptosystem is based on the digital signature schemes (DLP) in conic curve groups. We associate some hyperoperations to hyperconics and investigate their properties. At the end, a collection of canonical hypergroups connected to hyperconics is proposed.

Keywords: hypergroup; hyperring; hyperfield; (hyper)conics; quadratic forms

MSC: 20N20; 14H52; 11G05

1. Introduction

In 1934, Marty initiated the notion of hypergroups as a generalization of groups and referred to its utility in solving some problems of groups, algebraic functions and rational fractions [1]. To review this theory one can study the books of Corsini [2], Davvaz and Leoreanu-Fotea [3], Corsini and Leoreanu [4], Vougiouklis [5] and in papers of Hoskova and Chvalina [6] and Hoskova-Mayerova and Antampoufis [7]. In recent years, the connection of hyperstructures theory with various fields has been entered into a new phase. For this we advise the researchers to see the following papers. (i) For connecting it to number theory, incidence geometry, and geometry in characteristic one [8–10]. (ii) For connecting it to tropical geometry, quadratic forms [11,12] and real algebraic geometry [13,14]. (iii) For relating it to some other objects see [15–19]. M. Krasner introduced the concept of the hyperfield and hyperring in Algebra [20,21]. The theory which was developed for the hyperrings is generalizing and extending the ring theory [22–25]. There are different types of hyperrings [22,25,26]. In the most general case a triplet $(R, +, \cdot)$ is a hyperring if $(R, +)$ is a hypergroup, $(R, \cdot)$ is a semihypergroup and the multiplication is bilaterally distributive with regards to the addition [3]. If $(R, \cdot)$ is a semigroup instead of semihypergroup, then the hyperring is called additive. A special type of additive hyperring is the Krasner's hyperring and hyperfield [20,21,24,27,28]. The construction of different classes of hyperrings can be found in [29–33]. There are different kinds of curves that basically are used in cryptography [34,35]. An elliptical curve is a curve of the form $y^2 = p(x)$, where $p(x)$ is a cubic polynomial with no-repeat roots over the field F. This kind of curves are considered and extended over Krasner's hyperfields in [13]. Now let $g(x,y) = ax^2 + bxy + cy^2 + dx + ey + f \in F[x,y]$ and

$g(x,y) = 0$ be the quadratic equation of two variables in field of F, if $a = c = 0$ and $b \neq 0$ then the equation $g(x,y) = 0$ is called homographic transformation. In [14] Vahedi et. al extended this particular quadratic equation on Krasner's quotient hyperfield $\frac{F}{G}$. The motivation of this paper goes in the same direction of [14]. If in the general form of the equation of quadratic form one suppose that $ae \neq 0$ and $b = 0$ then initiate an important quadratic equation which is called a conic. Notice that the conditions which are considered for the coefficients of the equations of a conic curve and a homographic curve are completely different. Until now the study of conic curves has been on fields. At the recent works the authors have investigated some main classes of curves; elliptic curves and homographics over Krasner's hyperfields (see [13,14]). In the present work, we study the conic curves over some quotients of Krasner's hyperfields.

2. Preliminaries

In the following, we recall some basic notions of Pell conics and hyperstructures theory that these topics can be found in the books [2,36,37]. Moreover, we fix here the notations that are used in this paper.

2.1. Conics

According to [36] a conic is a plane affine curve of degree 2. Irreducible conics C come in three types: we say that C is a hyperbola, a parabola, or an ellipse according as the number of points at infinity on (the projective closure of) C equals $2, 1$, or 0. Over an algebraically closed field, every irreducible conic is a hyperbola. Let d be a square free integer nonequal to 1 and put

$$\Delta = \begin{cases} d & \text{if } d \equiv 1 \pmod{4}, \\ 4d & \text{if } d \equiv 2,3 \pmod{4}. \end{cases}$$

The conic $C : Q_0(x,y) = 1$ associated to the principal quadratic form of discriminant Δ,

$$Q_0(x,y) = \begin{cases} x^2 + xy + \frac{1-d}{4}y^2 & \text{if } d \equiv 1 \pmod{4}, \\ x^2 - dy^2 & \text{if } d \equiv 2,3 \pmod{4}, \end{cases}$$

is called the *Pell conic* of discriminant. Pell conics are irreducible nonsingular affine curves with a distinguished integral point $N = (1, 0)$. The problem corresponding to the determination of E(Q) is finding the integral points on a Pell conic. The idea that certain sets of points on curves can be given a group structure is relatively modern. For elliptic curves, the group structure became well known only in the 1920s; implicitly it can be found in the work of Clebsch, and Juel, in a rarely cited article, wrote down the group law for elliptic curves defined over $\mathbb{R}$ and $\mathbb{C}$ at the end of the 19th century. The group law on Pell conics defined over a field F. For two rational points $p, q \in Q(F)$, draw the line through $\mathcal{O}$ parallel to line p, q, and denote its second point of intersection with $p * q$ which is the sum of two p, q, where $\mathcal{O}$ is an arbitrary point in pell conic perchance in infinity, is identity element of group. In the Figure 1 the operation is picturised on the conic section $Q_\mathbb{R}(f_{1,1})$.

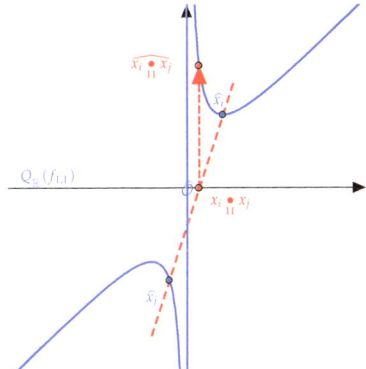

Figure 1. Conic section $Q_\mathbb{R}(f_{1,1})$.

Example 1. *Consider $f_{1,1}(x) = x^{-1} + x$ over finite field $F = \mathbb{Z}_7$. Then we have a Caley table of points (Table 1):*

Table 1. Conic group $(Q_{\mathbb{Z}_7}(f_{1,1}), \bullet_{a,b})$.

$\bullet_{11}$	$(0,\infty)$	$(1,2)$	$(2,-1)$	$(3,1)$	$(-3,-1)$	$(-2,1)$	$(-1,-2)$	(∞,∞)
$(0,\infty)$	$(0,\infty)$	$(1,2)$	$(2,-1)$	$(3,1)$	$(-3,-1)$	$(-2,1)$	$(-1,-2)$	(∞,∞)
$(1,2)$	$(1,2)$	$(-1,-2)$	$(-3,-1)$	$(-2,1)$	$(3,1)$	$(2,-1)$	$(0,\infty)$	(∞,∞)
$(2,-1)$	$(2,-1)$	$(-3,-1)$	$(1,2)$	$(-1,-2)$	(∞,∞)	$(0,\infty)$	$(-2,1)$	$(3,1)$
$(3,1)$	$(3,1)$	$(-2,1)$	$(-1,-2)$	$(1,2)$	$(0,\infty)$	(∞,∞)	$(-3,-1)$	$(2,-1)$
$(-3,-1)$	$(-3,-1)$	$(3,1)$	(∞,∞)	$(0,\infty)$	$(-1,-2)$	$(1,2)$	$(2,-1)$	$(-2,1)$
$(-2,1)$	$(-2,1)$	$(2,-1)$	$(0,\infty)$	(∞,∞)	$(1,2)$	$(-1,-2)$	$(3,1)$	$(-3,-1)$
$(-1,-2)$	$(-1,-2)$	$(0,\infty)$	$(-2,1)$	$(-3,-1)$	$(2,-1)$	$(3,+1)$	(∞,∞)	$(1,2)$
(∞,∞)	(∞,∞)	$(-1,-2)$	$(3,1)$	$(2,-1)$	$(-2,+1)$	$(-3,-1)$	$(1,2)$	$(0,\infty)$

The associativity of the group law is induced from a special case of Pascal's Theorem. In the following, we recall Pascal's Theorem which is a very special case of Bezout's Theorem.

Theorem 1 ([38] Pascal's Theorem). *For any conic and any six points $p_1, p_2, ..., p_6$ on it, the opposite sides of the resulting hexagram, extended if necessary, intersect at points lying on some straight line. More specifically, let $L(p,q)$ denote the line through the points p and q. Then the points $L(p_1,p_2) \cap L(p_4,p_5), L(p_2,p_3) \cap L(p_5,p_6)$, and $L(p_3,p_4) \cap L(p_6,p_1)$ lie on a straight line, called the Pascal line of the hexagon Figure 2.*

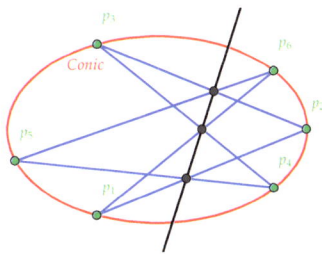

Figure 2. Pascal line of the hexagon.

2.2. Krasner's Hyperrings and Hyperfields

Let H be a non-empty set and $\mathcal{P}^*(H)$ denotes the set of all non-empty subsets of H. Any function $\cdot$ from the cartesian product $H \times H$ into $\mathcal{P}^*(H)$ is called a hyperoperation on H. The image of the pair $(a,b) \in H \times H$ under the hyperoperation $\cdot$ in $\mathcal{P}^*(H)$ is denoted by $a \cdot b$. The hyperoperation can be extended in a natural way to subsets of H as follows: for non-empty subsets A, B of H, define $A \cdot B = \bigcup_{a \in A, b \in B} a \cdot b$. The notation $a \cdot A$ is applied for $\{a\} \cdot A$ and also $A \cdot a$ for $A \cdot \{a\}$. Generally, we mean $H^k = H \times H... \times H$ (k times), for all $k \in \mathbb{N}$ and also the singleton $\{a\}$ is identified with its element a. The hyperstructure $(H, \cdot)$ is called a semihypergroup if $x \cdot (y \cdot z) = (x \cdot y) \cdot z$ for all $x, y, z \in H$, which means that

$$\bigcup_{u \in x \cdot y} u \cdot z = \bigcup_{v \in y \cdot z} x \cdot v.$$

A semihypergroup $(H, \cdot)$ is called a hypergroup if the reproduction law holds: $x \cdot H = H \cdot x = H$, for all $x \in H$.

Definition 1. [2] *Let $(H, \cdot)$ be a hypergroup and K be a non-empty subset of H. We say that $(K, \cdot)$ is a subhypergroup of H and it denotes $K \leqslant H$, if for all $x \in K$ we have $K \cdot x = K = x \cdot K$.*

Let $(H, \cdot)$ be a hypergroup, an element e_r (resp. e_l) of H is called a right identity (resp. left identity e_l) if for all $a \in H$, $x \in a \cdot e_r$ (resp. $a \in e_l \cdot a$). An element e is called an identity if, for all $a \in H$, $a \in a \cdot e \cap e \cdot a$. A right identity e_r (resp. left identity e_l) of H is called a scalar right identity (resp. scalar left identity) if for all $a \in H$, $a = a \cdot e_r$ ($a = e_l \cdot x$). An element e is called a scalar identity if for all $a \in H$, $a = a \cdot e = e \cdot x$. An element $a' \in H$ is called a right inverse (resp. left inverse) of a in H if $e_r \in a \cdot a'$, for some right identities e_r in H ($e_l \in a' \cdot a$). An element $a' \in H$ is called an inverse of $a \in H$ if $e \in a' \cdot a \cap a \cdot a'$, for some identities in H. We denote the set of all right inverses, left inverses and inverses of $a \in H$ by $i_r(a)$, $i_l(a)$, and $i(a)$, respectively.

Definition 2. [2] *A hypergroup $(H, \cdot)$ is called reversible, if the following conditions hold:*

(i) *At least H has one identity e;*
(ii) *every element x of H has one inverse, that is $i(x) \neq \emptyset$;*
(iv) *$x \in y \cdot z$ implies that $y \in x \cdot z'$ and $z \in y' \cdot x$, where $z' \in i(z)$ and $y' \in i(y)$.*

Definition 3. [2,23] *A hypergroup $(H, +)$ is called canonical, if the following conditions hold:*

(i) *for every $x, y, z \in H$, $x + (y + z) = (x + y) + z$,*
(ii) *for every $x, y \in H$, $x + y = y + x$,*
(iii) *there exists $0 \in H$ such that $0 + x = \{x\}$ for every $x \in H$,*
(iv) *for every $x \in H$ there exists a unique element $x' \in H$ such that $0 \in x + x'$; (we shall write $-x$ for x' and we call it the opposite of x.)*
(v) *$z \in x + y$ implies $y \in z - x$ and $x \in z - y$;*

Definition 4. [2] *Suppose that $(H, \cdot)$ and $(K, \circ)$ are two hypergroups. A function $f : H \rightarrow K$ is called a homomorphism if $f(x \cdot y) \subseteq f(x) \circ f(y)$, for all x and y in H. We say that f is a good homomorphism if for all x and y in H, $f(x \cdot y) = f(x) \circ f(y)$.*

The more general hyperstructure that satisfies the ring-like conditions is the hyperring. The notion of the hyperring and hyperfield was introduced in Algebra by M. Krasner in 1956 [21]. According to the current terminology, these initial hypercompositional structures are additive hyperrings and hyperfields whose additive part is a canonical hypergroup. Nowadays such hypercompositional structures are called Krasner's hyperrings and hyperfields.

Definition 5. [20] *A Krasner's hyperring is an algebraic structure $(R, +, \cdot)$ which satisfies the following axioms:*

(1) *$(R, +)$ is a canonical hypergroup,*

(2) $(R, \cdot)$ is a semigroup having zero as a bilaterally absorbing element, i.e., $x \cdot 0 = 0 \cdot x = 0$.
(3) The multiplication is distributive with respect to the hyperoperation $+$.

A Krasner's hyperring is called commutative if the multiplicative semigroup is a commutative monoid. A Krasner's hyperring is called a *Krasner's hyperfield*, if $(R - \{0\}, \cdot)$ is a commutative group. In [20] Krasner presented a class of hyperrings which is constructed from rings. He proved that if R is a ring and G is a normal subgroup of R's multiplicative semigroup, then the multiplicative classes $\bar{x} = xG, x \in R$, form a partition of R. He also proved that the product of two such classes, as subsets of R, is a class $mod\ G$ as well, while their sum is a union of such classes. Next, he proved that the set $\bar{R} = \frac{R}{G}$ of these classes becomes a hyperring, when:

(i) $xG \oplus yG = \{zG | z \in xG + yG\}$, and
(ii) $xG \odot yG = xyG$,

Moreover, he observed that if R is a field, then $\frac{R}{G}$ is a hyperfield. Krasner named these hypercompositional structures quotient hyperring and quotient hyperfield, respectively. At the same time, he raised the question if there exist non-quotient hyperrings and hyperfields [20]. Massouros in [27] generalized Krasner's construction using not normal multiplicative subgroups, and proved the existence of non-quotient hyperrings and hyperfields. Since the paper deals only with Krasner's hyperfields we will write simply quotient hyperfields instead of Krasner's quotient hyperfields.

3. Hyperconic

The notion of hyperconics on a quotient hyperfield will be studied in this section. By the use of hyperconic $Q_{\bar{F}}(f_{\bar{A},\bar{B}})$, we present some hyperoperations as a generalization group operations on fields. We investigate some attributes of the associated hypergroups from the hyperconics and the associated H_v-groups on the hyperconics.

Let $g(x,y) = ax^2 + bxy + cy^2 + dx + ey + f \in F[x,y]$ and $g(x,y) = 0$ be the quadratic equation of two variables in field of F. If $c = 0$ and equation $g(x,y) = 0$ still stay in quadratic and two variables or the other word $c = 0$ and $(a,b) \neq 0 \neq (e,b)$, then it can be calculated as an explicit function y in terms of x, also with a change of variables, can be expressed in the form of $Y = AX^2 + BX$ or $AX^{-1} + BX$, where $A, B \in F$.

For this purpose if $a, e \neq 0 = b$ set $x = X$ and $y = Y - fe^{-1}$ then $Y = AX^2 + BX$, where $A = -ae^{-1}, B = -de^{-1}$. If $b \neq 0 = a$ set $x = X - eb^{-1}$ and $y = Y - db^{-1}$ then $Y = AX^{-1}$, where $A = edb^{-2} - fb^{-1}$. If $b \neq 0 \neq a$ set $x = X - eb^{-1}$ and $y = Y + 2aeb^{-2}\alpha_F - db^{-1}$ that

$$\alpha_F = \begin{cases} 0, & \text{if } char(F) = 2 \\ 1, & \text{if } char(F) \neq 2, \end{cases}$$

then $Y = AX^{-1} + BX$, where $A = -ae^2b^{-3} + edb^{-2} - fb^{-1}$ and $B = -ab^{-1}$. Reduced quadratic equation of two variables $ax^2 + bxy + dx + ey + f = 0$ in field of F can be generalized in quotient hyperfield $\bar{F}$.

Definition 6. Let $\bar{F}$ be the quotient hyperfield and $(\bar{A}, \bar{B}) \in \bar{F}^2$ and $f_{\bar{A},\bar{B}}(\bar{x})$ be equal to $\bar{A}\bar{x}^{-1} \oplus \bar{B}\bar{x}$ or $\bar{A}\bar{x}^2 \oplus \bar{B}\bar{x}$. Then the relation $\bar{y} \in f_{\bar{A},\bar{B}}(\bar{x})$, is called generalized reduced two variable quadratic equation in $\bar{F}^2$. Moreover the set $Q(f_{\bar{A},\bar{B}}, \bar{F}) = \{(\bar{x}, \bar{y}) \in \bar{F}^2 | \bar{y} \in f_{\bar{A},\bar{B}}(\bar{x})\}$ is called conic hypersection, and if $\bar{A} \neq 0$, $Q(f_{\bar{A},\bar{B}}, \bar{F})$ is named non-degenerate conic hypersection. For all $a \in \bar{A}$ and $b \in \bar{B}$, $Q(f_{a,b}, F) = \{(x,y) \in F^2 | y = f_{a,b}(x)\}$ is conic section and for $a \neq 0$ is non-degenerate conic section, in which $f_{a,b}(z) = az^2 + bz$ or $az^{-1} + bz$ corresponding to $f_{\bar{A},\bar{B}}$. It is also said to $Q(f_{\bar{A},\bar{B}}, F) = \bigcup_{(\bar{x},\bar{y}) \in Q(f_{\bar{A},\bar{B}}, \bar{F})} \bar{x} \times \bar{y}$, conic hypersection, as a subset of F^2, and $Q(f_{a,b}, \bar{F}) = \overline{Q(f_{a,b}, F)} = \{\overline{(x,y)} | (x,y) \in Q(f_{a,b}, F)\}$, where

$\overline{(x,y)} = (\bar{x}, \bar{y})$ for all $(x,y) \in Q(f_{\bar{A},\bar{B}}, F)$.

Theorem 2. *Using the above notions we have* $Q(f_{\bar{A},\bar{B}}, F) = \bigcup_{a \in \bar{A}, b \in \bar{B}} Q(f_{a,b}, F)$.

Proof. Let $(x,y) \in Q(f_{\bar{A},\bar{B}}, F)$ and without losing of generality $f(x) = Ax^2 + Bx$. Then

$$(\bar{x}, \bar{y}) \in Q(f_{\bar{A},\bar{B}}, \bar{F}) \iff \bar{y} \in \bar{A}\bar{x}^2 \oplus \bar{B}\bar{x}$$
$$\iff \bar{y} \in \overline{Ax^2 + Bx}$$
$$\iff \bar{y} = \overline{ax^2 + bx}, \text{ for some } (a,b) \in \bar{A} \times \bar{B}$$
$$\iff y = agx^2 + bgx \text{ for some } g \in G$$
$$\iff y = a'x^2 + b'x, \text{ where } a' = ag, \ b' = bg$$
$$\iff (x,y) \in Q(f_{a',b'}, F), \text{ for some } (a',b') \in \bar{A} \times \bar{B}$$
$$\iff (x,y) \in \bigcup_{a \in \bar{A}, b \in \bar{B}} Q(f_{a,b}, F).$$

Consequently, $Q(f_{\bar{A},\bar{B}}, F) = \bigcup_{a \in \bar{A}, b \in \bar{B}} Q(f_{a,b}, F)$. □

Example 2. *Let $F = \mathbb{Z}_5$ be the field of order 5, $G = \{\pm 1\} \leqslant F^*$ and $f_{\bar{1},0}(\bar{x}) = \bar{x}^2$. Then we have $\bar{F} = \{\bar{0}, \bar{1}, \bar{2}\}$, $Q(f_{\bar{1},0}, \bar{F}) = \overline{Q(f_{1,0}, F)} \cup \overline{Q(f_{(-1),0}, F)}$, where*

$$Q(f_{1,0}, F) = \{(0,0), (1,1), (-1,1), (2,-1), (-2,-1)\},$$

$$Q(f_{(-1),0}, F) = \{(0,0), (1,-1), (-1,-1), (2,1), (-2,1)\},$$

and $Q(f_{1,0}, \bar{F}) = \overline{Q(f_{1,0}, F)} = \{(\bar{0}, \bar{0}), (\bar{1}, \bar{1}), (\bar{2}, \bar{1})\} = \overline{Q(f_{(-1),0}, F)} = Q(f_{(-1),0}, \bar{F})$. In this case $Q(f_{\bar{1},0}, \bar{F})$ is a non-degenerate conic hypersection because $\bar{A} = \bar{1} \neq \bar{0}$.

Definition 7. *Let F be a field, $x \in F$ and G be a subgroup in F^*. We take*

$$\mathcal{O} = \begin{cases} 0^{-1}, & \text{if } f_{a,b}(z) = az^2 + bz \\ 0, & \text{if } f_{a,b}(z) = az^{-1} + bz \end{cases}$$

$$\mathcal{G}_x(f_{a,b}) = \begin{cases} \{x\}, & \text{if } G = \{1\} \\ \{z \in F | f_{a,b}(z) = f_{a,b}(x)\}, & \text{if } G \neq \{1\}, f_{a,b}(z) = az^2 + bz \\ \{-x, x\}, & \text{if } G \neq \{1\}, f_{a,b}(z) = az^{-1} + bz. \end{cases}$$

Obviously, 0^{-1} is an element outside of F. We denote $0^{-1} = \frac{1}{0}$ by ∞, where $\infty \notin F$, and $\bar{\infty} = \infty$. Suppose that $\mathcal{G}_\infty(f_{a,b}) = \{\infty\}$, $f_{a,b}(\infty) = \infty$, $f_{a,b}(\mathcal{O}) = \infty$, for all $a \in \bar{A}, b \in \bar{B}$, also $\hat{X} = \{\hat{x} | x \in X\}$ where $\hat{x} = (x, f_{a,b}(x))$ and $X \subseteq F \cup \{\mathcal{O}\}$. Moreover, $\mathcal{O} \cdot \mathcal{O} = \mathcal{O} = \mathcal{O} + \mathcal{O}$, $x \cdot \mathcal{O} = \begin{cases} \mathcal{O}, & \text{if } x \neq 0 \\ 0, & \text{if } x = 0 \end{cases}$ and

$$x + \mathcal{O} = \begin{cases} \mathcal{O}, & \text{if } \mathcal{O} = 0^{-1} \\ x, & \text{if } \mathcal{O} = 0 \end{cases}, \text{ for all } x \text{ in field of } (F, +, \cdot).$$

Remark 1. *It should be noted that associativity by adding $\mathcal{O}$ to field of $(F, +, \cdot)$ for two operations of "+" and "·" remains preserved.*

Definition 8. Let $Q(f_{\tilde{A},\tilde{B}}, \tilde{F})$ be a non-degenerate conic hypersection, $F_\infty = F \cup \{\infty\}$ and

$$Q_F(f_{a,b}) = \{\hat{x} : x \in F_\infty, \hat{x} \notin L_0\},$$

$$Q_F(f_{\tilde{A},\tilde{B}}) = \bigcup_{a \in \tilde{A}, b \in \tilde{B}} Q_F(f_{a,b}),$$

where $L_0 = \{(x,0) | x \in F_{\mathcal{O}}\}$. For all $\hat{x}_i, \hat{x}_i \in Q_F(f_{a,b})$

$$\widehat{x_i \bullet_{ab} x_j} = (x_i \bullet_{ab} x_j, f_{a,b}(x_i \bullet_{ab} x_j)) \text{ in which } \{(x_i \bullet_{ab} x_j, 0)\} = L_0 \cap L_{a,b}(\hat{x}_i, \hat{x}_j),$$

and

$$L_{a,b}(\hat{x}_i, \hat{x}_j) = \begin{cases} \{(x,y) \in F^2 | y - f_{a,b}(x_i) = \frac{f_{a,b}(x_j) - f_{a,b}(x_i)}{(x_j - x_i)}(x - x_i)\}, & x_i \neq x_j, \mathcal{O} \notin \{x_i, x_j\} \\ \{(x,y) \in F^2 | y - f_{a,b}(x_i) = f'_{a,b}(x_i)(x - x_i)\}, & x_i = x_j \notin \{\mathcal{O}\} \\ \{(x,y) \in F^2 | \mathcal{O} \neq x \in \{x_i, x_j\}\} \cup \{\hat{\mathcal{O}}\}, & x_i \neq x_j, \mathcal{O} \in \{x_i, x_j\} \\ \{(\mathcal{O}, y) | y \in F_\infty = F \cup \{\infty\}\}, & (x_i, x_j) = (\mathcal{O}, \mathcal{O}), \end{cases}$$

and $f'_{a,b}$ is meant by formal derivative $f_{a,b}$.

We denote $\overline{Q_F(f_{a,b})}$ by $Q_{\bar{F}}(f_{a,b})$ and $\overline{Q_F(f_{\tilde{A},\tilde{B}})}$ by $Q_{\bar{F}}(f_{\tilde{A},\tilde{B}})$ also take $\bar{\mathcal{O}} = \{\mathcal{O}\} = \mathcal{O}, \overline{f(\mathcal{O})} = \{f(\mathcal{O})\} = f(\mathcal{O})$ and, $\overline{(\mathcal{O}, f(\mathcal{O}))} = (\bar{\mathcal{O}}, \overline{f(\mathcal{O})}) = (\mathcal{O}, f(\mathcal{O}))$. Moreover, $\mathcal{O} \odot \mathcal{O} = \mathcal{O}$ and $\mathcal{O} \oplus \mathcal{O} = \mathcal{O}$ also, for all $\bar{x}$ in hyperfield of $(\bar{F}, \oplus, \odot)$, $\bar{x} \odot \mathcal{O} = \begin{cases} \mathcal{O}, & \text{if } x \neq 0 \\ \bar{0}, & \text{if } x = 0 \end{cases}$ $\bar{x} \oplus \mathcal{O} = \begin{cases} \mathcal{O}, & \text{if } \mathcal{O} = 0^{-1} \\ \bar{x}, & \text{if } \mathcal{O} = 0 \end{cases}$ and agree to $L_0 \cap L(\hat{x}_i, \hat{x}_j) = \{(\infty, 0)\}$ if $f_{a,b}(x_i) = f_{a,b}(x_j)$. In addition say to $L_{a,b}(\hat{x}_i, \hat{x}_j)$ the line passing from $\hat{x}_i, \hat{x}_j$, Intuitively each line passing from $(\mathcal{O}, \infty)$ is called vertical line, and every vertical line pass through $(\mathcal{O}, \infty)$. $\hat{\mathcal{O}}$ is playing an asymptotic extension role for function $f_{a,b}$

Remark 2. By adding $\mathcal{O}$ to hyperfield of $(F, \oplus, \odot)$ associativity for two hyperoperations of "$\oplus$" and "$\odot$" remains preserved.

Suppose that $\hat{x} \in Q(f_{a,b}, F)$ and $\tilde{x} = \begin{cases} \{x\}, & f_{a,b}(x) = ax^2 + bx \\ \{x, -x\}, & f_{a,b}(x) = ax^{-1} + bx \end{cases}$ Hence, we the following proposition

Proposition 1. if $|Q_F(\tilde{f}_{a_1,b_1}) \cap Q_F(\tilde{f}_{a_2,b_2})| \geq 2$ then $Q(f_{a_1,b_1}, F) = Q(f_{a_2,b_2}, F)$.

Proof. Let $\{\tilde{x}_1, \tilde{x}_2\} \subseteq Q_F(\tilde{f}_{a_1,b_1}) \cap Q_F(\tilde{f}_{a_2,b_2})$, $\tilde{x}_1 \neq \tilde{x}_2$ and $i, j = 1, 2$. Then

$$y_i = a_j x_i^2 + b_j x_i \implies x_1 \neq x_2 \implies \begin{cases} a_1 = a_2 = \dfrac{x_2 y_1 - x_1 y_2}{x_1^2 x_2 - x_2^2 x_1}, \\ b_1 = b_2 = \dfrac{-x_2^2 y_1 + x_1^2 y_2}{x_1^2 x_2 - x_2^2 x_1}, \end{cases}$$

$$y_i = a_j x_i^{-1} + b_j x_i \implies x_1 \neq \pm x_2 \implies \begin{cases} a_1 = a_2 = \dfrac{x_2 y_1 - x_1 y_2}{x_2 x_1^{-1} - x_1 x_2^{-1}}, \\ b_1 = b_2 = \dfrac{-x_2^{-1} y_1 + x_1^{-1} y_2}{x_2 x_1^{-1} - x_1 x_2^{-1}}. \end{cases}$$

Hence $Q(f_{a_1,b_1}, F) = Q(f_{a_2,b_2}, F)$, as we expected. □

Definition 9. Let $Q(f_{\bar{A},\bar{B}}, \bar{F})$ be a non-degenerate conic hypersection then it is named hyperconic and denoted to $Q_{\bar{F}}(f_{\bar{A},\bar{B}})$, if the following implication for all $a,c \in \bar{A}$ and $b,d \in \bar{B}$ holds:

$$Q_F(f_{a,b}) \cap Q_F(f_{c,d}) \neq \begin{cases} \{\hat{\mathcal{O}}\}, & f_{a,b}(x) = ax^2 + bx \\ \{\hat{\mathcal{O}}, \hat{\infty}\}, & f_{a,b}(z) = ax^{-1} + bx \end{cases} \implies Q_F(f_{a,b}) = Q_F(f_{c,d}).$$

Proposition 2. Let $\hat{x}_i = (x_i, f(x_i))$ and $\hat{x}_j = (x_j, f(x_j))$ belong to $Q_F(f_{a,b})$, then

$$x_i \bullet_{ab} x_j = \begin{cases} \dfrac{x_i f_{a,b}(x_j) - x_j f_{a,b}(x_i)}{f_{a,b}(x_j) - f_{a,b}(x_i)} & x_i \neq x_j, \mathcal{O} \notin \{x_i, x_j\}, \\ x_i - \dfrac{f_{a,b}(x_i)}{f'_{a,b}(x_i)} & x_i = x_j \notin \{\mathcal{O}\}, \\ x_i & x_i \neq \mathcal{O} = x_j, \\ x_j & x_j \neq \mathcal{O} = x_i, \\ \mathcal{O} & (x_i, x_j) = (\mathcal{O}, \mathcal{O}). \end{cases}$$

Proof. The proof is straightforward for the first two cases. If $f_{a,b}(x_i) = f_{a,b}(x_j)$ then

$$x_i \bullet_{ab} x_j = \dfrac{x_i f_{a,b}(x_j) - x_j f_{a,b}(x_i)}{f_{a,b}(x_j) - f_{a,b}(x_i)} = \dfrac{x_i f_{a,b}(x_j) - x_j f_{a,b}(x_i)}{0} = \infty,$$

$$\{(x_i \bullet_{ab} x_j, 0)\} = L_o \cap L(\hat{x}_i, \hat{x}_j) = \{(\infty, 0)\} \implies x_i \bullet_{ab} x_j = \infty.$$

Suppose that $(x_i, x_j) \in Q_F^2(f_{a,b})$ by regarding Definition 8 if $x_i \neq \mathcal{O} = x_j$ then

$$\{(x_i \bullet_{ab} \mathcal{O}, 0)\} = L_0 \cap L_{a,b}(\hat{x}_i, \hat{\mathcal{O}}) = \{(x_i, 0)\} \implies x_i \bullet_{ab} \mathcal{O} = x_i,$$

if $x_j \neq \mathcal{O} = x_i$ then proof is similar to previous manner, ultimately if $x_i = x_j = \mathcal{O}$ then

$$\{(\mathcal{O} \bullet_{a,b} \mathcal{O}, 0)\} = L_0 \cap L(\hat{\mathcal{O}}, \hat{\mathcal{O}}) = \{(\mathcal{O}, 0)\} \implies \mathcal{O} \bullet_{a,b} \mathcal{O} = \mathcal{O}. \quad \square$$

Remark 3. $(Q_F(f_{a,b}), \bullet_{ab})$ is a conic group, for all $(a,b) \in \bar{A} \times \bar{B}$. Notice that $\bullet_{ab}$ is the group operation on the conic $Q_F(f_{a,b})$.

Example 3. Let $F = \mathbb{Z}_5$ the field of order 5, $G = \{\pm 1\} \leqslant F^*$ and $f_{\bar{1},\bar{0}}(\bar{x}) = \bar{x}^2$. Then we have $\bar{F} = \{\bar{0}, \bar{1}, \bar{2}\}$, $Q_{\bar{F}}(f_{\bar{1},\bar{0}}) = \overline{Q_F(f_{\bar{1},\bar{0}})} \cup \overline{Q_F(f_{(-1),\bar{0}})}$, where $Q_F(f_{\bar{1},\bar{0}}) = \{\hat{\mathcal{O}}, (1,1), (-1,1), (2,-1), (-2,-1)\}$ and $\overline{Q_F(f_{\bar{1},\bar{0}})} = \{\hat{\mathcal{O}}, (\bar{1},\bar{1}), (\bar{2},\bar{1})\}$, $Q_F(f_{(-1),\bar{0}}) = \{\hat{\mathcal{O}}, (1,-1), (-1,-1), (2,1), (-2,1)\}$, and $\overline{Q_F(f_{(-1),\bar{0}})} = \{\hat{\mathcal{O}}, (\bar{1},\bar{1}), (\bar{2},\bar{1})\}$, in this case $Q_{\bar{F}}(f_{\bar{1},\bar{0}})$ is a hyperconic because $Q_F(f_{\bar{1},\bar{0}}) \cap Q_F(f_{(-1),\bar{0}}) = \hat{\mathcal{O}}$.

Definition 10. We introduce hyperoperation "$\circ$" on $Q_F(f_{\bar{A},\bar{B}})$ as follows:
Let $(x,y), (x',y') \in Q_F(f_{\bar{A},\bar{B}})$. If $(x,y) \in Q_F(f_{a,b})$ and $(x',y') \in Q_F(f_{a',b'})$ for some $a,a' \in \bar{A}$ and $b,b' \in \bar{B}$.

$$(x,y) \circ (x',y') = \begin{cases} \{\widehat{x_i \bullet_{ab} x_j} | (x_i, x_j) \in \mathcal{G}_x(f_{a,b}) \times \mathcal{G}_{x'}(f_{a',b'})\}, & \text{if } Q_F(f_{a,b}) = Q_F(f_{a',b'}) \\ Q_F(f_{a,b}) \cup Q_F(f_{a',b'}), & \text{otherwise.} \end{cases}$$

Theorem 3. $(Q_F(f_{\bar{A},\bar{B}}), \circ)$ is a hypergroup.

Proof. Suppose that $\{X, Y, Z\} \subseteq Q_F(f_{\bar{A},\bar{B}})$, by Bezout's Theorem $(x,y) \circ (x',y') \subseteq P^*(Q_F(f_{a,b}))$. Now let $X = (x,y) \in Q_F(f_{a,b})$, $Y = (x',y') \in Q_F(f_{a',b'})$, $Z = (x'',y'') \in Q_F(f_{a'',b''})$, where $J = \{(a,b), (a',b'), (a'',b'')\} \subseteq \bar{A} \times \bar{B}$. If $(x,y) = (x_1, y_1)$ and $(x',y') = (x'_1, y'_1)$ then $x = x_1$

and $x' = x'_1$. Because $\mathcal{G}_x(f_{a,b}) = \mathcal{G}_{x_1}(f_{a,b})$ and $\mathcal{G}_{x'}(f_{a,b}) = \mathcal{G}_{x'_1}(f_{a,b})$ we have $\mathcal{G}_x(f_{a,b}) \times \mathcal{G}_{x'}(f_{a,b}) = \mathcal{G}_{x_1}(f_{a,b}) \times \mathcal{G}_{x'_1}(f_{a,b})$ thus

$$\{\widehat{z \bullet_{ab} w} | (z,w) \in \mathcal{G}_x(f_{a,b}) \times \mathcal{G}_{x'}(f_{a,b})\} = \{\widehat{z \bullet_{ab} w} | (z,w) \in \mathcal{G}_{x_1}(f_{a,b}) \times \mathcal{G}_{x'_1}(f_{a,b})\}$$

and that is $(x,y) \circ (x',y') = (x_1,y_1) \circ (x'_1,y'_1)$, consequently "$\circ$" is well defined. If $X = (\mathcal{O}, \infty)$ or $Y = (\mathcal{O}, \infty)$ or $Z = (\mathcal{O}, \infty)$, associativity is evident. If this property is not met, the following cases may occur:

Case1. If $|J| = 1$. In this case we have $Q_F(f_{a,b}) = Q_F(f_{a',b'}) = Q_F(f_{a'',b''})$.

$$[(x,y) \circ (x',y')] \circ (x'',y'') = \left\{ \widehat{(x_i \bullet_{ab} x'_j)} | (x_i, x'_j) \in \mathcal{G}_x(f_{a,b}) \times \mathcal{G}_{x'}(f_{a,b}) \right\} \circ (x'',y'')$$

$$= \left\{ \widehat{(x_i \bullet_{ab} x'_j) \bullet_{ab} x''_k} | (x_i, x'_j, x''_k) \in \mathcal{G}_x(f_{a,b}) \times \mathcal{G}_{x'}(f_{a,b}) \times \mathcal{G}_{x''}(f_{a,b}) \right\}.$$

Similarly

$$(x,y) \circ [(x',y') \circ (x'',y'')] = \left\{ \widehat{x_i \bullet_{ab} (x'_j \bullet_{ab} x''_k)} | (x_i, x'_j, x''_k) \in \mathcal{G}_x(f_{a,b}) \times \mathcal{G}_{x'}(f_{a,b}) \times \mathcal{G}_{x''}(f_{a,b}) \right\}.$$

On the other hand we have

$$L(\hat{x}_i, \hat{x}'_j) \cap L(\widehat{x_i \bullet_{ab} x'_j}, \hat{\mathcal{O}}) = \{(x_i \bullet_{ab} x_j, 0)\} \subseteq L_0,$$

$$L(\hat{x}'_j, \hat{x}''_k) \cap L(\hat{\mathcal{O}}, \widehat{x'_j \bullet_{ab} x''_k}) = \{(x_j \bullet_{ab} x_k, 0)\} \subseteq L_0.$$

Therefore by Pascal's Theorem we have

$$L(\hat{x}''_k, \widehat{x_i \bullet_{ab} x'_j}) \cap L(\hat{x}'_j \bullet_{ab} \hat{x}''_k, \hat{x}_i) \subseteq L_0,$$

and in addition

$$\{((x_i \bullet_{ab} x'_j) \bullet_{ab} x''_k, 0)\} = L_0 \cap L(\widehat{x_i \bullet_{ab} x'_j}, \widehat{x''_k}),$$

$$\{(x_i \bullet_{ab} (x'_j \bullet_{ab} x''_k), 0)\} = L_0 \cap L(\hat{x}_i, \widehat{x'_j \bullet_{ab} x''_k}),$$

$$L_0 \cap L(\widehat{x_i \bullet_{ab} x'_j}, \widehat{x''_k}) = L(\hat{x}''_k, \widehat{x_i \bullet_{ab} x'_j}) \cap L(\widehat{x'_j \bullet_{ab} x''_k}, \hat{x}_i) = L_0 \cap L(\hat{x}_i, \widehat{x'_j \bullet_{ab} x''_k}).$$

On the other word

$$(x_i \bullet_{ab} x'_j) \bullet_{ab} x''_k = x_i \bullet_{ab} (x'_j \bullet_{ab} x''_k).$$

So

$$\left(\widehat{(x_i \bullet_{ab} x'_j) \bullet_{ab} x''_k}\right) = \left(\widehat{x_i \bullet_{ab} (x'_j \bullet_{ab} x''_k)}\right) \text{ for all } (x_i, x'_j, x''_k) \in \mathcal{G}_x(f_{a,b}) \times \mathcal{G}_{x'}(f_{a,b}) \times \mathcal{G}_{x''}(f_{a,b})$$

Case2. If $|J| = 2$. (i) If $Q_F(f_{a,b}) = Q_F(f_{a',b'}) \neq Q_F(f_{a'',b''})$. We have

$$[(x,y) \circ (x',y')] \circ (x'',y'') = [\{\widehat{z \bullet_{ab} w} | (z,w) \in \mathcal{G}_x(f_{a,b}) \times \mathcal{G}_{x'}(f_{a,b})\}] \circ (x'',y'')$$

$$= \bigcup_{(u,v) \in (x,y) \circ (x',y')} (u,v) \circ (x'',y'')$$

$$= Q_F(f_{a,b}) \cup Q_F(f_{a'',b''}).$$

Otherwise

$$(x,y) \circ [(x',y') \circ (x'',y'')] = (x,y) \circ (Q_F(f_{a',b'}) \cup Q_F(f_{a'',b''}))$$
$$= Q_F(f_{a',b'}) \cap Q_F(f_{a,b}) \cup Q_F(f_{a'',b''})$$
$$= Q_F(f_{a,b}) \cup Q_F(f_{a'',b''}).$$

(ii) If $Q_F(f_{a,b}) \neq Q_F(f_{a',b'}) = Q_F(f_{a'',b''})$. This case similar to (i).

(iii) If $Q_F(f_{a,b}) = Q_F(f_{a'',b''}) \neq Q_F(f_{a',b'})$. We have

$$[(x,y) \circ (x',y')] \circ (x'',y'') = (Q_F(f_{a,b}) \cup Q_F(f_{a',b'})) \circ (x'',y'')$$
$$= Q_F(f_{a,b}) \cup Q_F(f_{a',b'}) \cup Q_F(f_{a'',b''})$$
$$= Q_F(f_{a,b}) \cup Q_F(f_{a',b'}).$$

On the other hand

$$(x,y) \circ [(x',y') \circ (x'',y'')] = (x,y) \circ (Q_F(f_{a',b'}) \cup Q_F(f_{a'',b''}))$$
$$= Q_F(f_{a,b}) \cup Q_F(f_{a',b'}) \cup Q_F(f_{a'',b''})$$
$$= Q_F(f_{a,b}) \cup Q_F(f_{a',b'}).$$

Case3. If $|J| = 3$. In this case we have

$$[(x,y) \circ (x',y')] \circ (x'',y'') = (Q_F(f_{a,b}) \cup Q_F(f_{a',b'})) \circ (x'',y'')$$
$$= Q_F(f_{a,b}) \cup Q_F(f_{a',b'}) \cup Q_F(f_{a'',b''}).$$

On the other hand

$$(x,y) \circ [(x',y') \circ (x'',y'')] = (x,y) \circ (Q_F(f_{a',b'}) \cup Q_F(f_{a'',b''}))$$
$$= Q_F(f_{a,b}) \cup Q_F(f_{a',b'}) \cup Q_F(f_{a'',b''}).$$

To prove the validity of reproduction axiom for "$\circ$" let us consider two cases:

Case1. If $|\bar{A} \times \bar{B}| = 1$ then $\bar{F} = F$ and $Q_F(f_{\bar{A},\bar{B}}) = Q_F(f_{a,b})$, where $a \in \bar{A}, b \in \bar{B}$ also $(Q_F(f_{a,b}), \circ)$ is a conic group, hence there is nothing to prove.

Case2. If $|\bar{A} \times \bar{B}| > 1$, consider arbitrary element $\hat{x} \in Q_F(f_{a,b}) \subseteq Q_F(f_{\bar{A},\bar{B}})$, then

$$\hat{x} \circ Q_F(f_{\bar{A},\bar{B}}) = \left(\hat{x} \circ \bigcup_{a \neq i \in \bar{A}, b \neq j \in \bar{B}} Q_F(f_{i,j}) \right) \cup (\hat{x} \circ Q_F(f_{a,b})),$$
$$= \left(\bigcup_{a \neq i \in \bar{A}, b \neq j \in \bar{B}} \hat{x} \circ Q_F(f_{i,j}) \right) \cup Q_F(f_{a,b}),$$
$$= \left(\bigcup_{i \in \bar{A}, j \in \bar{B}} Q_F(f_{i,j}) \right) \cup Q_F(f_{a,b}),$$
$$= Q_F(f_{\bar{A},\bar{B}}).$$

Similarly, $Q_F(f_{\bar{A},\bar{B}}) \circ \hat{x} = Q_F(f_{\bar{A},\bar{B}})$ and reproduction axiom is established. Thus, $(Q_F(f_{\bar{A},\bar{B}}), \circ)$ is a hypergroup. □

Remark 4. *The hyperconic and the associated hypergroup are conic and conic group, respectively, if $G = \{1\}$.*

Example 4. Let $F = \mathbb{Z}_5$ be the field of order 5 and $G = \{\pm 1\} \leqslant F^*$. We have $\bar{F} = \{\bar{0}, \bar{1}, \bar{2}\}$. In addition, if we go back to Example 3 then $Q_{\bar{F}}(f_{\bar{1},\bar{0}}) = \overline{Q_F(f_{1,0})} \cup \overline{Q_F(f_{(-1),0})}$ is hyperconic, where
$Q_F(f_{1,0}) = \{\hat{\mathcal{O}}, (1,1), (-1,1), (2,-1), (-2,-1)\}$
$Q_F(f_{(-1),0}) = \{\hat{\mathcal{O}}, (1,-1), (-1,-1), (2,1), (-2,1)\}$, $Q_{\bar{F}}(f_{\bar{1},\bar{0}}) = \{\hat{\mathcal{O}}, (\bar{1}, \bar{1}), (\bar{2}, \bar{1})\}$. Now let $H = Q_F(f_{1,0})$ and $K = Q_F(f_{(-1),0})$. Then H and K are reversible subhypergroups of $Q_F(f_{\bar{1},\bar{0}})$, which are defined by the Caley Tables 2 and 3, respectively.

Table 2. Cayle table $(Q_{\mathbb{Z}_5}(f_{1,0}), \circ)$.

$\circ$	$\hat{\mathcal{O}}$	$(1,1)$	$(-1,1)$	$(2,-1)$	$(-2,-1)$
$\hat{\mathcal{O}}$	$\hat{\mathcal{O}}$	$(\pm 1, 1)$	$(\pm 1, 1)$	$(\pm 2, -1)$	$, (\pm 2, -1)$
$(1,1)$	$(\pm 1, 1)$	$\hat{\mathcal{O}}, (\pm 2, -1)$	$\hat{\mathcal{O}}, (\pm 2, -1)$	$(\pm 1, 1), (\pm 2, -1)$	$(\pm 1, 1), (\pm 2, -1)$
$(-1,1)$	$(\pm 1, 1)$	$\hat{\mathcal{O}}, (\pm 2, -1)$	$\hat{\mathcal{O}}, (\pm 2, -1)$	$(\pm 1, 1), (\pm 2, -1)$	$(\pm 1, 1), (\pm 2, -1)$
$(2,-1)$	$(\pm 2, -1)$	$(\pm 1, 1), (\pm 2, -1)$	$(\pm 1, 1), (\pm 2, -1)$	$(\pm 1, 1), \hat{\mathcal{O}}$	$(\pm 1, 1), \hat{\mathcal{O}}$
$(-2,-1)$	$(\pm 2, -1)$	$(\pm 1, 1), (\pm 2, -1)$	$(\pm 1, 1), (\pm 2, -1)$	$(\pm 1, 1), \hat{\mathcal{O}}$	$(\pm 1, 1), \hat{\mathcal{O}}$

Table 3. Caley table $(Q_{\mathbb{Z}_5}(f_{(-1),0}), \circ)$.

$\circ$	$\hat{\mathcal{O}}$	$(1,-1)$	$(-1,-1)$	$(2,1)$	$(-2,1)$
$\hat{\mathcal{O}}$	$\hat{\mathcal{O}}$	$(\pm 1, -1)$	$(\pm 1, -1)$	$(\pm 2, 1)$	$, (\pm 2, 1)$
$(1,-1)$	$(\pm 1, -1)$	$\hat{\mathcal{O}}, (\pm 2, 1)$	$\hat{\mathcal{O}}, (\pm 2, 1)$	$(\pm 1, -1), (\pm 2, 1)$	$(\pm 1, -1), (\pm 2, 1)$
$(-1,-1)$	$(\pm 1, -1)$	$\hat{\mathcal{O}}, (\pm 2, 1)$	$\hat{\mathcal{O}}, (\pm 2, 1)$	$(\pm 1, -1), (\pm 2, 1)$	$(\pm 1, -1), (\pm 2, 1)$
$(2,1)$	$(\pm 2, 1)$	$(\pm 1, -1), (\pm 2, 1)$	$(\pm 1, -1), (\pm 2, 1)$	$(\pm 1, -1), \hat{\mathcal{O}}$	$(\pm 1, -1), \hat{\mathcal{O}}$
$(-2,1)$	$(\pm 2, 1)$	$(\pm 1, -1), (\pm 2, 1)$	$(\pm 1, -1), (\pm 2, 1)$	$(\pm 1, -1), \hat{\mathcal{O}}$	$(\pm 1, -1), \hat{\mathcal{O}}$

Proposition 3. H is subhypergroup of $Q_F(f_{\bar{A},\bar{B}})$ if and only if $H = \bigcup_{(i,j) \in I \subseteq \bar{A} \times \bar{B}} Q_F(f_{i,j})$ or $H \leqslant Q_{i,j}(F)$, for some $(i,j) \in \bar{A} \times \bar{B}$.

Proof. ($\Rightarrow$). Let us assume that $H \nleqslant Q_F(f_{i,j})$ for every (i,j) in $\bar{A} \times \bar{B}$. Then in $\bar{A} \times \bar{B}$ exist $(i,j) \neq (s,t)$ such that $H \cap Q_F(f_{i,j}) \neq \emptyset \neq H \cap Q_F(f_{s,t})$. Now let $I = \{(i,j) \in \bar{A} \times \bar{B} | H \cap Q_F(f_{i,j}) \neq \emptyset\}$, thus we have $H \subseteq \bigcup_{(i,j) \in I} Q_F(f_{i,j}) \subseteq \bigcup_{(s,t),(i,j) \in I} (Q_F(f_{i,j}) \cap H) \circ (Q_F(f_{s,t}) \cap H) \subseteq H$. Accordingly, $H = \bigcup_{(i,j) \in I} Q_F(f_{i,j})$.
($\Leftarrow$). It is obvious. □

Proposition 4. Let H be a subhypergroup of $Q_F(f_{\bar{A},\bar{B}})$. Then H is reversible hypergroup if and only if $H \leqslant Q_F(f_{i,j})$, for some $(i,j) \in \bar{A} \times \bar{B}$.

Proof. ($\Leftarrow$). First we prove that $H \leqslant Q_F(f_{i,j})$ is a regular reversible hypergroup for all $(i,j) \in \bar{A} \times \bar{B}$. Let (x,y) and (x',y') are elements in $Q_F(f_{i,j})$.

Case1. If $x' \notin \mathcal{G}_x(f_{a,b})$, then

$$(x'',y'') \in (x,y) \circ (x',y') \Longrightarrow (x'',y'') = \widehat{z \bullet_{ij} w}, \text{ for some}(z,w) \in \mathcal{G}_x(f_{a,b}) \times \mathcal{G}_{x'}(f_{a,b})$$
$$\Longrightarrow x'' = z \bullet_{ij} w,$$
$$\Longrightarrow z = x'' \bullet_{ij} h \qquad \text{where, } w \bullet_{ij} h = \mathcal{O},$$
$$\Longrightarrow (z, f_{a,b}(z)) = \widehat{x'' \bullet_{ij} h} \text{ and } h \in \mathcal{G}_w(f_{a,b}) = \mathcal{G}_{x'}(f_{a,b})$$
$$\Longrightarrow (z, f_{a,b}(z)) \in (x'', f_{a,b}(x'')) \circ (x', f_{a,b}(x')).$$

Case2. If $x' \in \mathcal{G}_x(f_{a,b})$, then $z, w \in \mathcal{G}_x(f_{a,b})$ and $\hat{m} \in \hat{z} \circ \hat{w} = \{\widehat{x \bullet_{ab} x}, \widehat{n \bullet_{ab} n}, \hat{\mathcal{O}}\}$, where $x \bullet_{ab} n = \mathcal{O}$ then $\hat{z} \in \hat{m} \circ \hat{w}$ and $\hat{w} \in \hat{z} \circ \hat{m}$.

Case3. If $Y \in \infty \circ X = X \circ \infty$, then $\infty \in Y \circ X$ and $X \in \infty \circ Y$. Notice that $\infty \in X \circ X$, for all $X \in Q_F(f_{i,j})$ (i.e. every element is one of its inverses).

($\Rightarrow$). Assume that $(x, y) \in H \cap Q_F(f_{i,j})$ and $(x', y') \in H \cap Q_F(f_{s,t})$, in which $(i, j) \neq (s, t)$, $(x, y) \neq \hat{\mathcal{O}} \neq (x', y')$ and $(x'', y'') \in (x, y) \circ (x', y') \cap Q_F(f_{i,j})$. Then $(x', y') \in (z, w) \circ (x'', y'') \subseteq Q_F(f_{i,j})$, where $z \in \mathcal{G}_x(f_{a,b})$. Hence $Q_F(f_{i,j}) = Q_F(f_{s,t})$ and this means reversibility conditions do not hold. □

The class of H_v- groups is more general than the class of hypergroups which is introduced by Th. Vougiouklis [39]. The hyperstructure $(H, \circ)$ is called an H_v−group if $x \circ H = H = H \circ x$, and also the weak associativity condition holds, that is $x \circ (y \circ z) \cap (x \circ y) \circ z \neq \emptyset$ for all $x, y, z \in H$. In [13,14] the authors have investigated some hyperoperations denoted by $\bar{\circ}$ and $\diamond$ on some main classes of curves; elliptic curves and homographics over Krasner's hyperfields. In the following, we study them on hyperconic. Consider the following hyperoperation on the hyperconic; $(Q_{\bar{F}}(f_{\bar{A},\bar{B}}))$:

$$(\bar{x}, \bar{y})\bar{\circ}(\bar{x}', \bar{y}') = \{(\bar{v}, \bar{w}) | (v, w) \in (\bar{x} \times \bar{y}) \circ (\bar{x}' \times \bar{y}')\},$$

for all $(\bar{x}, \bar{y}), (\bar{x}', \bar{y}')$ in $Q_{\bar{F}}(f_{\bar{A},\bar{B}})$.

Proposition 5. $(Q_{\bar{F}}(f_{\bar{A},\bar{B}}), \bar{\circ})$ is an H_v-group.

Proof. The proof is straightforward. □

Proposition 6. If $\psi_{\bar{A},\bar{B}} : Q_F(f_{\bar{A},\bar{B}}) \longrightarrow Q_{\bar{F}}(f_{\bar{A},\bar{B}})$, $\psi_{\bar{A},\bar{B}}(x,y) = (\bar{x}, \bar{y})$, then $\psi_{\bar{A},\bar{B}}$ is an epimorphism of H_v-groups.

Proof. The base of the proof is similar to the proof of Proposition 3 in [14]. □

Example 5. Let $G = \{\pm 1\}$ be a subgroup of F^*, where $F = \mathbb{Z}_5$. Consider $f_{\bar{1},\bar{1}}(\bar{x}) = \bar{x}^2 \oplus \bar{x}$ on $\bar{F} = \{\bar{0}, \bar{1}, \bar{2}\}$. Consequently $Q_{\bar{F}}(f_{\bar{1},\bar{1}}) = \{\hat{\mathcal{O}}, (\bar{1}, \bar{2}), (\bar{2}, \bar{1}), (\bar{2}, \bar{2})\}$ is a hyperconic, a calculation gives us the Table 4 of H_v-group.

Table 4. Conic H_v-group, $(Q_{\mathbb{Z}_5}(f_{\bar{1},\bar{1}}), \bar{\circ})$.

$\bar{\circ}$	$\hat{\mathcal{O}}$	$(\bar{1},\bar{2})$	$(\bar{2},\bar{1})$	$(\bar{2},\bar{2})$
$\hat{\mathcal{O}}$	$\hat{\mathcal{O}}$	$(\bar{1},\bar{2}),(\bar{2},\bar{2})$	$(\bar{2},\bar{1})$	$(\bar{2},\bar{2}),(\bar{1},\bar{2})$
$(\bar{1},\bar{2})$	$(\bar{1},\bar{2}),(\bar{2},\bar{2})$	$(\bar{1},\bar{2}),(\bar{2},\bar{1}),(\bar{2},\bar{2})$	$(\bar{1},\bar{2}),(\bar{2},\bar{1}),(\bar{2},\bar{2})$	$(\bar{1},\bar{2}),(\bar{2},\bar{1}),(\bar{2},\bar{2})$
$(\bar{2},\bar{1})$	$(\bar{2},\bar{1})$	$(\bar{1},\bar{2}),(\bar{2},\bar{1}),(\bar{2},\bar{2})$	$Q_{\bar{F}}(f_{\bar{1},\bar{1}})$	$(\bar{1},\bar{2}),(\bar{2},\bar{1}),(\bar{2},\bar{2})$
$(\bar{2},\bar{2})$	$(\bar{1},\bar{2}),(\bar{2},\bar{2})$	$(\bar{1},\bar{2}),(\bar{2},\bar{1}),(\bar{2},\bar{2})$	$(\bar{1},\bar{2}),(\bar{2},\bar{1}),(\bar{2},\bar{2})$	$(\bar{1},\bar{2}),(\bar{2},\bar{1}),(\bar{2},\bar{2})$

Let A be a finite set called *alphabet* and let K be a non-empty subset of A, called *key-set* and also let "·" be a hyperoperation on A. In [40] Berardi et.al. utilized the hyperoperations with the following condition $k \cdot x = k \cdot y \Rightarrow x = y$, for all $(x, y) \in A^2$ and $k \in K$. Let the subhypergroup $(Q_F(f_{\hat{m},\hat{n}}), \circ)$ of $(Q_F(f_{m,n}), \circ)$ and $A = \{a_x | \hat{x} \in Q_F(f_{m,n})\}$, where $a_x = i(x)$. Notice that $i(x) = \widehat{\mathcal{G}_x(f_{a,b})}$, is the set of all inverses of $\hat{x}$, for all $\hat{x} \in Q_F(f_{m,n})$. We define the hyperoperation $\diamond$ on A as bellow:

$$a_u \diamond a_v = \{a_w | w \in u \circ v\}.$$

Theorem 4. $(A, \diamond)$ is a canonical hypergroup which satisfies Berardi's condition.

Proof. The proof is similar to the one of Theorem 4.1 in [13]. □

4. Conclusions

Conic curve cryptography (CCC) is rendering efficient digital signature schemes (CCDLP). They have a high level of security with small keys size. Let $g(x,y) = ax^2 + bxy + cy^2 + dx + ey + f \in F[x,y]$ and $g(x,y) = 0$ be the quadratic equation of two variables in field of F, if $a = c = 0$ and $b \neq 0$ then the equation $g(x,y) = 0$ is called homographic transformation. In [14] Vahedi et. al extended this particular quadratic equation on the quotient hyperfield $\frac{F}{G}$. Now suppose that $ae \neq 0$ and $b = 0$ in $g(x,y)$. Then the curve is called a conic. The motivation of this paper goes in the same direction of [14]. In fact, by a similar way the notion of conic on a field extended to hyperconic over a quotient hyperfield hyperfield, as picturized in Figure 3. Notice that as one can see the group structures of these two classes of curves have different applications, the associated hyperstructures can be different in applications. In the last part of the paper a canonical hypergroup which is assigned by $(Q_F(f_{n,m}), \diamond)$ is investigated.

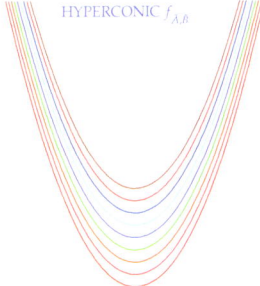

Figure 3. Hyperconic, $Q_\mathbb{R}(f_{\bar{A},\bar{B}})$.

Author Contributions: Conceptualization, methodology, investigation, resources, writing—original draft, writing—review and editing, all authors worked on equally; project administration and funding acquisition, S.H.-M. All authors have read and agreed to the published version of the manuscript.

Funding: Sarka Hoskova-Mayerova was supported within the project for development of basic and applied research developed in the long term by the departments of theoretical and applied bases FMT (Project code: DZRO K-217) supported by the Ministry of Defence in the Czech Republic.

Conflicts of Interest: Authors declare no conflict of interest. The funders had no role in the design of the study; in the collection, analysis, or interpretation of data; in the writing of the manuscript, or in the decision to publish the results.

References

1. Marty, F. Sur Une Generalization de Group. In Proceedings of the 8th Congres Des Mathematiciens Scandinaves, Stockholm, Sweden, 14–18 August 1934; pp. 45–49.
2. Corsini, P. *Prolegomena of Hypergroup Theory*; Aviani Editore: Tricesimo, Italy, 1993.
3. Davvaz, B.; Leoreanu-Fotea, V. *Hyperring Theory and Applications*; International Academic Press: Cambridge, MA, USA, 2007.
4. Corsini, P.; Leoreanu-Fotea, V. *Applications of Hyperstructure Theory*; Kluwer Academical Publications: Dordrecht, The Netherlands, 2003.
5. Vougiouklis, T. *Hyperstructures and Their Representations*; Hadronic Press: Palm Harbor, FL, USA, 1994.
6. Hoskova, S.; Chvalina, J. A survey of investigations of the Brno research group in the hyperstructure theory since the last AHA Congress. In Proceedings of the AHA 2008: 10th International Congress-Algebraic Hyperstructures and Applications, University of Defence, Brno, Czech Republic, 3–9 September 2008; pp. 71–83, ISBN 978-80-7231-688-5.

7. Antampoufis, N.; Hoskova-Mayerova, S. A Brief Survey on the two Different Approaches of Fundamental Equivalence Relations on Hyperstructures. *Ratio Math.* **2017**, *33*, 47–60, doi:10.23755/rm.v33i0.388.
8. Connes, A.; Consani, C. From monoids to hyperstructures: In search of an absolute arithmetic. In *Casimir Force, Casimir Operators and the Riemann Hypothesis: Mathematics for Innovation in Industry and Science*; van Dijk, G., Wakayama, M., Eds.; De Gruyter: Berlin, Germany, 2010; pp. 147–198.
9. Connes, A.; Consani, C. The hyperring of adele classes. *J. Number Theory* **2011**, *131*, 159–194.
10. Connes, A.; Consani, C. The universal thickening of the field of real numbers. In Proceedings of the Thirteenth Conference of the Canadian Number Theory Association, Ottawa, ON, Canada, 16–20 June 2014.
11. Viro, O. On basic concepts of tropical geometry. *Proc. Steklov Inst. Math.* **2011**, *273*, 252–282.
12. Gladki, P.; Marshall, M. Orderings and signatures of higher level on multirings and hyperfields. *J.-Theory-Theory Its Appl. Algebr. Geom. Topol.* **2012**, *10*, 489–518.
13. Vahedi, V.; Jafarpour, M.; Aghabozorgi, H.; Cristea, I. Extension of elliptic curves on Krasner hyperfields. *Comm. Algebra* **2019**, *47*, 4806–4823.
14. Vahedi, V.; Jafarpour, M.; Cristea, I. Hyperhomographies on Krasner Hyperfields. *Symmetry Class. Fuzzy Algebr. Hypercompos. Struct.* **2019**, *11*, 1442.
15. Freni, D. Strongly Transitive Geometric Spaces: Applications to Hypergroups and Semigroups Theory. *Comm. Algebra* **2004**, *32*, 969–988.
16. Izhakian, Z.; Knebusch, M.; Rowen, L. Layered tropical mathematics. *J. Algebra* **2014**, *416*, 200–273.
17. Izhakian, Z.; Rowen, L. Supertropical algebra. *Adv. Math.* **2010**, *225*, 2222–2286.
18. Lorscheid, O. The geometry of blueprints: Part i: Algebraic background and scheme theory. *Adv. Math.* **2012**, *229*, 1804–1846.
19. Lorscheid, O. Scheme theoretic tropicalization. *arXiv* **2015**, arXiv:1508.07949.
20. Krasner, M. A class of hyperrings and hyperfields. *Int. J. Math. Math. Sci.* **1983**, *6*, 307–311.
21. Krasner, M. Approximation des corps value complets de caracteristique p10 par ceux de caracteristique 0. In *Colloque d Algebre Superieure, Tenu a Bruxelles du 19 au 22 Decembre 1956*; Centre Belge de Recherches Mathematiques Etablissements Ceuterick: Louvain, Belgium; Gauthier-Villars: Paris, France, 1957; pp. 129–206.
22. Ameri, R.; Eyvazi, M.; Hoskova-Mayerova, S. Superring of Polynomials over a Hyperring. *Mathematics* **2019**, *7*, 902, doi:10.3390/math7100902.
23. Mittas, J.D. Hypergroupes canoniques. *Math. Balk.* **1972**, *2*, 165–179.
24. Massouros, C.G. A field theory problem relating to questions in hyperfield theory. In Proceedings of the International Conference on Numerical Analysis and Applied Mathematics, Halkidiki, Greece, 19–25 September 2011; Volume 1389, pp. 1852–1855.
25. Dakic, J.; Jancic-Rasovic, S.; Cristea, I. Weak Embeddable Hypernear-Rings. *Symmetry* **2019**, *11*, 96.
26. Massouros, C.G.; Massouros, G.G. On Join Hyperrings. In Proceedings of the 10th International Congress on Algebraic Hyperstructures and Applications, University of Defence, Brno, Czech Republic, 3–9 September 2008; pp. 203–215.
27. Massouros, C.G. On the theory of hyperrings and hyperfields. *Algebra Log.* **1985**, *24*, 728–742.
28. Massouros, G.G.; Massouros, C.G. Homomorphic Relations on Hyperringoids and Join Hyperrings. *Ratio Math.* **1999**, *13*, 61–70.
29. Ameri, R.; Kordi, A.; Sarka-Mayerova, S. Multiplicative hyperring of fractions and coprime hyperideals. *An. Stiintifice Univ. Ovidius Constanta Ser. Mat.* **2017**, *25*, 5–23. doi: 10.1515/auom-2017-0001.
30. Jancic-Rasovic, S. About the hyperring of polynomials. *Ital. J. Pure Appl. Math.* **2007**, *21*, 223–234.
31. Cristea, I.; Jancic-Rasovic, S. Operations on fuzzy relations: A tool to construct new hyperrings. *J. Mult-Val. Logic Soft Comput.* **2013**, *21*, 183–200.
32. Jancic-Rasovic, S.; Dasic, V. Some new classes of (m,n)-hyperrings. *Filomat* **2012**, *26*, 585–596.
33. Cristea, I. ; Jancic-Rasovic, S. Composition Hyperrings. *An. Stiintifice Univ. Ovidius Constanta Ser. Mat.* **2013**, *21*, 81–94.
34. Flaut, C.; Flaut, D.; Hoskova-Mayerova, S.; Vasile, R. From Old Ciphers to Modern Communications. *Adv. Mil. Technol.* **2019**, *14*, 79–88, doi:10.3849/aimt.01281.
35. Saeid, A.B.; Flaut, C.; Hoskova-Mayerova, S. Some connections between BCK-algebras and n-ary block codes. *Soft Comput.* **2017**, doi:10.1007/s00500-017-2788-z.

36. Hankerson, D.; Menezes, A.; Vanstone, S.A. *Guide to Elliptic Curve Cryptography*; Springer: Berlin/Heidelberg, Germany, 2004.
37. Koblitz, N. Introduction to elliptic curves and modular forms. In *Volume 97 of Graduate Texts in Mathematics*; Springer: New York, NY, USA, 1984.
38. Silverman, J.H.; Tate, I.T. *Rational Points on Elliptic Curves*; Springer: New York, NY, USA, 1992; p. 289
39. Vougiouklis, T. The fundamental relation in hyperrings, The general hyperfield. In Proceedings of the 4th International Congress on AHA, Xanthi, Greece, 27–30 June 1990; World Scientific: Singapore, 1991; pp. 209–217.
40. Berardi, L.; Eugeni, F.; Innamorati, S. Remarks on Hypergroupoids and Cryptography. *J. Combin. Inform. Syst Sci.* **1992**, *17*, 217–231.

© 2020 by the authors. Licensee MDPI, Basel, Switzerland. This article is an open access article distributed under the terms and conditions of the Creative Commons Attribution (CC BY) license (http://creativecommons.org/licenses/by/4.0/).

Article

On the Theory of Left/Right Almost Groups and Hypergroups with their Relevant Enumerations

Christos G. Massouros [1,*] and Naveed Yaqoob [2]

[1] Core Department, Euripus Campus, National and Kapodistrian University of Athens, GR 34400 Euboia, Greece
[2] Department of Mathematics and Statistics, Riphah International University, Sector I-14, Islamabad 44000, Pakistan; nayaqoob@ymail.com
* Correspondence: ChrMas@uoa.gr or Ch.Massouros@gmail.com

Abstract: This paper presents the study of algebraic structures equipped with the inverted associativity axiom. Initially, the definition of the left and the right almost-groups is introduced and afterwards, the study is focused on the more general structures, which are the left and the right almost-hypergroups and on their enumeration in the cases of order 2 and 3. The outcomes of these enumerations compared with the corresponding in the hypergroups reveal interesting results. Next, fundamental properties of the left and right almost-hypergroups are proved. Subsequently, the almost hypergroups are enriched with more axioms, like the transposition axiom and the weak commutativity. This creates new hypercompositional structures, such as the transposition left/right almost-hypergroups, the left/right almost commutative hypergroups, the join left/right almost hypergroups, etc. The algebraic properties of these new structures are analyzed and studied as well. Especially, the existence of neutral elements leads to the separation of their elements into attractive and non-attractive ones. If the existence of the neutral element is accompanied with the existence of symmetric elements as well, then the fortified transposition left/right almost-hypergroups and the transposition polysymmetrical left/right almost-hypergroups come into being.

Citation: Massouros, C.G.; Yaqoob, N. On the Theory of Left/Right Almost Groups and Hypergroups with their Relevant Enumerations. *Mathematics* **2021**, *9*, 1828. https://doi.org/10.3390/math9151828

Academic Editor: Askar Tuganbaev

Received: 14 June 2021
Accepted: 6 July 2021
Published: 3 August 2021

Publisher's Note: MDPI stays neutral with regard to jurisdictional claims in published maps and institutional affiliations.

Copyright: © 2021 by the authors. Licensee MDPI, Basel, Switzerland. This article is an open access article distributed under the terms and conditions of the Creative Commons Attribution (CC BY) license (https://creativecommons.org/licenses/by/4.0/).

Keywords: hypercompositional algebra; magma; left/right almost-group; left/right almost-hyper group; transposition axiom; Mathematica

1. Introduction

This paper is generally classified in the area of hypercompositional algebra. Hypercompositional algebra is the branch of abstract algebra which studies hypercompositional structures, i.e., structures equipped with one or more multi-valued operations. Multi-valued operations, also called hyperoperations or hypercompositions, are operations in which the result of the synthesis of two elements is multi-valued, rather than a single element. More precisely, a *hypercomposition* on a non-void set H is a function from $H \times H$ to the powerset $P(H)$ of H. Hypercompositional structures came into being through the notion of the *hypergroup*. The hypergroup was introduced in 1934 by Marty, in order to study problems in non-commutative algebra, such as cosets determined by non-invariant subgroups [1–3].

In [4] the *magma* is defined as an ordered pair $(H, \perp)$ where H is a non-void set and "$\perp$" is a law of synthesis, which is either a composition or a hypercomposition. Per this definition, if A and B are subsets of H, then:

$$A \perp B = \{a \perp b \in H \mid a \in A, b \in B\}, \text{ if } \perp \text{ is a composition}$$

and

$$A \perp B = \bigcup_{(a,b) \in A \times B} (a \perp b), \text{ if } \perp \text{ is a hypercomposition}$$

In particular if $A = \emptyset$ or $B = \emptyset$, then $A \perp B = \emptyset$ and vice versa. $A \perp b$ and $a \perp B$ have the same meaning as $A \perp \{b\}$ and $\{a\} \perp B$. In general, the singleton $\{a\}$ is identified with its member a. Sometimes it is convenient to use the relational notation $A \approx B$ instead of $A \cap B \neq \emptyset$. Then, as the singleton $\{a\}$ is identified with its member a, the notation $a \approx A$ or $A \approx a$ is used as a substitute for $a \in A$ or Aa. The relation $\approx$ may be considered as a weak generalization of equality, since, if A and B are singletons and $A \approx B$, then $A = B$. Thus, $a \approx b \perp c$ means either $a = b \perp c$, when the synthesis is a composition or $a \in b \perp c$, when the synthesis is a hypercomposition.

It is possible that the result of the synthesis of a pair of elements in a magma is the void set, when the law of synthesis is a hypercomposition. Then, the structure is called a *partial hypergroupoid*, otherwise, it is called a *hypergroupoid*.

Every law of synthesis in a magma induces two new laws of synthesis. If the law of synthesis is written multiplicatively, then the two induced laws are:

$$a/b = \{x \in E \mid a \approx xb\}$$

and

$$b \backslash a = \{x \in E \mid a \approx bx\}$$

Thus $x \approx a/b$ if and only if $a \approx xb$ and $x \approx b \backslash a$ if and only if $a \approx bx$. In the case of a multiplicative magma, the two induced laws are named *inverse laws* and they are called the *right division* and the *left division*, respectively [4]. It is obvious that, if the law of synthesis is commutative, then the right division and left division coincide.

A law of synthesis on a set H is called *associative* if the property,

$$(x \perp y) \perp z = x \perp (y \perp z)$$

is valid, for all elements x, y, z in H. A magma whose law of synthesis is associative, is called an *associative magma* [4].

Definition 1. *An associative magma in which the law of synthesis is a composition is called a semigroup, while it is called a semihypergroup if the law of synthesis is a hypercomposition and $ab \neq \emptyset$ for each pair of its elements.*

A law of synthesis $(x, y) \to x \perp y$ on a set H is called *reproductive* if the equality,

$$x \perp H = H \perp x = H$$

is valid for all elements x in H. A magma whose law of synthesis is reproductive is called a *reproductive magma* [4].

Definition 2. *A reproductive magma in which the law of synthesis is a composition is called a quasigroup, while it is called a quasi-hypergroup if the law of synthesis is a hypercomposition and $ab \neq \emptyset$ for each pair of its elements.*

Definition 3. [4] *An associative and reproductive magma is called a group, if the law of synthesis is a composition, while it is called a hypergroup if the law of synthesis is a hypercomposition.*

The above unified definition of the group and the hypergroup is presented in [4], where it is also proved analytically that it is equivalent to the well-known dominant definition of the group.

A composition or a hypercomposition on a non-void set H is called *left inverted associative* if:

$$(a \perp b) \perp c = (c \perp b) \perp a, \text{ for every } a, b, c \in H,$$

while it is called *right inverted associative* if

$$a \perp (b \perp c) = c \perp (b \perp a), \text{ for every } a, b, c \in H.$$

The notion of the inverted associativity was initially conceived by Kazim and Naseeruddin [5] who endowed a groupoid with the left inverted associativity, thus defining the LA-semigroup. A magma equipped with left inverted assisiativity is called a *left inverted associative magma*, while if it is equipped with right inverted assisiativity is called a *right inverted associative magma*.

Recall that if $(E, \perp)$ is a magma, then the law of synthesis:

$$(x, y) \to x \perp^{op} y = y \perp x$$

is called the *opposite* of "$\perp$". The magma $(E, \perp^{op})$ is called the *opposite magma* of $(E, \perp)$ [4].

Theorem 1. *If $(H, \perp)$ is a left inverted associative magma, then $(H, \perp^{op})$ is a right inverted associative magma.*

Proof.
$$a \perp^{op} (b \perp^{op} c) = (b \perp^{op} c) \perp a = (c \perp b) \perp a =$$
$$= (a \perp b) \perp c = c \perp^{op} (a \perp b) = c \perp^{op} (b \perp^{op} a) \quad \square$$

As it is detailed in [4] the group and the hypergroup satisfy exactly the same axioms, i.e., the *reproductive axiom* and the *associative axiom*. This led to the unified definition of the group and the hypergroup that was repeated as Definition 3 in this paper. Using the same approach, the definition of the *left/right almost-group* and the *left/right almost-hypergroup* is:

Definition 4. (FIRST DEFINITION OF THE LEFT/RIGHT ALMOST GROUP/HYPER GROUP) *A reproductive magma which satisfies the axiom of the left inverted associativity is called a left almost-group (LA-group), if the law of synthesis on the magma is a composition, while it is called a left almost-hypergroup (LA-hypergroup) if the law of synthesis is a hypercomposition. A reproductive, right inverted associative magma, is called a right almost-group (RA-group) or a right almost-hypergroup (RA-hypergroup) if the law of synthesis is a composition or a hypercomposition respectively.*

Remark 1. *Obviously if the law of synthesis is commutative, then the LA- or RA- groups and hypergroups are groups and hypergroups respectively, indeed:*

$$(a \perp b) \perp c = (c \perp b) \perp a = a \perp (c \perp b) = u \perp (b \perp c)$$

As shown in Theorem 11 in [4], when the law of synthesis is a composition, then the reproductive axiom is valid if and only if the inverse laws are compositions. Hence another definition can be given for the left/right almost-group:

Definition 5. (SECOND DEFINITION OF THE LEFT/RIGHT ALMOST-GROUP) *A magma which satisfies the axiom of the left/right inverted associativity is called a left/right almost-group (LA-group/RA-group), if the law of synthesis on the magma is a composition, and the two induced laws of synthesis are compositions as well.*

Thus, if the law of synthesis on a magma is written multiplicatively, then the magma is a left/right almost-group if and only if it satisfies the axiom of the left/right inverted associativity and both the right and the left division, a/b, $b \backslash a$ respectively, result in a single element, for every pair of elements a, b in the magma.

In a similar way, because of Theorem 14 in [4], if $a/b \neq \varnothing$ and $b \backslash a \neq \varnothing$, for all pairs of elements a, b of a magma, then the magma is reproductive. Therefore, a second definition of the left/right almost-hypergroup can be given:

Definition 6. (SECOND DEFINITION OF THE LEFT/RIGHT ALMOST-HYPERGROUP) *A magma which satisfies the axiom of the left/right inverted associativity is called a left/right almost-hypergroup (LA-hypergroup/RA-hypergroup), if the law of synthesis on the magma is a hypercomposition and the result of each one of the two inverse hypercompositions is nonvoid for all pairs of elements of the magma.*

Example 1. *This example proves the existence of non-trivial left almost-groups (Table 1) and right almost-groups (Table 2).*

Table 1. Left almost-group.

∘	1	2	3
1	1	2	3
2	3	1	2
3	2	3	1

Table 2. Right almost-group.

∘	1	2	3
1	1	3	2
2	2	1	3
3	3	2	1

Example 2. *This example presents a non-trivial left almost-hypergroup (Table 3) and a non-trivial right almost-hypergroup (Table 4).*

Table 3. Left almost-hypergroup.

∘	1	2	3
1	{1}	{1}	{1,2,3}
2	{1}	{1}	{2,3}
3	{2,3}	{2,3}	{1,2,3}

Table 4. Right almost-hypergroup.

∘	1	2	3
1	{2}	{3}	{1,3}
2	{1,2}	{1,3}	{1,2,3}
3	{1,3}	{1,2,3}	{1,2,3}

Remark 2. *It is noted that in the groups, the unified definition which uses only the reproductive axiom and the associative axiom, leads to the existence of a bilaterally neutral element and consequently to the existence of a symmetric element for each one of the group's elements, as it is proved in Theorem 2 in [4]. The same doesn't hold in the case of the left or right almost-groups. Indeed, if e is a neutral element in a left almost-group, then:*

$$(a \perp e) \perp c = a \perp c \text{ and } (a \perp e) \perp c = (c \perp e) \perp a = c \perp a$$

Hence the composition is commutative and therefore the left almost-group is a group.

A direct consequence of Theorem 1 is the following theorem:

Theorem 2. *If $(H, \perp)$ is a left almost-group or hypergroup, then $(H, \perp^{op})$ is a right almost-group or hypergroup respectively.*

Corollary 1. *The transpose of the Cayley table of a left almost-group or hypergroup, is the Cayley table of a right almost-group or hypergroup respectively and vice versa.*

Example 3. *The transpose of the Cayley Table 5 which describes a LA-hypergroup is Table 6, which describes a RA-hypergroup.*

Table 5. Left almost-hypergroup.

∘	1	2	3
1	{1}	{1,2}	{1,3}
2	{1,3}	{1,2,3}	{3}
3	{1,2}	{2}	{1,2,3}

Table 6. Right almost-hypergroup.

∘	1	2	3
1	{1}	{1,3}	{1,2}
2	{1,2}	{1,2,3}	{2}
3	{1,3}	{3}	{1,2,3}

Corollary 2. *The cardinal number of the left almost-groups which are defined over a set H is equal to the cardinal number of the right almost-groups defined over H.*

Corollary 3. *The cardinal number of the left almost-hypergroups which are defined over a set H is equal to the cardinal number of the right almost-hypergroups defined over H.*

Hypercompositional structures with inverted associativity were studied by many authors (e.g., Yaqoob et al. [6–10], Hila et al. [11,12], etc.).

Special notation: In the following, in addition to the typical algebraic notations, we use Krasner's notation for the complement and difference. So, we denote with A··B the set of elements that are in the set A, but not in the set B.

2. Hypercompositional Structures with Inverted Associativity

The hypergroup, was progressively enriched with additional axioms, thus creating a significant number of special hypergroups, e.g., [13–55]. Prenowitz enriched the hypergroup with an additional axiom, in order to use it in the study of geometry [56–61]. More precisely, he introduced the *transposition axiom*

$$a/b \cap c/d \neq \varnothing \text{ implies } ad \cap bc \neq \varnothing, \text{ for all } a,b,c,d \in H,$$

into a commutative hypergroup of idempotent elements, i.e., a commutative hypergroup H, all the elements of which satisfy the properties $aa = a$ and $a/a = a$. He named this new hypergroup *join space*. Prenowitz was followed by others, such as Jantosciak [62,63], Barlotti and Strambach [64], Freni [65,66], Massouros [67–71], Dramalidis [72], etc. For the sake of terminology unification, the commutative hypergroups which satisfy the transposition axiom are called *join hypergroups*. It has been proved that the join hypergroups also comprise a useful tool in the study of languages and automata [35–38,73,74]. Later on, Jantosciak generalized the above axiom in an arbitrary hypergroup as follows:

$$b\backslash a \cap c/d \neq \varnothing \text{ implies } ad \cap bc \neq \varnothing, \text{ for all } a,b,c,d \in H.$$

He named this particular hypergroup *transposition hypergroup* [33].

A *quasicanonical hypergroup* or *polygroup* [26–29] is a transposition hypergroup H containing a scalar identity, that is, there exists an element e such that $ea = ae = a$ for each a in H. A *canonical hypergroup* [17–19] is a commutative polygroup. A canonical hypergroup may also be characterized as a join hypergroup with a scalar identity [23]. The following Proposition connects the canonical hypergroups with the RA-hypergroups.

Proposition 1. *Let $(H, \cdot)$ be a canonical hypergroup and "/" the induced hypercomposition which follows from "·". Then $(H, /)$ is a right almost-hypergroup.*

Proof. In a canonical hypergroup $(H, \cdot)$ each element $x \in H$ has a unique inverse, which is denoted by x^{-1}. Moreover $e/x = x^{-1}$ and $y/x = yx^{-1}$ [4,32,34]. Thus:

$$a/(b/c) = a/\left(bc^{-1}\right) = a\left(bc^{-1}\right)^{-1} = a\left(cb^{-1}\right) = (ac)b^{-1} = (ca)b^{-1} = c\left(ab^{-1}\right) =$$
$$= c\left(ba^{-1}\right)^{-1} = c/ba^{-1} = c/(b/a)$$

Hence, the right inverted associativity is valid. Moreover, in any hypergroup holds [4,75]:

$$a/H = H/a = H$$

Thus, the reproductive axiom is valid and therefore $(H, /)$ is a RA-hypergroup. □

Subsequently, the left and right almost-hypergroups can be enriched with additional axioms. The first axiom to be used for this purpose is the transposition axiom, as it has been introduced into many hypercompositional structures and has given very interesting and useful properties (e.g., see [75,76]).

Definition 7. *If a left almost-hypergroup $(H, \cdot)$ satisfies the transposition axiom, i.e.,*

$$b\backslash a \cap c/d \neq \varnothing \text{ implies } ad \cap bc \neq \varnothing, \text{ for all } a, b, c, d \in H$$

then it will be called transposition left almost-hypergroup. If a right almost-hypergroup satisfies the transposition axiom, then it will be called transposition right almost-hypergroup.

Example 4. *This example presents a transposition left almost-hypergroup (Table 7) and a transposition right almost-hypergroup (Table 8).*

Table 7. Transposition left almost-hypergroup.

∘	1	2	3
1	{1,2,3}	{1,2,3}	{1,2,3}
2	{1,3}	{1,2}	{1,3}
3	{2,3}	{2,3}	{1}

Table 8. Transposition right almost-hypergroup.

∘	1	2	3
1	{1,2}	{1,2,3}	{1,3}
2	{1,2}	{1,3}	{1,2,3}
3	{1,2,3}	{1,2}	{1,3}

In [4] the *reverse transposition axiom* was introduced:

$$ad \cap bc \neq \varnothing \text{ implies } b\backslash a \cap c/d \neq \varnothing, \text{ for all } a, b, c, d \in H$$

Thus, the following hypercompositional structure can be defined:

Definition 8. *If a left (right) almost-hypergroup $(H, \cdot)$ satisfies the reverse transposition axiom, then it will be called a reverse transposition left (right) almost-hypergroup.*

Definition 9. *A hypercomposition on a non-void set H is called left inverted weakly associative if*

$$(ab)c \cap (cb)a \neq \varnothing, \text{ for every } a, b, c \in H,$$

while it is called right inverted weakly associative if

$$a(bc) \cap c(ba) \neq \varnothing, \text{ for every } a, b, c \in H.$$

Definition 10. *A quasi-hypergroup equipped with a hypercomposition which is left inverted weakly associative is called a weak left almost-hypergroup (WLA-hypergroup), while it is called a weak right almost-hypergroup (WRA-hypergroup) if the hypercomposition is right inverted weakly associative.*

Recall that a quasi-hypergroup which satisfies the weak associativity is called H_V-group [77].

Proposition 2. *A commutative WLA-hypergroup (or WRA-hypergroup) is a commutative H_V-group.*

Proof. Suppose that H is a commutative WLA-hypergroup, then:

$$(ab)c \cap (cb)a \neq \varnothing \Leftrightarrow (ab)c \cap a(cb) \neq \varnothing \Leftrightarrow (ab)c \cap a(bc) \neq \varnothing.$$

Hence H is an H_V-group. □

Proposition 3. *Let $(H, \cdot)$ be a left almost-hypergroup. An arbitrary subset I_{ab} of H is associated to each pair of elements $(a, b) \in H^2$ and the following hypercomposition is introduced into H:*

$$a * b = ab \cup I_{ab}.$$

Then $(H, *)$ is a WLA-hypergroup.

Proof. Since xH and Hx are subsets of $x*H$ and $H*x$ respectively, it follows that the reproductive axiom holds. On the other hand:

$$(a*b)*c = (ab \cup I_{ab})*c = (ab)*c \cup I_{ab}*c = \left[\bigcup_{r \in ab}(rc \cup I_{rc})\right] \cup \left[\bigcup_{s \in I_{ab}}(sc \cup I_{sc})\right] =$$

$$= (ab)c \cup \left[\bigcup_{r \in ab} I_{rc}\right] \cup \left[\bigcup_{s \in I_{ab}}(sc \cup I_{sc})\right]$$

and

$$(c*b)*a = (cb \cup I_{cb})*a = (cb)*a \cup I_{cb}*a = \left[\bigcup_{r \in cb}(ra \cup I_{ra})\right] \cup \left[\bigcup_{s \in I_{cb}}(sa \cup I_{sa})\right] =$$

$$= (cb)a \cup \left[\bigcup_{r \in cb} I_{ra}\right] \cup \left[\bigcup_{s \in I_{cb}}(sa \cup I_{sa})\right]$$

Since $(ab)c = (cb)a$, it follows that $a*(b*c) \cap (a*b)*c \neq \varnothing$. □

Proposition 4. *If $\bigcap\limits_{a,b \in H} I_{ab} \neq \varnothing$, then $(H, *)$ is a transposition WLA-hypergroup.*

It is obvious that if the composition or the hypercomposition is commutative, then the inverted associativity coincides with the associativity. Thus, a commutative LA-hypergroup is simply a commutative hypergroup. However, in the hypercompositions there exists a property that does not appear in the compositions. This is the weak commutativity. A hypercomposition on a non-void set H is called *weakly commutative* if

$$ab \cap ba \neq \varnothing, \text{ for all } a, b \in H.$$

Definition 11. *A left almost commutative hypergroup (LAC-hypergroup) is a left almost-hypergroup which satisfies the weak commutativity. A right almost commutative hypergroup (RAC-hypergroup) is a right almost-hypergroup which satisfies the weak commutativity. A LAC-hypergroup (resp. RAC-hypergroup) which satisfies the transposition axiom will be called join left almost-hypergroup (resp. reverse join left almost-hypergroup). A LAC-hypergroup (resp. RAC-hypergroup) which satisfies the reverse transposition axiom will be called reverse join left almost-hypergroup (resp. reverse join right almost-hypergroup).*

Example 5. *This example presents a join left almost-hypergroup (Table 9) and a join right almost-hypergroup (Table 10).*

Table 9. Join left almost-hypergroup.

∘	1	2	3
1	{2,3}	{1,2,3}	{1,3}
2	{1,2}	{1,3}	{1,2,3}
3	{1,3}	{2,3}	{1,2}

Table 10. Join right almost-hypergroup.

∘	1	2	3
1	{1,3}	{1,2}	{1,2,3}
2	{1,2,3}	{1,3}	{1,2}
3	{1,3}	{1,2}	{2,3}

A weak left almost-hypergroup which satisfies the weak commutativity will be named a *weak left almost commutative hypergroup* (WLAC-hypergroup). Analogous is the definition of the *weak right almost commutative hypergroup* (WRAC-hypergroup),

Proposition 5. *If $(H, \cdot)$ is a LAC-hypergroup, then $(H, *)$ is WLAC-hypergroup.*

Proposition 6. *Let $(H, \cdot)$ be a left almost-hypergroup and $I_{ab} = I_{ba}$ for all $a, b \in H$, then $(H, *)$ is a WLAC-hypergroup.*

Proposition 7.

i. If $(H, \cdot)$ is a LAC-hypergroup and $\bigcap\limits_{a,b \in H} I_{ab} \neq \varnothing$, then $(H, *)$ is a weak join left almost-hypergroup.

ii. If $(H, \cdot)$ is a LA-hypergroup, $I_{ab} = I_{ba}$ for all $a, b \in H$ and $\bigcap\limits_{a,b \in H} I_{ab} \neq \varnothing$, then $(H, *)$ is a weak join left almost-hypergroup.

Corollary 4. *If H is a left almost-hypergroup and w is an arbitrary element of H, then H endowed with the hypercomposition*

$$x * y = xy \cup \{x, y, w\}$$

is a weak join left almost-hypergroup.

Remark 3. *Analogous propositions to the above 3–7, hold for the right almost-hypergroups as well.*

3. Enumeration and Structure Results

The enumeration of hypercompositional structures is the subject of several papers (e.g., [78–87]). In [78] a symbolic manipulation package is developed which enumerates the hypergroups of order 3, separates them into isomorphism classes and calculates their cardinality. Following analogous techniques, in this paper, a package is developed (see Appendix A) which, when combined with the package in [78], enumerates the left almost-hypergroups and the right almost-hypergroups with 3 elements, classifies them in isomorphism classes and computes their cardinality.

For the purpose of the package, the set $H = \{1, 2, 3\}$ is used as the set with three elements. The laws of synthesis in H are defined through the Cayley Table 11:

Table 11. General form of a 3-element magma's Cayley table

∘	1	2	3
1	a_{11}	a_{12}	a_{13}
2	a_{21}	a_{22}	a_{23}
3	a_{31}	a_{32}	a_{33}

where the intersection of row i with column j, i.e., a_{ij}, is the result of $i \circ j$. As in the case of hypergroups, in the left almost-hypergroups or right almost-hypergroups, the result of the hypercomposition of any two elements is non-void (see Theorem 4 below). Thus, the cardinality of the set of all possible magmas with 3 elements which are not partial hypergroupoids, is $7^9 = 40\,353\,607$. As it is mentioned above in the commutative case the left inverted associativity and the right inverted associativity coincide with the associativity. Among the 40 353 607 magmas only 2520 are commutative hypergroups, which are the trivial cases of left almost-hypergroups and right almost-hypergroups. Thus, the package focuses on the non-trivial cases, that is on the non-commutative magmas. The enumeration reveals that there exist 65 955 reproductive non-commutative magmas which satisfy the left inverted associativity only and obviously the same number of reproductive non-commutative magmas which satisfy the right inverted associativity only. That is, there exist 65 955 non-trivial left almost-hypergroups and 65 955 non-trivial right almost-hypergroups. Moreover, there exist 7036 reproductive magmas which satisfy both the left and the right inverted associativity i.e., there exist 7036 both left and right almost-hypergroups. Furthermore, there are 16 044 reproductive magmas that satisfy the left inverted associativity, the right inverted associativity and the associativity. This means that there exist 16 044 structures which are simultaneously left almost-hypergroups, right almost-hypergroups and hypergroups. Finally, there do not exist any reproductive magmas which satisfies both the left (or right) inverted associativity and the associativity.

The following examples present the worth mentioning cases in which a hypercompositional structure is (a) both left and right almost-hypergroup and (b) simultaneously left almost-hypergroup, right almost-hypergroup and non-commutative hypergroup.

Example 6. *The hypercompositional structure described in Cayley Table 12 is both left and right almost-hypergroup.*

Table 12. A LA- and RA-hypergroup.

∘	1	2	3
1	{1}	{1,2}	{1,2,3}
2	{1,2}	{3}	{1,3}
3	{1,2,3}	{1,2,3}	{1,2,3}

Example 7. *The hypercompositional structure described in Cayley Table 13 is simultaneously left almost-hypergroup, right almost-hypergroup and non-commutative hypergroup.*

Table 13. A non-commutative hypergroup which is also left and right almost-hypergroup.

∘	1	2	3
1	{1}	{1}	{1,2,3}
2	{1}	{1}	{2,3}
3	{1,2,3}	{1,2,3}	{1,2,3}

A magma though, with three elements, can be isomorphic to another magma, which results from a permutation of these three elements. The isomorphic structures which appear in this way, can be considered as members of the same class. These classes can be constructed and enumerated, with the use of the techniques and methods which are developed in [78]. Having done so, the following conclusions occurred:

- The 65 955 non-trivial left almost-hypergroups are partitioned in 11 067 isomorphism classes. 10 920 of them consist of 6 members, 142 have 3 members, 4 have 2 members and the last one is a one-member class. The same are valid for the 65 955 non-trivial right almost-hypergroups.
- The 7036 both left and right almost-hypergroups are partitioned in 1174 isomorphism classes. 1172 of them consist of 6 members, while the other 2 have 2 members.
- The 16 044 noncommutative structures which are simultaneously left almost-hyper groups, right almost-hypergroups and hypergroups are partitioned in 2733 isomorphism classes. 2617 of them consist of 6 members, 110 have 3 members and the last 6 have 2 members.

The above results, combined with the ones of [78] for the hypergroups, are summarized and presented in the following Table 14:

Table 14. Classification of the LA-hypergroups and the RA-hypergroups with three elements.

	Total Number	Isomorphism Classes	Classes with 1 Member (Rigid)	Classes with 2 Members	Classes with 3 Members	Classes with 4 or 5 Members	Classes with 6 Members
non-trivial left almost-hypergroups	65 955	11 067	1	4	142	0	10 920
non-trivial right almost-hypergroups	65 955	11 067	1	4	142	0	10 920
non-trivial both left and right almost-hypergroups	7036	1174	0	2	0	0	1172
simultaneously left almost-hypergroups, right almost-hypergroups and non-commutative hypergroups	16 044	2733	0	6	110	0	2617
non-commutative hypergroups (satisfying the associativity only)	4628	800	0	1	56	0	723
simultaneously left almost-hypergroups and non-commutative hypergroups	0						
simultaneously right almost-hypergroups and non-commutative hypergroups	0						
commutative hypergroups (trivial left almost-hypergroups and trivial right almost-hypergroups)	2520	466	6	3	78	0	399
hypergroups	23 192	3999	6	10	244	0	3739

In the above table we observe that there is only one class of left almost-hypergroups and only one class of right almost-hypergroups which contains a single member. The member of this class is of particular interest, since its automorphism group is of order 1. Such hypercompositional structures are called *rigid*. Additionally, observe that there are 6 rigid hypergroups all of which are commutative. A study and enumeration of these rigid hypergroups, as well as other rigid hypercompositional structures is given in [81]. The following Table 15 presents the unique rigid left almost-hypergroup, while Table 16 describes the unique rigid right almost-hypergroup with 3 elements.

Table 15. The rigid left almost-hypergroup of three elements.

∘	1	2	3
1	{1,2,3}	{2,3}	{2,3}
2	{1,3}	{1,2,3}	{1,3}
3	{1,2}	{1,2}	{1,2,3}

Table 16. The rigid right almost-hypergroup of three elements.

∘	1	2	3
1	{1,2,3}	{1,3}	{1,2}
2	{2,3}	{1,2,3}	{1,2}
3	{2,3}	{1,3}	{1,2,3}

More generally, the next theorem is valid:

Theorem 3. *A non-void set H equipped with the hypercomposition:*

$$xy = \begin{cases} H, & \text{if } x = y \\ H \cdot \cdot \{x\}, & \text{if } x \neq y \end{cases}$$

becomes a rigid left almost-hypergroup, while equipped with the hypercomposition:

$$xy = \begin{cases} H, & \text{if } x = y \\ H \cdot \cdot \{y\}, & \text{if } x \neq y \end{cases}$$

becomes a rigid right almost-hypergroup. Both of them satisfy the transposition axiom.

Remark 4. *There exist 81 hypergroupoids of two elements, 3 of which are LA-hypergroups, 3 are RA-hypergroups, 4 are commutative hypergroups and 4 are non-commutative hypergroups. The enumeration of the two-element crisp and fuzzy hypergroups can be found in [82].*

4. Algebraic Properties

In hypercompositional algebra it is dominant that a hypercomposition on a set E is a mapping of $E \times E$ to the non-void subsets of E. In [4], it is shown that this restriction is not necessary, since it can be proved that the result of the hypercomposition is non void in many hypercompositional structures. Such is the hypergroup (see [4], Theorem 12, and [75], Property 1.1), the weakly associative magma ([4], Proposition 5) and consequently the H_V-group ([76], Proposition 3.1). The next theorem shows that in the case of the left/right almost hypergroups the result of the hypercomposition is non-void as well. The proof is similar to the one in [4], Theorem 12, but not the same due to the validity of the inverted associativity in these structures, instead of the associativity.

Theorem 4. *In a left almost-hypergroup or in a right almost-hypergroup H, the result of the hypercomposition of any two elements is non-void.*

Proof. Suppose that H is a left almost-hypergroup and $cb = \varnothing$, for some c, b in H. Because of the reproductivity the equalities $H = Hc = cH$ and $H = Hb = bH$ are valid. Then, the left inverted associativity gives:

$$H = Hc = (Hb)c = (cb)H = \varnothing H = \varnothing,$$

which is absurd. Next assume that H is a right almost-hypergroup and $bc = \varnothing$, for some b, c in H. The right inverted associativity gives:

$$H = cH = c(bH) = H(bc) = H\varnothing = \varnothing,$$

which is absurd. □

Proposition 8. *A left almost-hypergroup H is a hypergroup if and only if $a(bc) = (cb)a$ holds for all a, b, c in H.*

Proof. Let H be a hypergroup. Then, the associativity $(ab)c = a(bc)$ holds for all a, b, c in H. Moreover because of the assumption, $a(bc) = (cb)a$ is valid for all a, b, c in H. Therefore $(ab)c = (cb)a$, thus H is a left almost-hypergroup. Conversely now, suppose that $a(bc) = (cb)a$ holds for all a, b, c in H. Since H is a left almost-hypergroup, the sequence of the equalities $a(bc) = (cb)a = (ab)c$ is valid. Consequently, H is a hypergroup. □

Theorem 5. *In any left or right almost-hypergroup H the following are valid:*
i. $a/b \neq \varnothing$ and $b\backslash a \neq \varnothing$, for all a, b in H,
ii. $H = H/a = a/H$ and $H = a\backslash H = H\backslash a$, for all a in H,
iii. *the non-empty result of the induced hypercompositions is equivalent to the reproductive axiom.*

Proof. (i) Because of the reproduction, $Hb = H$ for all $b \in H$. Hence, for every $a \in H$ there exists $x \in H$, such that $a \in xb$. Thus, $x \in a/b$ and, therefore, $a/b \neq \varnothing$. Similarly, $b\backslash a \neq \varnothing$.

(ii) Because of Theorem 4, the result of the hypercomposition in H is always a non-empty set. Thus, for every $x \in H$ there exists $y \in H$, such that $y \in xa$, which implies that $x \in y/a$. Hence, $H \subseteq H/a$. Moreover, $H/a \subseteq H$. Therefore, $H = H/a$. Next, let $x \in H$. Since $H = xH$, there exists $y \in H$ such that $a \in xy$, which implies that $x \in a/y$. Hence, $H \subseteq a/H$. Moreover, $a/H \subseteq H$. Therefore, $H = a/H$.

(iii) Suppose that $x/a \neq \varnothing$, for all $a, x \in H$. Thus, there exists $y \in H$, such that $x \in ya$. Therefore, $x \in Ha$, for all $x \in H$, and so $H \subseteq Ha$. Next, since $Ha \subseteq H$ for all $a \in H$, it follows that $H = Ha$. Per duality, $H = aH$. Conversely now, per Theorem 4, the reproductivity implies that $a/b \neq \varnothing$ and $b\backslash a \neq \varnothing$, for all a, b in H. □

Proposition 9. *In any left or right almost-hypergroup H the following are valid:*
i. $a(b/c) \cup a/(c/b) \subseteq (ab)/c$ and
ii. $(c\backslash b)a \cup (b\backslash c)\backslash a \subseteq c\backslash(ba)$
for all a, b, c in H.

Proof. (i) Let $x \in a(b/c)$. Then, $b/c \cap a\backslash x \neq \varnothing$ (1). Next, if $x \in a/(c/b)$, then $a \in x(c/b)$ or $c/b \cap x\backslash a \neq \varnothing$ (2). From both (1) and (2) it follows that $xc \cap ab \neq \varnothing$. So, there exists $z \in ab$, such that $z \in xc$ which implies that $x \in z/c$. Hence, $x \in (ab)/c$. Thus (i) is valid. Similar is the proof of (ii). □

Corollary 5. *If A, B, C are non-empty subsets of any left or right almost-hypergroup H, then the following are valid:*
i. $A(B/C) \cup A/(C/B) \subseteq (AB)/C$ and
ii. $(C\backslash B)A \cup (B\backslash C)\backslash A \subseteq C\backslash(BA)$.

Proposition 10. *Let a, b, c, d be arbitrary elements of any left or right almost-hypergroup H. Then the following are valid:*
i. $(b\backslash a)(c/d) \subseteq (b\backslash ac)/d \cap b\backslash(ac/d)$,
ii. $(b\backslash a)/(d/c) \subseteq (b\backslash ac)/d$,
iii. $(c\backslash d)\backslash(a/b) \subseteq d\backslash(ac/b)$.

Proof. (i) Let $x \in (b\backslash a)(c/d)$. Then, because of Proposition 9.i, there exists $y \in b\backslash a$, such that $x \in y(c/d) \subseteq (yc)/d$. Thus $xd \cap yc \neq \varnothing$ or $xd \cap (b\backslash a)c \neq \varnothing$. Because of Proposition 9.ii, it holds that: $(b\backslash a)c \subseteq b\backslash ac$. Therefore $x \in (b\backslash ac)/d$ (1). Next, since $x \in (b\backslash a)(c/d)$, there exists $z \in c/d$, such that $x \in (b\backslash a)z$. Because of Proposition 9.ii, the inclusion relation $(b\backslash a)z \subseteq b\backslash(az)$ holds. Thus, $bx \cap az \neq \varnothing$ or equivalently $bx \cap a(c/d) \neq \varnothing$ or, because of Proposition 9.i, $bx \cap ac/d \neq \varnothing$. Therefore, $x \in b\backslash(ac/d)$ (2). Now (1) and (2) give (i).

(ii) Suppose that $x \in (b\backslash a)/(d/c)$. Then there exists $y \in b\backslash a$ such that $x \in y/(d/c) \subseteq (yc)/d$. Thus, $xd \cap yc \neq \emptyset$ or equivalently $xd \cap (b\backslash a)c \neq \emptyset$. But according to Proposition 9.ii, the inclusion $(b\backslash a)c \subseteq b\backslash ac$ holds. Hence, $x \in (b\backslash ac)/d$.

(iii) can be proved in a similar manner. □

Corollary 6. *If A, B, C, D are non-empty subsets of any left or right almost-hypergroup H, then the following are valid:*

i. $(B\backslash A)(C/D) \subseteq (B\backslash AC)/D \cap B\backslash (AC/D)$,
ii. $(B\backslash A)/(D/C) \subseteq (B\backslash AC)/D$,
iii. $(C\backslash D)\backslash(A/B) \subseteq D\backslash(AC/B)$.

Proposition 11. *Let a, b be elements of a left or right almost-hypergroup H, then:*

i. $b \in (a/b)\backslash a$ and
ii. $b \in a/(b\backslash a)$.

Proof. (i) Let $x \in a/b$. Then $a \in xb$. Hence, $b \in x\backslash a$. Thus, $b \in (a/b)\backslash a$. Therefore, (i) is valid. The proof of (ii) is similar. □

Corollary 7. *If A, B are non-empty subsets of any left or right almost-hypergroup H, then:*

i. $B \subseteq (A/B)\backslash A$ and
ii. $B \subseteq A/(B\backslash A)$.

Remark 5. *The above properties are consequences of the reproductive axiom and therefore are valid in both, the left and the right almost-hypergroups as well as in the hypergroups [39].*

Proposition 12.

i. *In any left almost-hypergroup the following property is valid:*

$$(a/b)/c = (bc)\backslash a \quad (\text{mixed left inverted associativity})$$

ii. *In any right almost-hypergroup the following property is valid:*

$$c\backslash(b\backslash a) = a/(cb) \quad (\text{mixed right inverted associativity}).$$

Proof. (i) Let $x \in (a/b)/c$. Then the following sequence of equivalent statements is valid:

$$x \in (a/b)/c \Leftrightarrow xc \cap a/b \neq \emptyset \Leftrightarrow a \in (xc)b \Leftrightarrow a \in (bc)x \Leftrightarrow x \in (bc)\backslash a.$$

Similar is the proof of (ii). □

Corollary 8.

i. *If A, B, C are non-empty subsets of a left almost-hypergroup H, then:*

$$(A/B)/C = (BC)\backslash A.$$

ii. *If A, B, C are non-empty subsets of a right almost-hypergroup H, then:*

$$C\backslash(B\backslash A) = A/(CB).$$

Proposition 13.

i. *In any left almost-hypergroup the right inverted associativity of the induced hypercompositions is valid:*

$$b\backslash(a/c) = c\backslash(a/b).$$

ii. In any right almost-hypergroup the left inverted associativity of the induced hypercompositions is valid:
$$(b\backslash a)/c = (c\backslash a)/b.$$

Proof. For (i) it holds that:
$$b\backslash(a/c) = \{x \in H \mid bx \cap a/c \neq \varnothing\} = \{x \in H \mid a \in (bx)c\} =$$
$$= \{x \in H \mid a \in (bx)c\} = \{x \in H \mid a/b \cap cx \neq \varnothing\} = c\backslash(a/b).$$

Regarding (ii) it is true that:
$$(b\backslash a)/c = \{x \in H \mid b\backslash a \cap xc \neq \varnothing\} = \{x \in H \mid a \in b(xc)\} =$$
$$= \{x \in H \mid a \in c(xb)\} = \{x \in H \mid c\backslash a \cap xb \neq \varnothing\} = (c\backslash a)/b.$$

This completes the proof. □

Corollary 9.

i. If A, B, C are non-empty subsets of a left almost-hypergroup H, then:
$$B\backslash(A/C) = C\backslash(A/B).$$

ii. If A, B, C are non-empty subsets of a right almost-hypergroup H, then:
$$(B\backslash A)/C = (C\backslash A)/B.$$

5. Identities and Symmetric Elements

Let H be a non-void set endowed with a hypercomposition. An element e of H is called *right identity*, if $x \in x \cdot e$ for all x in H. If $x \in e \cdot x$ for all x in H, then e is called *left identity*, while e is called *identity* if it is both right and left identity. An element e of H is called *right scalar identity*, if $x = x \cdot e$ for all x in H. If $x = e \cdot x$ for all x in H, then e is called *left scalar identity*, while e is called *scalar identity* if it is both right and left scalar identity [4,13,18,88]. When a left (resp. right) scalar identity exists in H, then it is unique.

Example 8. *The Cayley Tables 17 and 18 describe left/right almost-hypergroups with left/right scalar identity.*

Table 17. Left almost-hypergroup with left scalar identity.

∘	1	2	3
1	{1}	{2}	{3}
2	{3}	{1,2,3}	{1,3}
3	{2}	{1,2}	{1,2,3}

Table 18. Right almost-hypergroup with right scalar identity.

∘	1	2	3
1	{1}	{3}	{2}
2	{2}	{1,2,3}	{1,3}
3	{3}	{1,2}	{1,2,3}

If the equality $e = ee$ is valid for an identity e, then e is called *idempotent identity* [4,39,88]. e is a *right strong identity*, if $x \in x \cdot e \subseteq \{e, x\}$ for all x in H while e is a *left strong identity*, if $x \in e \cdot x \subseteq \{e, x\}$. e is a *strong identity*, if it is right and left strong [4,32,34,39,88–90]. Note that the strong identity needs not be unique [32,34,39]. Both the scalar identity and the strong identity are idempotent identities.

Theorem 6. *If e is a strong identity in a left almost-hypergroup H, then*

$$x/e = e \backslash x = x, \text{ for all } x \in H - \{e\}$$

Proof. Suppose that $y \in x/e$. Then $x \in ye \subseteq \{y, e\}$. Consequently $y = x$. □

Let e be an identity element in H and x an element in H. Then, x will be called *right e-attractive*, if $e \in e \cdot x$, while it will be called *left e-attractive* if $e \in x \cdot e$. If x is both left and right e-attractive, then it will be called *e-attractive*. When there is no likelihood of confusion, e can be omitted. See [32] for the origin of the terminology.

Proposition 14. *In a left almost-hypergroup with idempotent identity e, $e \backslash e$ is the set of right e-attractive elements of H and e/e is the set of left e-attractive elements of H.*

Proof. Suppose that x is a right attractive element of H. Then $e \in ex$. Thus $x \in e \backslash e$. Additionally, if $x \in e \backslash e$, then $e \in ex$. Hence $e \backslash e$ consists of all the right attractive elements of H. The rest follows in a similar way. □

When the identity is strong and x is an attractive element, then $e \cdot x = x \cdot e = \{e, x\}$, while, if x is non attractive, then $e \cdot x = x \cdot e = x$ is valid. In the case of a strong identity, the non-attractive elements are called *canonical*. See [32] for the origin of the terminology

Proposition 15. *If x is not an e-attractive element in a left almost-hypergroup with idempotent identity e, then xe consists of elements that are not e-attractive.*

Proof. Suppose that $y \in xe$ and assume that y is attractive. Then

$$ye \subseteq (xe)e = (ee)x = ex.$$

Since $e \in ey$, it follows that $e \in ex$, which is absurd. □

Proposition 16. *If x is a right (resp. left) e-attractive element in a transposition left almost-hypergroup with idempotent identity e, then all the elements of xe are right (resp. left) e-attractive.*

Proof. Let $y \in xe$, then $x \in y/e$. Moreover, $x \in e \backslash e$. Thus $e \backslash e \cap y/e \neq \emptyset$, which implies that $ee \cap ey \neq \emptyset$. Therefore, $e \in ey$. □

Proposition 17. *If x is a right (resp. left) attractive element in a transposition left almost-hypergroup with idempotent identity e, then its right (resp. left) inverses are also right (resp. left) attractive elements.*

Proof. Since $e \in ex$, it follows that $x \in e\backslash e$. Moreover, if x' is a right inverse of x, then $e \in xx'$. Therefore, $x \in e/x'$. Consequently, $e\backslash e \approx e/x'$. Transposition gives $ee \approx ex'$ and since e is idempotent, $e \in ex'$. Thus, x' is right attractive. □

Corollary 10. *If x is not a right (resp. left) attractive element in a transposition left almost-hypergroup with idempotent identity e, then its right (resp. left) inverses are also not right (resp. left) attractive elements.*

Proposition 18. *If a left almost-hypergroup (resp. right almost-hypergroup) has a right scalar identity (resp. left scalar identity), then it is a hypergroup.*

Proof. Suppose that e is a right scalar identity in a left almost-hypergroup H. Then for any two elements b, c in H we have:

$$bc = (be)c = (ce)b = cb.$$

Hence H is commutative and therefore H is a hypergroup. Similar is the proof for the right almost-hypergroups. □

Proposition 19. *If a left almost-hypergroup or a right almost-hypergroup H has a strong identity e and if it consists only of attractive elements, then $x, y \in xy$.*

Proof. $(ex)y = \{e, x\}y = \{e, y\} \cup xy$ and $(yx)e = yx \cup \{e\}$. Since $(ex)y = (yx)e$ it derives that $yx \cup \{e\} = xy \cup \{e, y\}$. Similarly, $xy \cup \{e\} = yx \cup \{e, x\}$. Therefore:

$$xy \cup \{e\} = yx \cup \{e, x\} = xy \cup \{e, y\} \cup \{e, x\} = xy \cup \{e, x, y\}.$$

Hence, $x, y \in xy$. □

Corollary 11. *If a left almost-hypergroup or a right almost-hypergroup H has a strong identity e and if it consists of attractive elements only, then*
i. $x \in x/y$ and $x \in y\backslash x$, for all $x, y \in H$
ii. $x/x = x\backslash x = H$, for all $x, y \in H$

Proposition 20. *If a left almost-hypergroup or a right almost-hypergroup H has a strong identity e and if the relation $e \in bc$ implies that $e \in cb$, for all $b, c \in H$, then H is a hypergroup.*

Proof. For any two elements in H we have:

$$cb \subseteq \{e, c\}b = (ec)b = (bc)e = \{e\} \cup bc. \text{ (i)}$$

If $e \in bc$ then $cb \subseteq bc$. Moreover,

$$bc \subseteq \{e, b\}c = (eb)c = (cb)e = \{e\} \cup cb. \text{ (ii)}$$

According to the assumption if $e \in bc$, then $e \in cb$. Hence $bc \subseteq cb$. Thus $cb = bc$. Next, if $e \notin bc$ then $e \notin cb$ and (i) implies that $cb \subseteq bc$ while (ii) implies that $bc \subseteq cb$. Thus $cb = bc$. Consequently H is commutative and therefore H is a hypergroup. □

Proposition 21. *If a left almost-hypergroup or a right almost-hypergroup H has a strong identity e and if A is the set of its attractive elements, then*

$$A/A \subseteq A \text{ and } A\backslash A \subseteq A.$$

Proof. Since A is the set of the attractive elements, $e/e = A$ and $e\backslash e = A$ is valid. Therefore:

$$A/A = (e\backslash e)/(e/e) \subseteq (e\backslash ee)/e = (e\backslash e)/e$$

But $e\backslash e = e/e$, thus

$$(e\backslash e)/e = (e/e)/e = ee\backslash e = e\backslash e = A$$

Consequently $A/A \subseteq A$. Similarly, $A\backslash A \subseteq A$. □

An element x' is called *right e-inverse* or *right e-symmetric* of x, if there exists a right identity $x' \neq e$ such that $e \in x \cdot x'$. The definition of the *left e-inverse* or *left e-symmetric* is analogous to the above, while x' is called *e-inverse* or *e-symmetric* of x, if it is both right and left inverse with regard to the same identity e. If e is an identity in a left almost-hypergroup H, then the set of the left inverses of $x \in H$, with regard to e, will be denoted by $S_{el}(x)$, while $S_{er}(x)$ will denote the set of the right inverses of $x \in H$ with regard to e. The intersection $S_{el}(x) \cap S_{er}(x)$ will be denoted by $S_e(x)$.

Proposition 22. *If e is an identity in a left almost-hypergroup H, then*

$$S_{er}(x) = (e/x) \cdots \{e\} \text{ and } S_{er}(x) = (x\backslash e) \cdots \{e\}.$$

Proof. $y \in e/x$, if and only if $e \in yx$. This means that either $y \in S_{el}(x)$ or $y = e$, if x is right attractive. Hence, $e/x \subseteq \{e\} \cup S_{el}(x)$. The rest follows in a similar way. □

Corollary 12. *If $S_{el}(x) \cap S_{er}(x) \neq \emptyset$, $x \in H$, then $x\backslash e \cap e/x \neq \emptyset$.*

Proposition 23. *If H is a transposition left almost-hypergroup with an identity e and $z \in xy$, then:*
  i. $ey \cap x'z \neq \emptyset$, for all $x' \in S_{el}(x)$,
  ii. $xe \cap zy' \neq \emptyset$, for all $y' \in S_{er}(y)$.

Proof. $z \in xy$ implies that $x \in z/y$ and that $y \in x\backslash z$. Let $x' \in S_{el}(x)$ and $y' \in S_{er}(y)$. Then $e \in x'x$ and $e \in yy'$. Thus $x \in x'\backslash e$ and $y \in e\backslash y'$. Therefore $x'\backslash e \cap z/y \neq \emptyset$ and $x\backslash z \cap e/y' \neq \emptyset$. Hence, because of the transposition, $ey \cap x'z \neq \emptyset$ and $xe \cap zy' \neq \emptyset$. □

Proposition 24. *Let H be a transposition left almost-hypergroup with a strong identity e and let x, y, z be elements in H such that $x, y, z \neq e$ and $z \in xy$. Then:*
  i. *if $S_{el}(x) \cap S_{el}(z) = \emptyset$, then $y \in x'z$, for all $x' \in S_{el}(x)$,*
  ii. *if $S_{er}(y) \cap S_{er}(z) = \emptyset$, then $x \in zy'$, for all $y' \in S_{er}(y)$.*

Proof. (i) According to Proposition 23, $z \in xy$ implies that $ey \cap x'z \neq \emptyset$ for all $x' \in S_l(x)$. Since e is strong $ey = \{e, y\}$. Hence $\{e, y\} \cap x'z \neq \emptyset$. But $S_{el}(x)$ and $S_{el}(z)$ are disjoint. Thus $e \notin x'z$, therefore $y \in x'z$. Analogous is the proof of (ii). □

6. Substructures of the Left/Right Almost-Hypergroups

There is a big variety of substructures in the hypergroups, which is much bigger than the one in the groups. Analogous is the variety of the substructures which are revealed here in the case of the left/right almost-hypergroups. For the consistency of the terminology [4,13,33,55,91–99], the terms semisub-left/right almost-hypergroup, sub-left/right almost-hypergroup, etc. will be used in exactly the same way as the prefixes sub- and semisub- are used in the cases of the groups and the hypergroups, e.g., the terms subgroup, subhypergroup are used instead of hypersubgroup, etc. The following research is inspired by the methods and techniques used in [93–97].

Let H be a left almost-hypergroup. Then,

Definition 12. *A non-empty subset K of H is called semisub-left-almost-hypergroup (semisub-LA-hypergroup) when it is stable under the hypercomposition, i.e., when it has the property $xy \subseteq K$ for all $x, y \in K$.*

Definition 13. *A semisub-LA-hypergroup K of H is a sub-left-almost-hypergroup (sub-LA-hypergroup) of H if it satisfies the reproductivity, i.e., if the equality $xK = Kx = K$ is valid for all $x \in K$.*

Example 9. *In the left almost-hypergroup, which is described in the Cayley Table 19, $\{5,6,7\}$ is a semisub-LA-hypergroup while $\{1,2\}$, $\{3,4\}$, $\{1,2,3,4\}$, $\{1,2,5,6,7\}$ and $\{3,4,5,6,7\}$ are sub-LA-hypergroup.*

Table 19. Left almost-hypergroup.

	1	2	3	4	5	6	7
1	{1,2}	{1,2}	{1,2,3,4}	{1,2,3,4}	{1,2,5,6,7}	{1,2,5,6,7}	{1,2,5,6,7}
2	{1}	{1,2}	{1,2,3,4}	{1,2,3,4}	{1,2,5,6,7}	{1,2,5,6,7}	{1,2,5,6,7}
3	{1,2,3,4}	{1,2,3,4}	{3,4}	{4}	{3,4,5,6,7}	{3,4,5,6,7}	{3,4,5,6,7}
4	{1,2,3,4}	{1,2,3,4}	{3}	{3,4}	{3,4,5,6,7}	{3,4,5,6,7}	{3,4,5,6,7}
5	{1,2,5,6,7}	{1,2,5,6,7}	{3,4,5,6,7}	{3,4,5,6,7}	{5,7}	{7}	{6,7}
6	{1,2,5,6,7}	{1,2,5,6,7}	{3,4,5,6,7}	{3,4,5,6,7}	{6,7}	{7}	{7}
7	{1,2,5,6,7}	{1,2,5,6,7}	{3,4,5,6,7}	{3,4,5,6,7}	{6,7}	{6,7}	{6,7}

Proposition 25. *If a semisub-LA-hypergroup K of H is stable under the induced hypercompositions, then K is a sub-LA-hypergroup of H.*

Proof. We have to prove the reproductivity. Let $x \in K$. Obviously $xK \subseteq K$ and $Kx \subseteq K$. Next let y be an element of K. Then $x \backslash y = \{z \in H \mid y \in xz\}$ is a subset of H. Therefore, there exists an element $z \in K$ such that $y \in xz \subseteq xK$. Thus $K \subseteq xK$. Similarly, $y/x \subseteq K$ yields $K \subseteq Kx$. □

Proposition 26. *If K is a sub-LA-hypergroup of H, then $H \cdot \cdot K \subseteq (H \cdot \cdot K)s$ and $H \cdot \cdot K \subseteq s(H \cdot \cdot K)$, for all $s \in K$.*

Proof. Let r be an element in $H \cdot \cdot K$ which does not belong to $(H \cdot \cdot K)s$. Because of the reproductive axiom, $r \in Hs$ and since $r \notin (H \cdot \cdot K)s$, r must be a member of Ks. Thus, $r \in Ks \subseteq KK = K$. This contradicts the assumption and so $H \cdot \cdot K \subseteq (H \cdot \cdot K)s$. The second inclusion follows similarly. □

Proposition 27. *If K is a sub-LA-hypergroup of H, $A \subseteq K$ and $B \subseteq H$, then*

i. $A(B \cap K) \subseteq AB \cap K$ and
ii. $(B \cap K)A \subseteq BA \cap K$.

Proof. Let $t \in A(B \cap K)$. Then $t \in ax$, with $a \in A$ and $x \in B \cap K$. Since x lies in $B \cap K$, it derives that $x \in B$ and $x \in K$. Hence $ax \subseteq aB$ and $ax \subseteq aK = K$. Thus $ax \subseteq AB \cap K$ and therefore $t \in AB \cap K$. The second inclusion follows similarly. □

The next definition introduces the notion of the closed sub-LA/RA-hypergroups. The significance of the closed subhypergroups which is mentioned in [4,13,33,55,91–96,99] remains exactly the same in the case of the sub-LA/RA-hypergroups.

Definition 14. *A sub-LA-hypergroup K of H is called right closed in H if it is stable under the right induced hypercomposition, that is if $a/b \subseteq K$ for any two elements a and b in K. Similarly, K is called left closed if $b \backslash a \subseteq K$, for all $a, b \in K$. K is closed when it is both right and left closed.*

Proposition 28.
i.   *K is a right closed sub-LA-hypergroup of H, if and only if $xK \cap K = \emptyset$, for every $x \in H \cdot \cdot K$*
ii. *K is a left closed sub-LA-hypergroup of H, if and only if $Kx \cap K = \emptyset$, for every $x \in H \cdot \cdot K$*
iii. *K is a closed sub-LA-hypergroup of H, if and only if $xK \cap K = \emptyset$ and $Kx \cap K = \emptyset$, for every $x \in H \cdot \cdot K$*

Proof. (i) Suppose that $xK \cap K \neq \emptyset$ for some $x \in H$. So there exist $y, z \in K$ such that $y \in xz$ or equivalently $x \in y/z$. Since K is right closed $y/z \subseteq K$, therefore $x \in K$. Conversely now. Let $y, z \in K$. Then for every $x \in y/z$ we have that $y \in xz$ or equivalently $xK \cap K \neq \emptyset$. Consequently $x \in K$, thus $y/z \subseteq K$. The proof of (ii) is similar, while (iii) derives directly from (i) and (ii). □

Proposition 29.
i. *A sub-LA-hypergroup K of H is right closed in H, if and only if*

$$(H \cdot \cdot K)s = H \cdot \cdot K, \text{ for all } s \in K.$$

ii. *A sub-LA-hypergroup K of H is left closed in H, if and only if*

$$s(H \cdot \cdot K) = H \cdot \cdot K, \text{ for all } s \in K.$$

iii. *A sub-LA-hypergroup K of H is closed in H, if and only if*

$$s(H \cdot \cdot K) = (H \cdot \cdot K)s = H \cdot \cdot K, \text{ for all } s \in K.$$

Proof. (i) Let K be right closed in H. Suppose that z lies in $H \cdot \cdot K$ and assume that $zs \cap K \neq \emptyset$. Then, there exists an element y in K such that $y \in zs$, or equivalently, $z \in y/s$. Therefore $z \in K$, which is absurd. Hence $(H \cdot \cdot K)s \subseteq H \cdot \cdot K$. Next, because of Proposition 26, $H \cdot \cdot K \subset (H \cdot \cdot K)s$ and therefore $H \cdot \cdot K = (H \cdot \cdot K)s$. Conversely now. Suppose that $(H \cdot \cdot K)s = H \cdot \cdot K$ for all $s \in K$. Then $(H \cdot \cdot K)s \cap K = \emptyset$ for all $s \in K$. Hence $x \notin rs$ and so $r \notin x/s$ for all $x, s \in K$ and $r \in H \cdot \cdot K$. Therefore $x/s \cap (H \cdot \cdot K) = \emptyset$ which implies that $x/s \subseteq K$. Thus K is right closed in H. (ii) follows in a similar way and (iii) is an obvious consequence of (i) and (ii). □

Proposition 30.
i. *If K is a sub-LA-hypergroup of H and*

$$Kx \cap (H \cdot \cdot K)x = \emptyset$$

for all $x \in H$, then K is right closed in H.

ii. *If K is a sub-LA-hypergroup of H and*

$$xK \cap x(H \cdot \cdot K) = \emptyset$$

for all $x \in H$, then K is left closed in H.

Proof. (i) From the reproductive axiom we have:

$$K \cup (H \cdot \cdot K) = H = Hx = Kx \cup (H \cdot \cdot K)x$$

According to the hypothesis $Kx \cap (H \cdot \cdot K)x = \emptyset$, which implies that $K \cap (H \cdot \cdot K)x = \emptyset$ when $x \in K$. Therefore, $H = K \cup (H \cdot \cdot K)x$ is a union of disjoint sets. Thus $(H \cdot \cdot K)x = H \cdot \cdot K$. So, per Proposition 29, K is right closed in H. Similar is the proof of (ii). □

Proposition 31.

i. If K is a right closed sub-LA-hypergroup in H, $A \subseteq K$ and $B \subseteq H$, then

$$(B \cap K)A = BA \cap K.$$

ii. If K is a left closed sub-LA-hypergroup in H, $A \subseteq K$ and $B \subseteq H$, then

$$A(B \cap K) = AB \cap K.$$

Proof. (i) Let $t \in BA \cap K$. Since K is right closed, for any element y in $B \cdot \cdot K$, it is valid that $yA \cap K \subseteq yK \cap K = \emptyset$. Hence $t \in (B \cap K)A \cap K$. But $(B \cap K)A \subseteq KK = K$. Thus $t \in (B \cap K)A$. Therefore $BA \cap K \subseteq (B \cap K)A$. Next the inclusion becomes equality because of Proposition 27. (ii) derives in a similar way. □

Proposition 32.

i. If K is a right closed sub-LA-hypergroup in H, $A \subseteq K$ and $B \subseteq H$, then

$$(B \cap K)/A = (B/A) \cap K.$$

ii. If K is a left closed sub-LA-hypergroup in H, $A \subseteq K$ and $B \subseteq H$, then

$$(B \cap K) \backslash A = B \backslash A \cap K.$$

Proof. (i) Since $B \cap K \subseteq B$, it derives that $(B \cap K)/A \subseteq B/A$. Moreover $A \subseteq K$ and $B \cap K \subseteq K$, thus $(B \cap K)/A \subseteq K$. Hence $(B \cap K)/A \subseteq (B/A) \cap K$. For the reverse inclusion now suppose that $x \in (B/A) \cap K$. Then, there exist $a \in A$ and $b \in B$ such that $x \in b/a$ or equivalently $b \in ax$. Since $ax \subseteq K$ it derives that $b \in K$ and so $b \in B \cap K$. Therefore $b/a \subseteq (B \cap K)/A$. Thus $x \in (B \cap K)/A$. Hence $(B/A) \cap K \subseteq (B \cap K)/A$. (ii) derives in a similar way. □

Although the non-void intersection of two sub-LA-hypergroups is stable under the hypercomposition, it is usually not a sub-LA-hypergroup since the reproductive axiom is not always valid in it.

Proposition 33. *The non-void intersection of any two closed sub-LA-hypergroups of H is a closed sub-LA-hypergroup of H.*

Proof. Let K, M be two closed LA-subhypergroups of H and suppose that x, y are two elements in $K \cap M$. Then $xy \subseteq K$ and $xy \subseteq M$. Therefore $xy \subseteq K \cap M$. Next, since K, M are closed LA-subhypergroups of H, x/y and $y \backslash x$ are subsets of $K \cap M$. Thus, because of Proposition 25, $K \cap M$ is a closed LA-subhypergroup of H. □

Corollary 13. *The set of the closed sub-LA-hypergroups of H which are containing a non-void subset of H, is a complete lattice.*

Proposition 34. *If K is a closed sub-LA-hypergroups of H and $x \in K$, then:*

$$x/K = K/x = K = K\backslash x = x\backslash K$$

Proof. Since K is closed $x/K \subseteq K$ and $K/x \subseteq K$. Let $y \in K$. Because of the reproductivity $x \in yK$ or equivalently $y \in x/K$. Therefore $x/K = K$. Moreover since K is a LA-subhypergroups of H, $yx \subseteq K$. Thus $y \in K/x$. So $K/x = K$. The equalities $K = K\backslash x = x\backslash K$ follow in a similar way. □

Corollary 14. *In any left almost-hypergroup, K is a closed sub-LA-hypergroup if and only if:*

$$K/K = K\backslash K = K$$

Definition 15. *A sub-LA-hypergroup M of H is called right invertible if $x/y \cap M \neq \varnothing$ implies $y/x \cap M \neq \varnothing$, while it is called left invertible if $y\backslash x \cap M \neq \varnothing$ implies $x\backslash y \cap M \neq \varnothing$. If M is right and left invertible, then it is called invertible.*

Direct consequence of the above definition is the following proposition:

Proposition 35.

i. *M is a right invertible sub-LA-hypergroup of H, if and only if:*

$$x \in My \Rightarrow y \in Mx, \; x,y \in H$$

ii. *M is a left invertible sub-LA-hypergroup of H, if and only if:*

$$x \in yM \Rightarrow y \in xM, \; x,y \in H$$

Proposition 36. *If K is an invertible sub-LA-hypergroup of H, then K is closed.*

Proof. Let $x \in K/K$. Then $K \cap xK \neq \varnothing$ and $x\backslash K \cap K \neq \varnothing$. Since K is invertible $K\backslash x \cap K \neq \varnothing$. Thus $x \in KK$. But K is a LA-subhypergroup of H, so $KK = K$. Therefore $x \in K$. Hence $K/K \subseteq K$. Similarly, $K\backslash K \subseteq K$ and so the Proposition. □

Definition 16. *If H has an identity e, then a sub-LA-hypergroup K of H is called symmetric if $x \in K$ implies $S_{el}(x) \cup S_{er}(x) \subset K$.*

Proposition 37. *To any pair of symmetric sub-LA-hypergroups K and M of a LA-hypergroup H there exists a least symmetric sub-LA-hypergroup $K \vee M$ containing them both.*

Proof. Let U be the set of all symmetric subhypergroups R of H which contain both K and M. The intersection of all these symmetric subhypergroups R of H is a symmetric subhypergroup with the desired property. □

7. Fortification in Transposition Left Almost-Hypergroups

The transposition left almost-hypergroups can be fortified through the introduction of neutral elements. Next, we will present two such hypercompositional structures.

Definition 17. *A transposition LA-hypergroup H, is left fortified if it contains an element e which satisfies the axioms,*

i. *e is a left identity and $ee = e$*
ii. *for every $x \in H \cdots \{e\}$ there exists a unique $y \in H \cdots \{e\}$ such that $e \in xy$ and $e \in yx$*

213

For $x \in H \cdot \cdot \{e\}$ the notation x^{-1} is used for the unique element of $H \cdot \cdot \{e\}$ that satisfies axiom (ii). Clearly $(x^{-1})^{-1} = x$. The next results are obvious.

Proposition 38.

i   If $x \neq e$, then $e \backslash x = x$.
ii $e \backslash e = H$.

Proposition 39. Let $x \in H \cdot \cdot \{e\}$. Then $e \in xy$ or $e \in yx$ implies $y \in \{x^{-1}, e\}$.

Now the role of the identity in the transposition left almost-hypergroups can be clarified.

Proposition 40. *The identity e of the transposition left almost-hypergroup is left strong.*

Proof. It suffices to prove that $ex \subseteq \{e, x\}$. For $x = e$ the inclusion is valid. Let $x \neq e$. Suppose that $y \in ex$. Then $x \in e \backslash y$. But $e \in xx^{-1}$, hence $x \in e/x^{-1}$. Thus, $(e \backslash y) \cap (e/x^{-1}) \neq \emptyset$. The transposition axiom gives $e = ee = yx^{-1}$. By the previous proposition, $y \in \{e, x\}$. Therefore, the proposition holds. □

Proposition 41. *The identity e of the transposition left almost-hypergroup is unique.*

Proof. Suppose that u is an identity distinct from e. Then, there would exist the inverse of e, i.e., an element v distinct from u such that $u \in ev$, which is absurd because $ev \subseteq \{e, v\}$. □

Proposition 42. *If H consists of attractive elements only and $e \neq x$ then*

$$e/x = ex^{-1} = \{e, x^{-1}\} = x \backslash e.$$

Proof. Since e is a left strong identity, $ex^{-1} = \{e, x^{-1}\}$ is valid. Moreover

$$e/x = \{z \in H \mid e \in zx\} = \{e, x^{-1}\} \text{ and } x \backslash e = \{z \in H \mid e \in xz\} = \{e, x^{-1}\}. \square$$

Corollary 15. *If A is a non-empty subset of H and $e \notin A$ then:*

$$e/A = eA^{-1} = A^{-1} \cup \{e\} = A \backslash e$$

Definition 18. *A transposition polysymmetrical left almost-hypergroup is a transposition left almost-hypergroup P that contains a left idempotent identity e which satisfies the axiom:*

for every $x \in P \cdot \cdot \{e\}$ there exists at least one element $x' \in P \cdot \cdot \{e\}$, a symmetric of x, such that $e \in x\, x'$ which furthermore satisfies $e \in x'x$.

The set of the symmetric elements of x is denoted by $S(x)$ and it is called the *symmetric set* of x.

Example 10. *Cayley Table 20 describes a transposition polysymmetrical left almost-hypergroup in which, the element 1 is a left idempotent identity.*

Table 20. Transposition polysymmetrical left almost-hypergroup.

∘	1	2	3
1	{1}	{1,2}	{1,3}
2	{1,3}	{1,2,3}	{1,3}
3	{1,2}	{1,2}	{1,2,3}

Proposition 43. $e \backslash x$ *always contains the element* x.

Corollary 16. *If* X *is non-empty, then* $X \subseteq e \backslash X$.

Proposition 44. *Let* $x \neq e$. *Then*
i. $S(x) \cup \{e\} = e/x$ *and* $S(x) = x \backslash e$, *if* x *is attractive*
ii. $eS(x) = e/x$, *if* e *is left strong identity and* x *is attractive*
iii. $S(x) = x \backslash e = e/x$, *if* x *is non attractive*.

Corollary 17. *Let* X *be a non-empty set and* $e \notin X$. *Then*
i. $S(X) \cup \{e\} = e/X$ *and* $S(X) = X \backslash e$, *if* X *contains an attractive element*
ii. $eS(X) = e/X$, *if* e *is left strong identity and* X *contains an attractive element*
iii. $S(X) = X \backslash e = e/X$, *if* X *consists of non-attractive elements*.

Proposition 45. *For every element* x *of a transposition polysymmetrical left almost-hypergroup it holds* $ex \subseteq \{e\} \cup S(S(x))$, *while for every* $x' \in S(x)$ *it holds:* $(ex) \cap S(x') \neq \varnothing$.

Proof. Let $y \neq e$ and $y \in ex$, then $x \in e \backslash y$. Moreover, for every $x' \in S(x)$ it holds $x \in e/x'$. Consequently $e/x' \cap e \backslash y \neq \varnothing$, so, per transposition axiom, $ee \cap yx' \neq \varnothing$, that is $e \in yx'$ and thus $y \in S(x') \subseteq S(S(x))$. □

Proposition 46. *If* $x \neq e$, *is a right attractive element of a transposition polysymmetrical left almost-hypergroup, then* $S(x)$ *consists of left attractive elements*.

Proof. Let $e \in ex$. Then $x \in e \backslash e$. Moreover, if x' is an arbitrary element from $S(x)$, then $e \in xx'$. Therefore $x \subset e/x'$. Consequently $e \backslash e \cap e/x' \neq \varnothing$. Per transposition axiom $ee \cap x'e \neq \varnothing$. So $e \in x'e$. Thus x' is an attractive element. □

Corollary 18. *If* x *is a non-attractive element, then* $S(x)$ *consists of non-attractive elements only*.

Corollary 19. *If* $x \neq e$, *is an attractive element of a transposition polysymmetrical left almost-hypergroup, then* $S(x)$ *consists of attractive elements only*.

Proposition 47.
i. *If* x *is a right attractive element of a transposition polysymmetrical left almost-hypergroup, then all the elements of* xe *are right attractive*.
ii. *If* x *is a left attractive element of a transposition polysymmetrical left almost-hypergroup, then all the elements of* ex *are left attractive*.

Proof. Assuming that x is a right attractive element we have that $e \in ex$, which implies that $x \in e \backslash e$. Additionally, if z is an element in xe, then $x \in z/e$. Thus $(z/e) \cap (e \backslash e) \neq \varnothing$. Per transposition axiom, $(ee) \cap (ez) \neq \varnothing$ and therefore $e \in ez$, i.e., z is right attractive. Similar is the proof of (ii). □

8. Conclusions and Open Problems

In [70,93] and later in [4], with more details, it was proved that the group can be defined with the use of two axioms only: the associativity and the reproductivity. Likewise, the left/right almost-group is defined here and their existence is proved via examples. The study of this structure reveals a very interesting research area in abstract algebra.

This paper focuses on the study of the more general structure, i.e., of the left/right almost-hypergroup. The enumeration of these structures showed that they appear more frequently than the hypergroups. Indeed, in the case of the hypergroupoids with three elements, there exists one hypergroup in every 1740 hypergroupoids, while there is one non-trivial purely left almost-hypergroup in every 612 hypergroupoids. The same holds for the non-trivial purely right almost-hypergroups, as it is proved in this paper that the cardinal number of the left almost-hypergroups is equal to the cardinal number of the right almost-hypergroups, over a set E. Moreover, there is one non-trivial left and right almost-hypergroup in every 5735 hypergroupoids. Considering the trivial cases as well, i.e., left and right almost-hypergroups which are also non-commutative hypergroups, there exists one left almost-hypergroup in every 453 hypergroupoids. This frequency, which is nearly 4 times higher than that of the hypergroups, justifies a more thorough study of these structures.

Subsequently, these structures were equipped with more axioms, the first one of which is the transposition axiom:

$$b\backslash a \cap c/d \neq \varnothing \text{ implies } ad \cap bc \neq \varnothing, \text{ for all } a,b,c,d \in H$$

The transposition left/right almost-hypergroup is studied here.

In [4] though, the reverse transposition axiom was introduced:

$$ad \cap bc \neq \varnothing \text{ implies } b\backslash a \cap c/d \neq \varnothing, \text{ for all } a,b,c,d \in H$$

The study of the reverse transposition left/right almost-hypergroup is an open problem.

Additionally, open problems for algebraic research are the studies of the properties of all the structures which are introduced in this paper (weak left/right almost-hypergroup, left/right almost commutative hypergroup, join left/right almost hypergroup, reverse join left/right almost hypergroup, weak left/right almost commutative hypergroup) as well as their enumerations. Especially, for the enumeration problems, it is worth mentioning those, which are associated with the rigid hypercompositional structures, that is hypercompositional structures whose automorphism group is of order 1. The conjecture is that there exists only one rigid left almost-hypergroup (and one rigid right almost-hypergroup), while it is known that there exist six such hypergroups, five of which are transposition hypergroups [81].

Author Contributions: Conceptualization, investigation, C.G.M. and N.Y.; software, validation, writing—original draft preparation, writing—review and editing, funding acquisition, C.G.M. Both authors have read and agreed to the published version of the manuscript.

Funding: This research received no external funding. The APC was funded by the MDPI journal Mathematics and the National and Kapodistrian University of Athens.

Institutional Review Board Statement: Not applicable.

Informed Consent Statement: Not applicable.

Data Availability Statement: Not applicable.

Conflicts of Interest: The authors declare no conflict of interest.

Appendix A

The following is the Mathematica [100] package that implements the axioms of left inverted associativity, right inverted associativity, associativity and reproductivity for testing whether a magma is a LA-hypergroup, a RA-hypergroup, a LRA-hypergroup or a non-commutative hypergroup.

```
BeginPackage["LRHtest`"];
Clear["LRHtest`*"];


LRHtest::usage = "LRHtest[groupoid] returns
                1: for Left Asso
                2: for Right Asso
                3: for Left+Right Asso
                4: for Hypergroup
                5: for Hypergroup + Left Asso
                6: for Hypergroup + Right Asso
                7: for Hypergroup + Left+Right Asso"


Begin["`Private`"];
Clear["LRHtest`Private`*"];

LRHtest[groupoid_List] := Module[{r}, r = 0;
   If[groupoid != Transpose[groupoid] && ReproductivityTest[groupoid],
     If[LeftAs[groupoid], r = 1];
     If[RightAs[groupoid], r = r + 2];
     If[Asso[groupoid], r = r + 4]]; Return[r]];


LeftAs[groupoid_List] :=
  Not[MemberQ[
    Flatten[Table[
      Union[Flatten[
        Union[Extract[groupoid,
          Distribute[{groupoid[[i, j]], {k}}, List]]]] ==
      Union[Flatten[
        Union[Extract[groupoid,
          Distribute[{groupoid[[k, j]], {i}}, List]]]]], {i, 1,
      Length[groupoid]}, {j, 1, Length[groupoid]}, {k, 1,
      Length[groupoid]}], 2], False]];

RightAs[groupoid_List] :=
  Not[MemberQ[
    Flatten[Table[
      Union[Flatten[
        Union[Extract[groupoid,
          Distribute[{{i}, groupoid[[j, k]]}, List]]]] ==
      Union[Flatten[
        Union[Extract[groupoid,
          Distribute[{{k}, groupoid[[j, i]]}, List]]]]], {i, 1,
      Length[groupoid]}, {j, 1, Length[groupoid]}, {k, 1,
      Length[groupoid]}], 2], False]];

Asso[groupoid1_List] :=
```

```
   Not[MemberQ[
     Flatten[Table[
       Union[Flatten[
         Union[Extract[groupoid1,
           Distribute[{groupoid1[[i, j]], {k}}, List]]]] ==
       Union[Flatten[
         Union[Extract[groupoid1,
           Distribute[{{i}, groupoid1[[j, k]]}, List]]]], {i, 1,
       Length[groupoid1]}, {j, 1, Length[groupoid1]}, {k, 1,
       Length[groupoid1]}], 2], False]];

ReproductivityTest[groupoid_List] :=
  Min[Table[
      Length[Union[Flatten[Transpose[groupoid][[j]]]]], {j, 1,
      Length[groupoid]}]] == Length[groupoid] &&
    Min[Table[
      Length[Union[Flatten[groupoid[[j]]]]], {j, 1,
      Length[groupoid]}]] == Length[groupoid];

End[];
EndPackage[];

Use of the package:
for checking a magma, for instance the following one:

{1}       {1}       {2,3}
{1}       {1}       {2,3}
{1,2,3}   {2,3}     {1,3}

write in Mathematica:

In[1]:=LRHtest[{{{1}, {1}, {2, 3}}, {{1}, {1}, {2, 3}}, {{1, 2, 3}, {2,3}, {1, 3}}}]

And the output is:

Out[1]=2

where number 2 corresponds to «RA-hypergroup»
```

References

1. Marty, F. Sur une Généralisation de la notion de Groupe. Huitième Congrès des Mathématiciens Scand. *Stockholm* **1934**, 45–49.
2. Marty, F. Rôle de la notion d'hypergroupe dans l'étude des groupes non abéliens. *Comptes Rendus Acad. Sci. Paris* **1935**, *201*, 636–638.
3. Marty, F. Sur les groupes et hypergroupes attachés à une fraction rationnelle. *Ann. Sci. l'École Norm. Supérieure* **1936**, *53*, 83–123. [CrossRef]
4. Massouros, C.; Massouros, G. An Overview of the Foundations of the Hypergroup Theory. *Mathematics* **2021**, *9*, 1014. [CrossRef]
5. Kazim, M.A.; Naseeruddin, M. On almost semigroups. *Portugaliae Math.* **1977**, *36*, 41–47.
6. Aslam, M.; Aroob, T.; Yaqoob, N. On cubic Γ-hyperideals in left almost Γ-semihypergroups. *Ann. Fuzzy Math. Inform.* **2013**, *5*, 169–182.
7. Gulistan, M.; Yaqoob, N.; Shahzad, M. A Note on Hv-LA-Semigroups. *UPB. Sci. Bull.* **2015**, *77*, 93–106.
8. Yaqoob, N.; Aslam, M. Faisal, On soft Γ-hyperideals over left almost Γ-semihypergroups. *J. Adv. Res. Dyn. Control. Syst.* **2012**, *4*, 1–12.
9. Yaqoob, N.; Corsini, P.; Yousafzai, F. On intra-regular left almost semi-hypergroups with pure left identity. *J. Math.* **2013**. [CrossRef]

10. Yaqoob, N.; Cristea, I.; Gulistan, M. Left almost polygroups. *Ital. J. Pure Appl. Math.* **2018**, *39*, 465–474.
11. Amjad, V.; Hila, K.; Yousafzai, F. Generalized hyperideals in locally associative left almost semihypergroups. *N. Y. J. Math.* **2014**, *20*, 1063–1076.
12. Hila, K.; Dine, J. On hyperideals in left almost semihypergroups. *ISRN Algebra* **2011**. [CrossRef]
13. Corsini, P. *Prolegomena of Hypergroup Theory*; Aviani Editore: Tricesimo, Udine, Italy, 1993.
14. Corsini, P.; Leoreanu, V. *Applications of Hyperstructures Theory*; Kluwer Academic Publishers: Dordrecht, The Netherlands, 2003.
15. Krasner, M. Hypergroupes moduliformes et extramoduliformes. *Comptes Rendus Acad. Sci. Paris* **1944**, *219*, 473–476.
16. Mittas, J. Hypergroupes et hyperanneaux polysymetriques. *Comptes Rendus Acad. Sci. Paris* **1970**, *271*, 920–923.
17. Mittas, J. Hypergroupes canoniques hypervalues. *Comptes Rendus Acad. Sci. Paris* **1970**, *271*, 4–7.
18. Mittas, J. Hypergroupes canoniques. *Math. Balk.* **1972**, *2*, 165–179.
19. Mittas, J. Hypergroupes canoniques values et hypervalues. Hypergroupes fortement et superieurement canoniques. *Bull. Greek Math. Soc.* **1982**, *23*, 55–88.
20. Mittas, J. Hypergroupes polysymetriques canoniques. In Proceedings of the Atti del Convegno su Ipergruppi, Altre Strutture Multivoche e loro Applicazioni, Udine, Italy, 15–18 October 1985; pp. 1–25.
21. Mittas, J. Generalized M-Polysymmetric Hypergroups. In Proceedings of the 10th International Congress on Algebraic Hyperstructures and Applications, Brno, Czech Republic, 3–9 September 2008; pp. 303–309.
22. Massouros, C.G.; Mittas, J. On the theory of generalized M-polysymmetric hypergroups. In Proceedings of the 10th International Congress, on Algebraic Hyperstructures and Applications, Brno, Czech Republic, 3–9 September 2008; pp. 217–228.
23. Massouros, C.G. Canonical and Join Hypergroups. *An. Științifice Univ. Alexandru Ioan Cuza* **1996**, *42*, 175–186.
24. Yatras, C. M-polysymmetrical hypergroups. *Riv. Mat. Pura Appl.* **1992**, *11*, 81–92.
25. Yatras, C. Types of Polysymmetrical Hypergroups. *AIP Conf. Proc.* **2018**, *1978*, 340004-1–340004-5. [CrossRef]
26. Bonansinga, P. Sugli ipergruppi quasicanonici. *Atti Soc. Pelor. Sci. Fis. Mat. Nat.* **1981**, *27*, 9–17.
27. Massouros, C.G. Quasicanonical hypergroups. In Proceedings of the 4th International Congress on Algebraic Hyperstructures and Applications, Xanthi, Greece, 27–30 June 1990; World Scientific: Singapore, 1991; pp. 129–136.
28. Comer, S. Polygroups derived from cogroups. *J. Algebra* **1984**, *89*, 397–405. [CrossRef]
29. Davvaz, B. *Polygroup Theory and Related Systems*; World Scientific: Singapore, 2013.
30. Massouros, G.G.; Zafiropoulos, F.A.; Massouros, C.G. Transposition polysymmetrical hypergroups. In Proceedings of the 8th International Congress on Algebraic Hyperstructures and Applications, Samothraki, Greece, 1–9 September 2002; Spanidis Press: Xanthi, Greece, 2003; pp. 191–202.
31. Massouros, G.G. Fortified join hypergroups and join hyperrings. *An. Științifice Univ. Alexandru Ioan Cuza Sect. I Mat.* **1995**, *XLI*, 37–44.
32. Massouros, G.G.; Massouros, C.G.; Mittas, J. Fortified join hypergroups. *Ann. Math. Blaise Pascal* **1996**, *3*, 155–169. [CrossRef]
33. Jantosciak, J. Transposition hypergroups, Noncommutative Join Spaces. *J. Algebra* **1997**, *187*, 97–119. [CrossRef]
34. Jantosciak, J.; Massouros, C.G. Strong Identities and fortification in Transposition hypergroups. *J. Discret. Math. Sci. Cryptogr.* **2003**, *6*, 169–193. [CrossRef]
35. Massouros, G.G. Hypercompositional structures in the theory of languages and automata. *An. Științifice Univ. Alexandru Ioan Cuza Sect. Inform.* **1994**, *III*, 65–73.
36. Massouros, G.G. Hypercompositional structures from the computer theory. *Ratio Math.* **1999**, *13*, 37–42.
37. Massouros, G.G. On the attached hypergroups of the order of an automaton. *J. Discret. Math. Sci. Cryptogr.* **2003**, *6*, 207–215. [CrossRef]
38. Massouros, C.G.; Massouros, G.G. Hypergroups associated with graphs and automata. *AIP Conf. Proc.* **2009**, *1168*, 164–167. [CrossRef]
39. Massouros, C.G.; Massouros, G.G. On the algebraic structure of transposition hypergroups with idempotent identity. *Iran. J. Math. Sci. Inform.* **2013**, *8*, 57–74.
40. Ameri, R. Topological transposition hypergroups. *Ital. J. Pure Appl. Math.* **2003**, *13*, 181–186.
41. De Salvo, M.; Lo Faro, G. On the n*-complete hypergroups. *Discret. Math.* **1999**, *208–209*, 177–188. [CrossRef]
42. De Salvo, M.; Freni, D.; Lo Faro, G. A new family of hypergroups and hypergroups of type U on the right of size five. *Far East J. Math. Sci.* **2007**, *26*, 393–418.
43. De Salvo, M.; Freni, D.; Lo Faro, G. Fully simple semihypergroups. *J. Algebra* **2014**, *399*, 358–377. [CrossRef]
44. De Salvo, M.; Fasino, D.; Freni, D.; Lo Faro, G. 1-hypergroups of small size. *Mathematics* **2021**, *9*, 108. [CrossRef]
45. Gutan, M. Sur une classe d'hypergroupes de type C. *Ann. Math. Blaise Pascal* **1994**, *1*, 1–19. [CrossRef]
46. Kankaras, M.; Cristea, I. Fuzzy reduced hypergroups. *Mathematics* **2020**, *8*, 263. [CrossRef]
47. Cristea, I. Complete hypergroups, 1-hypergroups and fuzzy sets. *An. St. Univ. Ovidius Constanta* **2000**, *10*, 25–38.
48. Angheluta, C.; Cristea, I. On Atanassov's intuitionistic fuzzy grade of complete hypergroups. *J. Mult. Val. Log. Soft Comput.* **2013**, *20*, 55–74.
49. Massouros, C.G.; Massouros, G.G. On certain fundamental properties of hypergroups and fuzzy hypergroups—Mimic fuzzy hypergroups. *Intern. J. Risk Theory* **2012**, *2*, 71–82.
50. Hoskova-Mayerova, S.; Al Tahan, M.; Davvaz, B. Fuzzy multi-hypergroups. *Mathematics* **2020**, *8*, 244. [CrossRef]
51. Al Tahan, M.; Hoskova-Mayerova, S.; Davvaz, B. Fuzzy multi-polygroups. *J. Intell. Fuzzy Syst.* **2020**, *38*, 2337–2345. [CrossRef]

52. Chvalina, J.; Novák, M.; Smetana, B.; Staněk, D. Sequences of Groups, Hypergroups and Automata of Linear Ordinary Differential Operators. *Mathematics* **2021**, *9*, 319. [CrossRef]
53. Novák, M.; Křehlík, S.; Cristea, I. Cyclicity in EL-hypergroups. *Symmetry* **2018**, *10*, 611. [CrossRef]
54. Novák, M.; Křehlík, Š. EL–hyperstructures revisited. *Soft. Comput.* **2018**, *22*, 7269–7280. [CrossRef]
55. Massouros, G.G. On the Hypergroup Theory. *Fuzzy Syst. A.I. Rep. Lett. Acad. Romana* **1995**, *IV*, 13–25.
56. Prenowitz, W. Projective Geometries as multigroups. *Am. J. Math.* **1943**, *65*, 235–256. [CrossRef]
57. Prenowitz, W. Descriptive Geometries as multigroups. *Trans. Am. Math. Soc.* **1946**, *59*, 333–380. [CrossRef]
58. Prenowitz, W. Spherical Geometries and mutigroups. *Can. J. Math.* **1950**, *2*, 100–119. [CrossRef]
59. Prenowitz, W. A Contemporary Approach to Classical Geometry. *Am. Math. Month.* **1961**, *68*, 1–67. [CrossRef]
60. Prenowitz, W.; Jantosciak, J. Geometries and Join Spaces. *J. Reine Angew. Math.* **1972**, *257*, 100–128.
61. Prenowitz, W.; Jantosciak, J. *Join Geometries: A Theory of Convex Sets and Linear Geometry*; Springer: Berlin/Heidelberg, Germany, 1979.
62. Jantosciak, J. Classical geometries as hypergroups. In Proceedings of the Atti del Convegno su Ipergruppi altre Structure Multivoche et loro Applicazioni, Udine, Italy, 15–18 October 1985; pp. 93–104.
63. Jantosciak, J. A brif survey of the theory of join spaces. In Proceedings of the 5th International Congress on Algebraic Hyperstructures and Applications, Iasi, Rumania, 4–10 July 1993; Hadronic Press: Palm Harbor, FL, USA, 1994; pp. 109–122.
64. Barlotti, A.; Strambach, K. Multigroups and the foundations of Geometry. *Rend. Circ. Mat. Palermo* **1991**, *XL*, 5–68. [CrossRef]
65. Freni, D. Sur les hypergroupes cambistes. *Rend. Ist. Lomb.* **1985**, *119*, 175–186.
66. Freni, D. Sur la théorie de la dimension dans les hypergroupes. *Acta Univ. Carol. Math. Phys.* **1986**, *27*, 67–80.
67. Massouros, C.G. Hypergroups and convexity. *Riv. Mat. Pura Appl.* **1989**, *4*, 7–26.
68. Mittas, J.; Massouros, C.G. Hypergroups defined from linear spaces. *Bull. Greek Math. Soc.* **1989**, *30*, 63–78.
69. Massouros, C.G. Hypergroups and Geometry. *Mem. Acad. Romana Math. Spec. Issue* **1996**, *XIX*, 185–191.
70. Massouros, C.G. On connections between vector spaces and hypercompositional structures. *Ital. J. Pure Appl. Math.* **2015**, *34*, 133–150.
71. Massouros, G.G.; Massouros, C.G. Hypercompositional algebra, computer science and geometry. *Mathematics* **2020**, *8*, 1338. [CrossRef]
72. Dramalidis, A. On geometrical hyperstructures of finite order. *Ratio Math.* **2011**, *21*, 43–58.
73. Massouros, G.G.; Mittas, J. Languages—Automata and hypercompositional structures. In Proceedings of the 5th International Congress on Algebraic Hyperstructures and Applications, Xanthi, Greece, 27–30 June 1990; World Scientific: Singapore, 1991; pp. 137–147.
74. Massouros, G.G. Automata and hypermoduloids. In Proceedings of the 5th International Congress on Algebraic Hyperstructures and Applications, Iasi, Rumania, 4–10 July 1993; Hadronic Press: Palm Harbor, FL, USA, 1994; pp. 251–265.
75. Massouros, C.G.; Massouros, G.G. The transposition axiom in hypercompositional structures. *Ratio Math.* **2011**, *21*, 75–90.
76. Massouros, C.G.; Dramalidis, A. Transposition Hv-groups. *ARS Comb.* **2012**, *106*, 143–160.
77. Vougiouklis, T. *Hyperstructures and Their Representations*; Hadronic Press: Palm Harbor, FL, USA, 1994.
78. Tsitouras, C.G.; Massouros, C.G. On enumeration of hypergroups of order 3. *Comput. Math. Appl.* **2010**, *59*, 519–523. [CrossRef]
79. Massouros, C.G.; Tsitouras, C.G. Enumeration of hypercompositional structures defined by binary relations. *Ital. J. Pure Appl. Math.* **2011**, *28*, 43–54.
80. Tsitouras, C.G.; Massouros, C.G. Enumeration of Rosenberg type hypercompositional structures defined by binary relations. *Eur. J. Comb.* **2012**, *33*, 1777–1786. [CrossRef]
81. Massouros, C.G. On the enumeration of rigid hypercompositional structures. *AIP Conf. Proc.* **2014**, *1648*, 740005–1–740005–6. [CrossRef]
82. Massouros, C.G.; Massouros, G.G. On 2-element Fuzzy and Mimic Fuzzy Hypergroups. *AIP Conf. Proc.* **2012**, *1479*, 2213–2216. [CrossRef]
83. Cristea, I.; Jafarpour, M.; Mousavi, S.; Soleymani, A. Enumeration of Rosenberg hypergroups. *Comput. Math. Appl.* **2010**, *60*, 2753–2763. [CrossRef]
84. Jafarpour, M.; Cristea, I.; Tavakoli, A. A method to compute the number of regular reversible Rosenberg hypergroup. *ARS Comb.* **2018**, *128*, 309–329.
85. Bayon, R.; Lygeros, N. Advanced results in enumeration of hyperstructures. *J. Algebra* **2008**, *320*, 821–835. [CrossRef]
86. Nordo, G. An algorithm on number of isomorphism classes of hypergroups of order 3. *Ital. J. Pure Appl. Math.* **1997**, *2*, 37–42.
87. Ameri, R.; Eyvazi, M.; Hoskova-Mayerova, S. Advanced results in enumeration of hyperfields. *AIMS Math.* **2020**, *5*, 6552–6579. [CrossRef]
88. Massouros, C.G.; Massouros, G.G. Identities in multivalued algebraic structures. *AIP Conf. Proc.* **2010**, *1281*, 2065–2068. [CrossRef]
89. Massouros, C.G.; Massouros, G.G. Transposition hypergroups with idempotent identity. *Int. J. Algebr. Hyperstruct. Appl.* **2014**, *1*, 15–27.
90. Massouros, C.G.; Massouros, G.G. Transposition polysymmetrical hypergroups with strong identity. *J. Basic Sci.* **2008**, *4*, 85–93.
91. Krasner, M. La loi de Jordan—Holder dans les hypergroupes et les suites génératrices des corps de nombres P—Adiqes, (I). *Duke Math. J.* **1940**, *6*, 120–140, [CrossRef], (II) *Duke Math. J.* **1940**, *7*, 121–135. [CrossRef]
92. Krasner, M.; Kuntzmann, J. Remarques sur les hypergroupes. *Comptes Rendus Acad. Sci. Paris* **1947**, *224*, 525–527.

93. Massouros, C.G. On the semi-subhypergroups of a hypergroup. *Int. J. Math. Math. Sci.* **1991**, *14*, 293–304. [CrossRef]
94. Massouros, G.G. The subhypergroups of the fortified join hypergroup. *Ital. J. Pure Appl. Math.* **1997**, *2*, 51–63.
95. Massouros, C.G.; Massouros, G.G. On subhypergroups of fortified transposition hypergroups. *AIP Conf. Proc.* **2013**, *1558*, 2055–2058. [CrossRef]
96. Massouros, C.G. Some properties of certain subhypergroups. *Ratio Math.* **2013**, *25*, 67–76.
97. Yatras, C. Subhypergroups of M-polysymmetrical hypergroups. In Proceedings of the 5th International Congress on Algebraic Hyperstructures and Applications, Iasi, Rumania, 4–10 July 1993; Hadronic Press: Palm Harbor, FL, USA, 1994; pp. 123–132.
98. De Salvo, M.; Freni, D.; LoFaro, G. Hypercyclic subhypergroups of finite fully simple semihypergroups. *J. Mult. Valued Log. Soft Comput.* **2017**, *29*, 595–617.
99. Sureau, Y. Sous-hypergroupe engendre par deux sous-hypergroupes et sous-hypergroupe ultra-clos d'un hypergroupe. *Comptes Rendus Acad. Sci. Paris* **1977**, *284*, 983–984.
100. Wolfram, S. *The Mathematica Book*, 5th ed.; Wolfram Media: Champaign, IL, USA, 2003.

MDPI
St. Alban-Anlage 66
4052 Basel
Switzerland
Tel. +41 61 683 77 34
Fax +41 61 302 89 18
www.mdpi.com

Mathematics Editorial Office
E-mail: mathematics@mdpi.com
www.mdpi.com/journal/mathematics

www.ingramcontent.com/pod-product-compliance
Lightning Source LLC
LaVergne TN
LVHW072332090526
838202LV00019B/2406